T0236526

Mathematics of the 19th Century

Edited by
A.N. Kolmogorov
A.P. Yushkevich

Translated from the Russian
by Roger Cooke

Geometry

Analytic Function Theory

Birkhäuser Verlag
Basel · Boston · Berlin

Editors' addresses

A.N. Kolmogorov †
Moscow State University
Faculty of Mathematics and Mechanics
117 234 Moscow
Russia

A.P. Yushkevich †
Institute of History of Science and Technology
Staropanski Pereulok 1/5
103 012 Moscow
Russia

Originally published as:
Matematika XIX veka: geometriya, teoriya analiticheskikh funktsii
© Izdatel'stvo «Nauka», Moskva 1981
Softcover reprint of the hardcover 1st edition 1981

All illustrations are taken from the original Russian edition.
ISBN 978-3-0348-9933-8   ISBN 978-3-0348-9173-8 (eBook)
DOI 10.1007/978-3-0348-9173-8

A CIP catalogue record for this book is available from the Library of Congress, Washington D.C., USA

Deutsche Bibliothek Cataloging-in-Publication Data
Mathematics of the 19th century / ed. by A. N. Kolmogorov ;
    Einheitssacht.: Mathematika XIX veka <engl.>
    Literaturangaben
NE: Kolmogorov, Andrej N. [Hrsg.]; EST
Geometry, analytic function theory. – 1996

© 1996 Birkhäuser Verlag, P.O.Box 133, CH-4010 Basel, Switzerland
Printed from the translator's camera-ready manuscript on acid-free paper produced from chlorine-free pulp. TCF ∞
Cover design: Markus Etterich, Basel

9 8 7 6 5 4 3 2 1

Contents

Preface

The general principles by which the editors and authors of the present edition have been guided were explained in the preface to the first volume of *Mathematics of the 19th Century*, which contains chapters on the history of mathematical logic, algebra, number theory, and probability theory (Nauka, Moscow 1978; English translation by Birkhäuser Verlag, Basel-Boston-Berlin 1992). Circumstances beyond the control of the editors necessitated certain changes in the sequence of historical exposition of individual disciplines. The second volume contains two chapters: history of geometry and history of analytic function theory (including elliptic and Abelian functions); the size of the two chapters naturally entailed dividing them into sections. The history of differential and integral calculus, as well as computational mathematics, which we had planned to include in the second volume, will form part of the third volume.

We remind our readers that the appendix of each volume contains a list of the most important literature and an index of names. The names of journals are given in abbreviated form and the volume and year of publication are indicated; if the actual year of publication differs from the nominal year, the latter is given in parentheses. The book *History of Mathematics from Ancient Times to the Early Nineteenth Century* [in Russian], which was published in the years 1970–1972, is cited in abbreviated form as HM (with volume and page number indicated). The first volume of the present series is cited as Bk. 1 (with page numbers).

The first chapter of this volume was read by Prof. P. K. Rashevskiĭ, and the second by Prof. E. D. Solomentsev; the entire volume was read by Prof. A. D. Solov'ev. The authors and the editors are grateful to all three for useful advice and corrections. We also wish to express our gratitude to Prof. Taton (France), Dr. E. Fellmann (Switzerland), and K. Jacobs (Federal Republic of Germany) who kindly provided us with a number of portraits.

The first volume of *Mathematics of the 19th Century* was published at the same time as the *Abrégé d'histoire des mathématiques, 1700–1900*, edited by J. Dieudonné (Hermann: Paris, 1978, T. 1, 2).

The editors announce with deep sorrow that after the manuscript of this book had been completed one of its authors passed away. Academician Alekseĭ Ivanovich Markushevich (2.4.1908–4.6.1979) of the Academy of Pedagogical Sciences of the

USSR was one of the editors of the first two volumes in the series. Articles and obituaries dedicated to the memory of this prominent scholar and activist in the field of popular education have appeared in many Soviet journals.

It should be remarked that Markushevich had communicated certain important conclusions of his research, relating in particular to the connections between the doctoral dissertation of B. Riemann and the papers of A. Cauchy and V. Puiset immediately preceding it, in a paper given at the International Congress of Mathematicians in Helsinki (1978) and published in expanded form in Issue XXV of *Историко-Математические Исследования* (1980). These conclusions were subsequently confirmed completely by the Zürich scholar E. Neuenschwander, who studied the papers of Riemann from the period of his doctoral dissertation, which are kept in the archives of the University of Göttingen.

Moscow, 3 June 1980

A. N. Kolmogorov
A. P. Yushkevich

Chapter 1
Geometry

INTRODUCTION

Although the main achievements of eighteenth-century mathematics were connected with the development of mathematical analysis, important discoveries were made in the course of the century in geometry also. The development of analysis was linked in the first instance with the development of analytic geometry. Plane analytic geometry, which had appeared in the work of Descartes and Fermat, was significantly advanced in the late seventeenth century and the first half of the eighteenth century in the work of Newton, Hermann, Stirling, Maupertuis, and Cramer, and assumed a form very close to its modern form in the second volume of Leonhard Euler's *Introductio in analysin infinitorum* (1748), and in Clairaut's book on curves of double curvature (1731). Analytic geometry was developed in three dimensions in the appendix to the second volume of Euler's *Introductio*; it was further developed in papers of Monge (1794–1805). In connection with the development of the concept of a function geometers made ever more extensive use of geometric transformations: Clairaut and Euler laid the foundations of the subject of affine transformations, d'Alembert and Euler founded the subject of conformal mappings, and Waring and Monge studied projective transformations from various points of view. Johann Bernoulli, Clairaut, and Euler solved a number of problems in the differential geometry of curves in space, including in particular the theory of geodesics on a surface. In his *Recherches sur la courbure des surfaces* (1767) Euler laid the foundations of the differential geometry of surfaces, which was further developed in the work of Monge. La Caille, Lambert, and especially Monge made significant advances in descriptive geometry. The first topological problems were solved in the papers of Euler on the Königsberg bridge problem and polyhedra. Spherical geometry and trigonometry received significant development, especially in the work of Euler and his students. Finally, Saccheri, Lambert, Bertrand, Legendre, and Gur'ev also advanced the theory of parallel lines, which led geometers directly to the discovery of non-Euclidean geometry. Kant and d'Alembert posed the problem of the geometry of multi-dimensional spaces, especially four-dimensional space.

As in other areas of mathematics, in the development of geometry in the nineteenth century one can trace the interaction of the two primary motives for the development of mathematics — the need to develop new methods of solving practical problems, and the internal logic of mathematical development. Thus, one of the important motives for developing differential geometry was the solution of practical problems of geodesy, mainly those that Gauss had encountered in his geodesic survey of the Duchy of Hannover. Further stimulus to the development of this theory came from geometric optics. The revival of projective geometry in the early nineteenth century was to a considerable degree due to Monge's *Géométrie descriptive* (1799), which solved problems of fortification, and also problems of the rapidly developing mechanical engineering. Vector calculus had as one of its sources the need to create the mathematical machinery for statics, kinematics, and dynamics; these same mechanical disciplines required the creation of the cross-product calculus. The creation of new branches of geometry required that other mathematical disciplines be developed as well: the rise of multiple integrals and the algebra of forms in several variables created the need for multi-dimensional geometry; the theory of multi-sheeted Riemann surfaces required the development of topology; the theory of Fuchsian groups of fractional-linear transformations of a complex variable led to the Poincaré model for hyperbolic geometry. The successes of group theory and its applications to the problem of solving algebraic equations by radicals played an exceptional role in the creation of the theory of transformation groups. The problem of determining the conditions for solvability of differential equations by quadratures became the motive for creating the theory of Lie groups, which are very important for geometry. On the other hand, the study of the mutual dependence of the axioms of geometry, in particular the Euclidean parallel postulate, led to the discovery of Lobachevskiĭ (hyperbolic) geometry, after which analogous problems led to the discovery of elliptic geometry and other geometries. The problem of providing a foundation for projective geometry without using a metric led to important discoveries. The attempt to solve problems of differential and non-Euclidean geometry also led to remarkable new geometric discoveries. At the same time the creation of new geometric systems posed from the outset the problems of applying them and, in certain cases, to experimental testing of the new theories — among these were the attempts of Lobachevskiĭ to determine experimentally which geometry holds in the real world and Riemann's application of his multi-dimensional geometry of curved spaces to the solution of a problem connected with the heat equation. In the final analysis the attempts of Zöllner to study the "fourth dimension" experimentally using "science in the spirit world," though fantastic, were an expression of the same trend. However, although nineteenth-century geometry continued many trends of the eighteenth century, the development of this subject in the nineteenth century led to significant qualitative differences between the new geometry and the old. The most important achievement of the new geometry was the creation of a variety of non-Euclidean and multi-dimensional geometries based on various transformation groups. As a result, by the late nineteenth century geometry, which at the beginning of the century had been the geometry of three-dimensional Euclidean space, had turned into a complicated system of geometric subjects suitable for applying geometric methods to many fields of mathematics and science.

1. ANALYTIC AND DIFFERENTIAL GEOMETRY

Analytic Geometry

Analytic geometry, which had arisen in the seventeenth century in the work of Descartes and Fermat and had received considerable development in the eighteenth century in the work of Euler, Monge, Lagrange, and others, assumed the form that it still has today, in Monge's *Applications de l'algèbre à la géométrie*, which we have already mentioned (cf. HM, Vol. 3, p. 181). There (p. 182) we have pointed out the significance for the development of analytic geometry of the book of S. F. Lacroix (1765–1843), *Traité élémentaire de la trigonométrie rectilinéaire et sphèrique et des applications de l'algèbre à la géométrie*, which was re-issued several times during the nineteenth century, and the appearance of the term "analytic geometry" in the early nineteenth century. We remark that, besides the *Essai de la géométrie analytique, appliquée aux courbes et surfaces du second ordre* of J. B. Biot (1805, cf. HM, Vol. 3, p. 183), this term occurred in the title *Eléments de géométrie analytique* (Paris 1801) of J. Garnier (1766–1840).

In contrast to the handbooks of Euler and Monge, the book of Lacroix contained the basic problems for a straight line in a form close to the modern form. These problems were also solved by Biot and other authors of French textbooks on analytic geometry and in the *Sammlung geometrischer Aufgaben* (Berlin 1807) of Meyer Hirsch (1765–1851). Many problems of analytic geometry that have become part of modern textbooks were first solved by Gabriel Lamé (1795–1870) in his *Examen des différentes méthodes employées pour résoudre les problèmes de géométrie* (Paris 1818). Here we find the equation of a plane "in segments," i.e., in the form $x/a + y/b + z/c = 1$; the analogous equation of a line first appears only in the *Sammlung mathematischer Aufsätze und Bemerkungen* (Berlin 1821, Bd. 1) of August Leopold Crelle (1780–1855), a German mathematician and member of the Berlin Academy of Sciences, who founded in 1826 the *Journal für die reine und angewandte Mathematik*, which we have referred to many times, and which is often called *Crelle's Journal*. Lamé's book gave a general criterion for three lines to meet in a single point, as well as the first exposition of an abbreviated notation very popular among nineteenth-century geometers, in which the left-hand sides of equations of lines or higher-order curves are denoted by the letters F_1 and F_2, the equations of an arbitrary line or curve of the pencil defined by the lines $F_1 = 0$ and $F_2 = 0$ are written $F_1 + \mu F_2 = 0$, and geometric theorems are proved by investigating the geometric meaning of the analytic relations thereby obtained. This method was subsequently extended to pencils of planes and surfaces of higher order, as well as to combinations of them: $F_1 + \lambda F_2 + \mu F_3 = 0$, and so on. This method lost its popularity only after the introduction of vector methods into analytic geometry in the first half of the twentieth century.

The normal equation of a line, $x \cos \alpha + y \sin \alpha - p = 0$, first appeared in the *Eléments d'analyse géométrique et d'analyse algébrique* (Paris 1809) by Simon

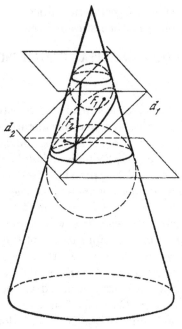

Fig. 1

L'Huilier (1750–1840), and the analogous normal equation for a plane and the parametric equations for a line in three-dimensional space appeared in the *Leçons sur les applications du calcul infinitésimal à la géométrie* (Paris 1826, Vol. 1) of Augustin-Louis Cauchy.

Following the French textbooks on analytic geometry the book *Analytische Geometrie* (Vienna 1823) of Joseph von Littrow (1781–1840) appeared in German. Littrow, who was a native of Bohemia, studied in Vienna and Prague and worked in Krakow and Kazan' (from 1809 to 1816) and in Budapest and Vienna. The *Elemente der analytischen Geometrie* (Leipzig 1839, Bd. 1–2) also appeared, written by J. A. Grunert (1797–1872), the editor of the journal *Grunerts Archiv*. The Russian book *Course of Analytic Geometry* appeared at the same time, written by Nikolaï Dmitrievich Brashman (1796–1866), also a native of Bohemia, who studied in Vienna and worked in the Petropavlovsk school in Petersburg during the years 1823–1825, at the University of Kazan' from 1825–1834, and at the University of Moscow after 1834. Also very popular and translated into many European languages (including Russian) were the English handbooks on analytic geometry by the Irish mathematician and theologian George Salmon (1819–1904). These were his *Conic Sections* (Dublin 1848) and his *Analytic Geometry of Three Dimensions* (Dublin 1862).

Among the particular epochs in the development of analytic geometry we note the results of the Belgian mathematician and engineer Germinal Pierre Dandelin (1794–1847), the author of a well-known method of approximate computation of the roots of algebraic equations (1826), which was later discovered independently by Lobachevskiĭ (1832) and developed in detail by the Swiss mathematician K. H. Gräffe (1837). In the paper "Sur quelques propriétés remarquables de la focale parabolique," (*Nouv. mém. Acad. Bruxelles*, 1882) Dandelin proved that the foci of a conic section can be defined as the points at which the cutting plane is tangent to spheres inscribed in the cone (in the case of a parabola, a single sphere), while the directrices of the conic section are the lines in which the cutting plane intersects the planes of the circles in which the spheres are tangent to the cone (Fig. 1). These facts make possible a particularly simple derivation of the focus and directrix properties of conic sections and different equations of these curves based on their geometric definition.

In contrast to the analytic geometry of the eighteenth century the subject of algebraic curves and surfaces of higher order ceased to belong to this branch of geometry in the nineteenth century; rather it became the so-called algebraic geometry. After the appearance of projective geometry, originally defined synthetically, analytic methods were extended to the projective plane and projective three-space in which problems of algebraic geometry also began to be solved. On the other hand, in the process of creating non-Euclidean geometry the analytic geometry of non-Euclidean space was also created. And analytic methods lay at the heart of the study of multi-dimensional spaces also. We shall study these varieties of analytic geometry below in the sections on projective, non-Euclidean, and multi-dimensional geometries, but at this point we continue the study of the history of analytic geometry of three-dimensional Euclidean space — the differential geometry of that space.

The Differential Geometry of Monge's Students

After Clairaut and Euler the central place in the development of differential geometry in the late eighteenth century was occupied by Monge (cf. HM, Vol. 3, pp. 184–186, 191–195). His works, especially his pedagogical activity in the Military Academy in Mézières (1768–1780) and at the École Polytechnique (1795–1809) attracted a large number of students and followers. Thus in these years and the first decades of the nineteenth century it was in France that geometric studies made the greatest progress. This research was conducted in both differential and projective geometry, whose foundations were laid by Monge's student Poncelet.

A characteristic feature of Monge's school of differential geometry is the direct geometric nature of the reasoning, in which the analytic machinery — the coordinate method and the results of the theory of differential equations — seems to be only reinforcement.

Of the students of Monge in the Mézières Academy we mention the names Meusnier (cf. HM, Vol. 3, pp. 194–195) and Tinseau (cf. HM, Vol. 3, pp. 180–181, 192–193).

Among the students of Monge at the École Polytechnique who developed his differential-geometric ideas one must mention Malus, Lancret, Rodrigues, and Dupin. We shall discuss the research of Malus below in connection with the theory of line congruences, of which he was one of the founders.

C. Dupin

M. Lancret (1774–1807) introduced both the curvature and the torsion of a space curve in his paper "Mémoire sur les courbes à double courbure," (*Mémoires présentés à l'Institut*, 1806), defining them as the infinitesimal angles of rotation of the normal and osculating planes, and calling them the first and second *flexions*. Subsequently these quantities, now in finite form, appear in the work of Cauchy, in his *Leçons sur l'application du calcul infinitésimal à la géométrie* (1826), after which their role was definitively established in the derivation formulas of Frenet and Serret (1847).

Olinde Rodrigues (1794–1851), also famous as a utopian socialist, was a student of H. C. Saint-Simon (1760–1825). In his paper "Recherches sur la théorie analytique de lignes et rayons de courbure des surfaces," (*Bull. soc. philomatique de Paris*, 1815) he obtained a number of results and formulas connected with lines of curvature, in particular the so-called *Rodrigues formulas*. Anticipating Gauss, he used a spherical mapping of a surface, studied the ratio of the areas of the corresponding surfaces, and arrived at a quantity later called the total or Gaussian curvature; he showed that it equals the product of the principal curvatures.

Charles Dupin (1784–1873), had a brilliant career in differential geometry. His works were long delayed in publication, since he was a naval officer and went on long sea voyages. His works were published in two books, *Développement de géométrie* (Paris 1813) and *Applications de géométrie et de mécanique* (Paris 1822). At the age of 16, by considering the envelope of the family of spheres tangent to three given spheres, he arrived at the concept of a remarkable surface — the cyclid (later named after him), both of whose families of lines of curvature are circles (published in 1804). Around 1807 he proved a beautiful theorem that came to be known by his name, asserting that the surfaces of a triorthogonal system intersect in lines of curvature. This made it possible to interpret the lines of curvature of an ellipsoid, which had been studied by Monge, as the intersection of the ellipsoid with families of second-order surfaces confocal with it. To study the curvatures of the normal sections of surfaces he introduced the indicatrix, which also now bears his name, and which makes it possible to visualize and study the behavior of the curvature of a normal section of a surface as the cutting plane is revolved about the normal. This approach provided a classification of the points of a surface that are not planar points into three types: elliptic, hyperbolic, and parabolic. The geometric meaning of umbilical points and asymptotic lines also became clear (the latter term was introduced by Dupin). He was also the first to introduce the concept of conjugate lines (with respect to an asymptotic grid), and he obtained a geometric proof of Monge's theorem that a surface consisting of umbilical points is a sphere.

Gauss' *Disquisitiones generales circa superficies curvas*

A decisive influence on the entire course of development of differential geometry was exerted by the publication of a remarkable paper of Gauss, "Disquisitiones generales circa superficies curvas" (Göttingen 1828), written in Latin, as was the custom in the seventeenth and eighteenth centuries. It was this paper, carefully polished and containing a wealth of new ideas, that gave this area of geometry more or less its present form and opened a large circle of new and important problems whose development provided work for geometers for many decades.

A government mandate — to conduct a precise measurement of an arc of a meridian from Göttingen to Alton, and then to produce a geodesic map of the Duchy of Hannover — forced Gauss to turn his attention to the problems and practice of geodesy. In the 1820's he laid down the principles of a new subject — higher geodesy (published in 1842 and 1847), and organized field geodesic measurements that took more than 15 years to complete. For about five of those years he was personally involved in the measurements and carried out an enormous volume of computational work. In the course of these studies he became involved in the study of surface theory and opened a new area of research — the intrinsic geometry of surfaces.

Gauss made the parametric representation of a surface and the corresponding expression for its element of length into the foundation of his *Disquisitiones*. He was the first to formulate clearly and explicitly the concept of intrinsic geometry of a surface, and he proved that the curvature could be measured by a quantity (the Gaussian curvature) that belongs to the intrinsic geometry, i.e., does not vary when the surface is deformed. He further developed the theory of geodesic lines, which also belong to the intrinsic geometry.

There are three ways of defining a surface:

1) by an implicit equation

$$w(x, y, z) = 0, \tag{1}$$

2) by a parametric representation

$$x = x(p, q), \quad y = y(p, q), \quad z = z(p, q), \tag{2}$$

3) by giving z as a function of x and y

$$z = z(x, y), \tag{3}$$

where the third method is a special case of both the first and second. Gauss used primarily the second of these methods, as the one corresponding most closely to the nature of a surface. Parametric representation had been used before Gauss in special cases by Euler (HM, Vol. 3, p. 190), but Gauss was the first to make systematic application of it.

If one of the parameters, say q, is given a fixed value c, Eqs. (2) become parametric equations of a line (a p-line). By varying the value c we obtain the family of p-lines (i.e., the lines $q = $ const.). The two families together form a curvilinear parametric grid, which serves as a coordinate grid in a portion of the surface in which at most one line of each family passes through each point $M(p, q)$ and two lines from different families intersect in only one point. Such a coordinate system (p, q) on the surface came to be known as *Gaussian curvilinear coordinates*. Gauss defined the derivatives $a = dx/dp$, $b = dy/dp$, $c = dz/dp$, and $a' = dx/dq$, $b' = dy/dq$, $c' = dz/dq$, which are nowadays regarded as the partial derivatives of the coordinates of the radius-vector of a point of the surface with respect to p and q; he also defined the functions $A = bc' - cb'$, $B = ca' - ac'$, and $C = ab' - ba'$, which are nowadays regarded as the coordinates of the cross product of these vectors and form a vector directed along the normal to the surface. Finally, he defined the second derivatives

$$\alpha = \frac{ddx}{dp^2}, \quad \beta = \frac{ddx}{dp\,dq}, \quad \gamma = \frac{ddx}{dq^2},$$

$$\alpha' = \frac{ddy}{dp^2}, \quad \beta' = \frac{ddy}{dp\,dq}, \quad \gamma' = \frac{ddy}{dq^2},$$

$$\alpha'' = \frac{ddz}{dp^2}, \quad \beta'' = \frac{ddz}{dp\,dq}, \quad \gamma'' = \frac{ddz}{dq^2}.$$

Using these functions Gauss constructed the two quadratic forms

$$dx^2 = E\,dp^2 + 2F\,dp\,dq + G\,dq^2 \tag{4}$$

and

$$D\,dp^2 + 2D'\,dp\,dq + D''\,dq^2, \tag{5}$$

where

$$E = aa + bb + cc, \quad F = aa' + bb' + cc', \quad G = a'a' + b'b' + c'c',$$
$$D = A\alpha + B\beta + C\gamma, \quad D' = A\alpha' + B\beta' + C\gamma', \quad D'' = A\alpha'' + B\beta'' + C\gamma''.$$

The first of these forms (now called the *first fundamental form* of the surface) expresses the square of the linear element ds^2 of the surface, i.e., the square of the distance ds between infinitely close points of the surface with coordinates (p, q) and $(p + dp, q + dq)$. This form plays a fundamental role in Gauss' investigations.

The second form differs only in the factor $AA + BB + CC = EG - FF$ from the second fundamental form of the modern theory of surfaces.

As early as 1816 Gauss was relying on the expression for the linear element in solving the problem of conformal mapping of two surfaces, i.e., a mapping that preserves similarity on the infinitesimal level. The requirement of conformality reduces to proportionality of the linear elements of these surfaces, i.e.,

$$E/E' = F/F' = G/G'.$$

This problem was solved by Gauss, who submitted the solution to a competition announced by the Copenhagen Academy in 1822. The paper was awarded the prize and published in 1825.

Another useful innovation due to Gauss was the use of spherical mapping in geometry, which was usually applied in astronomy. Every oriented line is assigned the point on the unit sphere having radius vector parallel to that line. Thus a region of the surface is mapped to a region on the sphere using the normals. Relying on this mapping Gauss introduced the concept of a measure of the curvature K (the Gaussian curvature of the surface at the given point) as the ratio of the areas of the corresponding infinitesimal regions on the sphere and on the surface. In other words K is the limit of the ratio of the areas of the corresponding regions on the sphere and the surface as the region on the surface is contracted to a point.

Gauss found the measure of the curvature for the third case of (3) in the form

$$K = \frac{z_{xx}z_{yy} - (z_{xy})^2}{(1 + z_x^2 + z_y^2)^2},$$

and he proved that it equals the product of the principal curvatures. We have noted above that the definition of the measure of the curvature and this last result had been found previously by Rodrigues in 1815 (although little used), but this result seems to have remained unknown to Gauss.

Gauss went on to compute an expression for K in the case of a general parametric definition of the surface (2) using the coefficients of the two forms (4) and (5) in the form

$$K = \frac{DD'' - D'D'}{EG - FF}.$$

Finally, after some highly artificial computations Gauss arrived at a remarkable result: the general expression he had found for K can be represented as

$$4(EG - FF)^2 K = E\left(\frac{dE}{dq}\frac{dG}{dq} - 2\frac{dF}{dp}\frac{dG}{dq} + \left(\frac{dG}{dp}\right)^2\right) +$$

$$+ F\left(\frac{dE}{dp}\frac{dG}{dq} - \frac{dE}{dq}\frac{dG}{dp} - 2\frac{dE}{dq}\frac{dF}{dq} + 4\frac{dF}{dp}\frac{dF}{dq} - 2\frac{dF}{dp}\frac{dG}{dq}\right) +$$

$$+ G\left(\frac{dE}{dp}\frac{dG}{dp} - 2\frac{dF}{dp}\frac{dF}{dq} + \left(\frac{dE}{dq}\right)^2\right)$$

$$- 2(EG - FF)\left(\frac{ddE}{dq^2} - 2\frac{ddF}{dp\,dq} + \frac{ddG}{dp^2}\right),$$

i.e., the "measure of the curvature" is a function of the coefficients of the first fundamental form alone and their derivatives. This formula, as Gauss pointed out, leads to his "great theorem" (*Theorema Egregium*): *If a curved surface is developed on any other surface, the measure of the curvature at each point remains invariant.*[1]

Subsequently it became clear that this theorem had been proved by Gauss already in 1816, but at that time he had managed to carry out the proof only in isometric coordinates, in which the linear element reduces to the form

$$ds^2 = m^2(dp^2 + dq^2).$$

From his *theorema egregium* Gauss drew the following corollary: *If one surface can be developed (in other words, laid down or isometrically mapped) on another, then the curvatures of the two surfaces must be equal at corresponding points.*

The only case of developing a surface that had been studied previously was the case of developing on a plane. This problem was solved by Euler (cf. HM, Vol. 3, p. 190), who found all such "developable" surfaces: cylinders, cones, and the surfaces formed by the tangents to a space curve. Gauss emphasized the importance of the new approach to the study of the properties of surfaces when a surface is regarded as a flexible but inextensible body one of whose dimensions is considered as vanishingly small. A flexible inextensible thin film of metal gives an idea of such an approach to the concept of a surface and to the problems of developing or deforming it.

The totality of properties of figures lying on a surface that are preserved when it is deformed constitute the so-called *intrinsic geometry* of the surface. A deformation is an isometric mapping, i.e., a mapping that preserves the length of lines; consequently the linear elements must be equal at corresponding points. Therefore if we assign equal curvilinear coordinates to the corresponding points, the coefficients of the first fundamental forms must also be equal:

$$E = E', \quad F = F', \quad G = G'.$$

_____

1) *On the Foundations of Geometry. A Collection of Classical Works on Hyperbolic Geometry and the Development of its Ideas* [in Russian]. Moscow 1956, p. 140.

It follows from this that angles and areas on the surface are also preserved (and Gauss gave the corresponding formulas for them), i.e., these concepts belong to the intrinsic geometry.

The *theorema egregium* established the new and unexpected fact that the Gaussian curvature also belongs to the intrinsic geometry and raised the possibility of considering questions of bending of surfaces. Gauss himself did not undertake the study of these questions; he left this problem to his successors.

The concept of a geodesic, i.e., a shortest line, also belongs to the intrinsic geometry, since geodesic lines remain geodesics under deformation. For that reason Gauss, in studying problems of intrinsic geometry, found the equation of the geodesic lines in curvilinear coordinates and studied their behavior further. He introduced the notion of a geodesic circle, i.e., the geometric locus of the endpoints of geodesic radii of constant length emanating from a single point, and he showed that it was orthogonal to its radii. He also considered semigeodesic coordinate systems on a surface analogous to the orthogonal Cartesian and polar coordinate systems, and he found that in such systems the linear element has the form

$$ds^2 = dr^2 + m^2 d\varphi^2,$$

which made possible a great simplification in the equations for geodesic lines. Using series expansions, he found approximate computational formulas, useful in applications, for certain geometric quantities, in particular for the angles of a rectilinear triangle obtained by straightening the sides of a geodesic triangle.

The derivation formulas obtained by Gauss in carrying out the intermediate computations, which expressed the second derivatives of the coordinates of a point of the surface in the curvilinear coordinates p and q on the surface in terms of the first derivatives of these coordinates and the direction cosines of the normal to the surface (the perpendicular to the tangent plane) were also of great significance for the subsequent development of the theory of surfaces. These formulas, together with the expression for the derivatives of the direction cosines of the normal, can be regarded as differential equations that define the first derivatives. In modern notation, replacing the coordinates of a point by the radius-vector $\mathbf{r}$ of the point whose rectangular coordinates equal those of the point and replacing the direction cosines of the normal by the unit normal vector $\mathbf{n}$ whose rectangular coordinates equal the direction cosines, we express the first and second derivatives of the coordinates as the vectors $\partial \mathbf{r}/\partial p$, $\partial \mathbf{r}/\partial q$, $\partial^2 \mathbf{r}/\partial p^2$, $\partial^2 \mathbf{r}/\partial p\,\partial q$, $\partial^2 \mathbf{r}/\partial q^2$, and the formulas of Gauss can be rewritten in the form

$$\frac{\partial^2 \mathbf{r}}{\partial p^2} = \left\{ \begin{matrix} pp \\ p \end{matrix} \right\} \frac{\partial \mathbf{r}}{\partial p} + \left\{ \begin{matrix} pp \\ q \end{matrix} \right\} \frac{\partial \mathbf{r}}{\partial q} + D\mathbf{n},$$

$$\frac{\partial^2 \mathbf{r}}{\partial p\,\partial q} = \left\{ \begin{matrix} pq \\ p \end{matrix} \right\} \frac{\partial \mathbf{r}}{\partial p} + \left\{ \begin{matrix} pq \\ q \end{matrix} \right\} \frac{\partial \mathbf{r}}{\partial q} + D'\mathbf{n},$$

$$\frac{\partial^2 \mathbf{r}}{\partial q^2} = \left\{ \begin{matrix} qq \\ p \end{matrix} \right\} \frac{\partial \mathbf{r}}{\partial p} + \left\{ \begin{matrix} qq \\ q \end{matrix} \right\} \frac{\partial \mathbf{r}}{\partial q} + D''\mathbf{n},$$

where $\left\{ \begin{matrix} pp \\ p \end{matrix} \right\}, \left\{ \begin{matrix} pp \\ q \end{matrix} \right\}, \ldots, \left\{ \begin{matrix} qq \\ q \end{matrix} \right\}$ are functions of the coefficients E, F, G introduced by Gauss and their partial derivatives with respect to p and q.

The remarkable formula found by Gauss for the sum of the angles of a geodesic triangle amounts to the statement that the excess over 180° of the sum of the angles of such a triangle in the case of a surface of positive curvature, or the deficiency in the case of a surface of negative curvature, equals the area of the spherical image of this triangle, called by Gauss the *total curvature* (curvatura integra) of the triangle,

$$A + B + C - \pi = \int K \, d\sigma.$$

This formula has a direct connection with Gauss' reflections and computations on non-Euclidean geometry, which he kept secret during his lifetime. In his geodesic measurements he actually found the angles of a huge spherical triangle formed by the summits of three mountains and then computed the angles of the rectilinear triangle having sides of these lengths, but he found no noticeable deviation from 180° in the sum of the angles of the latter.

Minding and the Formulation of the Problems of Intrinsic Geometry

Soon after the publication of Gauss' paper there began a gradually widening development of the problems of the intrinsic geometry of a surface. The first research extending Gauss' ideas on surface theory appeared in Germany. These were the papers of the young F. G. Minding (from 1830 on) and C. G. Jacobi (from 1836 on). But in general these ideas did not draw a large response among the German geometers, since the geometers grouped around Crelle's *Journal* were almost entirely absorbed by questions of projective and algebraic geometry (see below). Gauss' ideas were appreciated more widely, though considerably later, in France, where special attention had been paid to geometric questions since the time of Monge. There the soil had already been prepared for the acceptance of these ideas, and the results of Gauss were subjected to analysis, reworking, and further development in the work of Liouville (from 1847 on) and his students.

We shall first of all describe the contribution of C. G. Jacobi (Bk. 1, p. 69), whose geometric research was carried out under the direct influence of the work of Gauss. Although Jacobi's mathematical research was mainly in the theory of functions and theoretical mechanics, he discussed the problems of intrinsic geometry in his lectures at the University of Königsberg. In his paper "Demonstratio et amplificatio nova theorematis Gaussiani de curvatura integra trianguli in data superficies e lineis brevissimi formate" (*J. für Math.*, 1837) and in his paper "Note von der geodätischen Linie auf einem Ellipsoid und der verschiedenen Anwendungen einer merkwürdigen analytischen Substitution" (*J. für Math.*, 1839) Jacobi integrated the equations of the geodesics on a triaxial ellipsoid. He obtained an important result in global geometry about the behavior of geodesic lines on a surface in the paper "Zur Theorie der Variations-Rechnung und der Differential-Gleichungen," (*J. für Math.*, 1838), in which he introduced the concept of a conjugate point to a point lying on a geodesic line and formulated a condition for the geodesic line to be a shortest line. In his paper "Über einige merkwürdige Curventheoreme," (*Astron. Nachr.*, 1843) Jacobi proved an interesting theorem to the effect that the spherical image of the principal normals of a closed space curve divides the surface of the sphere into two equal parts.

Jacobi's contribution to the geometry of surfaces was not particularly significant. The intrinsic geometry of surfaces, especially the theory of deformation of surfaces, owes its fundamental results to Minding. His contemporaries regarded his work as completely inadequate, and at first passed over it in silence. In Germany Minding's work found no support for the reasons already mentioned, while the French geometers of Liouville's circle seldom cited him, even when they occasionally duplicated his results. We shall give a few details of his biography.

Ferdinand Gottlieb Minding (1806–1885) was the son of a lawyer. Graduating with distinction from the classical gymnasium in Hirschberg, he studied at the University of Halle for one year (1824), where he attended lectures on classical philology and physics. Then at the University of Berlin (1825–1828) he studied philosophy (with Hegel), philology, and natural science. He studied mathematics independently and defended a doctoral dissertation on integral calculus at the University of Halle after one year. In 1830, having passed the appropriate examinations, he received a license to teach as Privatdocent at the University of Berlin.

He began lecturing on algebra, mathematical analysis and mechanics, and from 1834 on he also taught at the Berlin Bauschule, all the while continuing his mathematical research. However his attempt to obtain a professorship at the university ended in failure after ten years of effort. An attempt to appoint Minding to the Berlin Academy of Sciences also failed, although his candidacy was put forward in 1842 by P. Lejeune-Dirichlet. Crelle, Dirksen, and Steiner were proposed for membership at the same time, and the higher-ranking senior candidates were given preference.

Naturally when Minding was offered (on the recommendation of C. G. Jacobi) a professorship of applied mathematics and mechanics at the University of Dorpat (now Tartu), he accepted and left Germany in 1844. His later career was spent in Russia, where he continued his successful research and teaching work for many years. One of his students was K. M. Peterson, who founded the Moscow school of geometry. Minding became a Russian citizen, along with his children. In 1865 he was elected a corresponding member of the Petersburg Academy of Sciences, and he became an honorary member in 1879.

We shall study Minding's work only in the theory of surfaces. Although he is the author of a considerable amount of valuable research in algebra and number theory, integration of algebraic functions, and the theory of differential equations (for which he was awarded the Demidov prize in 1861), his most important discoveries belong to geometry.

In his paper "Bemerkungen über die Abwickelung krummer Linien von Flächen," (*J. für Math.*, 1830), which was published almost immediately after Gauss' paper, the young Minding introduced a valuable complement to the concepts of intrinsic geometry. He proved that the quantity $1/\rho$ equal to the product of the

F. G. Minding

curvature $1/R$ of a curve and the cosine of the angle i formed by the osculating plane to the curve and the tangent plane to the surface, i.e.,

$$1/\rho = \cos i/R,$$

belongs to the intrinsic geometry of the surface. (In modern notation this is the projection of the curvature vector onto the tangent plane.) The quantity $1/\rho$ was later called the *geodesic curvature* by Bonnet (1848). The proof consisted of finding an expression for the geodesic curvature in terms of E, F, G, and their first-order derivatives.

Minding discovered the geodesic curvature by using the methods of calculus of variations to solve the isoperimetric problem: to find the shortest curve on a surface enclosing a given area. In his paper "Über die Curven des kürzesten Perimeters auf krummen Flächen," (*J. für Math.*, 1830) Minding found that if there exists a solution to this problem, then $\cos i/R$ is constant on the extremal, and he conjectured that such a curve must be a geodesic circle. For the case of surfaces of constant curvature he proved this conjecture in the paper just mentioned (it is true when the curvature is positive). The question was not finally decided until 1921, when A. Baule showed that geodesic circles in general have constant geodesic curvature only on surfaces of constant Gaussian curvature.

Subsequently in his paper "Beweis eines geometrischen Satzes," (*J. für Math.*, 1837), Minding proposed an interesting geometric interpretation of geodesic curvature. He showed that it can be defined as the curvature of the planar curve obtained from the given curve if one develops on the plane the developable surface that is the envelope of the family of planes tangent to the surface at the points of the given curve. This valuable idea of Minding of unrolling a line lying on a surface (more precisely a band of the surface) onto a plane was subsequently applied by Levi-Cività to introduce a fundamental concept — parallel translation of a vector along a curve lying on a surface (1917).

Minding's important results in the area of intrinsic geometry are related to the problem of deformation of surfaces, and were published in the years 1838–1840. In a paper bearing the title "Wie sich entscheiden lässt, ob zwei gegebene krumme Flächen auf einander abwickelbar sind, nebst Bemerkungen über die Flächen von unveränderlichen Krümmungsmasse," (*J. für Math.*, 1839) Minding derived definitive conditions under which one surface can be a deformation of another. He gave a nearly exhaustive treatment of the general question of criteria that are necessary and sufficient for developability of surfaces. Moreover he considered a number of fundamental special cases. However, there was a gap in his reasoning in regard to the possibility of singular solutions of a total differential equation. Consequently his requirements turned out to be excessively strict, i.e., certain possibilities were lost. A quarter of a century later this gap was discovered and corrected by O. Bonnet (1865).

Before obtaining the general conditions for deformability Minding first studied the deformation of general ruled surfaces into ruled surfaces in his paper "Über die Biegung gewisser Flächen" (*J. für Math.*, 1838) and showed that any ruled surface can be developed on another whose direction vectors form a circular cone. Then in his paper "Über die Biegung krummer Flächen," (*J. für Math.*, 1838) Minding studied the deformation of a certain family of surfaces of revolution containing the catenoid and was the first to point out that it can be developed on a rectilinear helicoid (an example that has become classical). He proved that arbitrary surfaces of revolution are deformable, stating (without giving an explicit reference) that "a closed convex surface, regarded as a whole, is undeformable, as is well-known,"[2] This conjecture (also found in the posthumous papers of Euler published in 1862) was not proved until 1899 by H. Liebmann for analytic surfaces and by A. V. Pogorelov in the general case (in his paper "Infinitesimal deformations of general convex surfaces," Khar'kov 1959).

The general conditions for developability of two surfaces, in other words the conditions for equivalence of the first fundamental forms of these surfaces, were obtained by Minding by successively introducing the quantities later called differential invariants (or parameters) by Beltrami, who applied them in his papers a quarter of a century later. As a result, based on the application of differential invariants, he noted that the question of developability of two given surfaces can be solved without using integration. In this same paper Minding proved a theo-

---

2) *J. für Math.*, **18** (1838), S. 368.

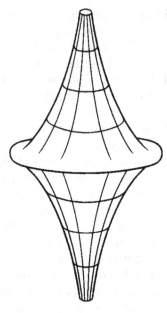

Fig. 2

rem of outstanding importance: *For two surfaces of constant Gaussian curvature,*
equality of the Gaussian curvatures is a sufficient condition for developability, and
the developing can be carried out in an infinite number of ways, depending on three
parameters. In the same paper he found the equations of the surfaces of revolution and heliacal surfaces of constant curvature, and in particular the surfaces of
constant negative curvature. One of these, the surface defined by the parametric
equations

$$r = \sqrt{x^2 + y^2} = \frac{1}{\cosh \varphi}, \quad z = \varphi - \tan \varphi,$$

can be obtained by revolving about the Oz-axis a tractrix, a curve having a cusp
on the Ox-axis which asymptotically approaches the Oz-axis as z tends to $+\infty$ and
$-\infty$ and is characterized by the property that at each of its points the segments
of the tangents from the point of tangency to the Oz-axis are equal (Fig. 2).
The surface of revolution described by Minding is called a *pseudosphere*, following
Beltrami.

Minding's article "Beiträge zur Theorie der kürzesten Linien auf krummen
Flächen," (*J. für Math.*, 1840) contained very valuable material. Minding found
the trigonometric relations in a triangle formed by geodesics (shortest lines) on
surfaces of constant Gaussian curvature K and remarked that these same formulas
can be obtained from the corresponding formulas of spherical trigonometry by replacing the radius of the sphere by $\sqrt{K}$, which is imaginary in the case of negative
Gaussian curvature K. This paper played an important role in Beltrami's interpretation of hyperbolic geometry (1868, see Sect. 4 below). Although Minding's

paper was published in the same journal as Lobachevskiĭ's "Imaginary Geometry," (1837), no mathematician before Beltrami noticed that these relations are the same for rectilinear triangles in hyperbolic geometry and for geodesic triangles in the intrinsic geometry of surfaces of constant negative curvature.

The French School of Differential Geometry

Research in mathematical physics, including the theory of elasticity and theoretical mechanics, which received a major development starting in the early decades of the nineteenth century, required researchers to create or advance a number of areas of mathematics, including geometry.

Typical representatives of this theoretical-applied trend in differential geometry were Lamé and Saint-Venant, who were mentioned above.

Gabriel Lamé, a graduate of the École Polytechnique, was invited to Russia as a young man and worked as a teacher and technical computer in Petersburg in the Civil Engineering Institute (1820–1832). Upon his return to France, he became a professor of the École Polytechnique (1832–1863) and was elected to the Paris Academy of Sciences (1843). Lamé's mathematical discoveries are closely connected with his research in the theory of elasticity and mathematical physics. In his paper "Mémoire sur les coordonnées curvilignes," (*J. math. pures et appl.*, 1840) Lamé was the first to apply curvilinear coordinates in space using an orthogonal system, so that he could write the element of length in the form

$$ds^2 = H^2 \, d\rho^2 + H_1^2 \, d\rho_1^2 + H_2^2 \, d\rho_2^2.$$

Earlier in his paper "Mémoire sur les surfaces isothermes dans les corps solides homogènes en équilibre de température," (*J. math. pures et appl.*, 1837) he introduced in rectangular coordinates the differential parameters (i.e., invariants) of a scalar field

$$\Delta_1 F = \sqrt{\left(\frac{\partial F}{\partial x}\right)^2 + \left(\frac{\partial F}{\partial y}\right)^2 + \left(\frac{\partial F}{\partial z}\right)^2}$$

and

$$\Delta_2 F = \frac{\partial^2 F}{\partial x^2} + \frac{\partial^2 F}{\partial y^2} + \frac{\partial^2 F}{\partial z^2}.$$

Lamé wrote a number of textbooks in mathematical physics, on special functions and their applications, where in particular he introduced the so-called Lamé functions in his *Leçons sur les coordonnées curvilignes et leurs diverses applications* (Paris 1859). These books enjoyed wide popularity and were used as textbooks even outside France.

Adhémard Jean-Claude Barré de Saint-Venant (1797–1886), although he studied mainly the theory of elasticity, in which he discovered the famous Saint-Venant equations, also wrote (in 1843) a detailed exposition, "Mémoire sur les lignes courbes non planes," (*J. Éc. Polyt.*, 1845), on the problem of equilibrium of a wire, which contained a discussion of the history of the problem and some new theorems. In particular he was the first to introduce the term *binormal*, which later caught on, for the perpendicular to the osculating plane. However the Frenet-Serret derivation formulas were unknown at the time.

The name of Liouville, which we have mentioned previously (Bk. 1, pp. 201–204) in connection with his work on transcendental numbers, is of particular interest in connection with the development of differential geometry in France in the 1840's. In his geometric work Liouville not only developed the topics of Monge, but also combined them with the ideas of Gauss, publishing a translation of Gauss' work together with some research of his own as an appendix to his re-edition of Monge's classical work *Application de l'analyse à la géométrie*, 5-ème éd. (Paris 1850). Liouville's annotations comprised one-third of the book. Among the members of Liouville's geometric school were the Sorbonne professors Ossian Bonnet (1819–1892), who was also a famous astronomer; the academician Joseph Alfred Serret (1819–1885), whom we have mentioned in connection with his work in algebra (Bk. 1, pp. 63, 66); the academician Victor Puiseux (1820–1883), famous for his work in analysis, which will be discussed below, and also as an astronomer and mechanic; the academician and professor of the Collège de France Joseph Bertrand (1822–1900), whose work in number theory and probability theory we have discussed in Book 1 (cf. pp. 185–188, 276); Professor Frédéric Frenet (1816–1900) of the University of Lions; and Professor Eugène Catalan (1814–1894) of the University of Liège (Belgium), who is also famous for his work in mathematical analysis. The work of Liouville and his geometric school involved primarily the problems of intrinsic geometry and developability of surfaces. We have already mentioned (rather frequently) that they either duplicated or changed only slightly the results of Minding without citing him. Thus Liouville gave a simpler form for Gauss' expression for the Gaussian curvature and found, duplicating a result of Minding, the equations for surfaces of revolution of both negative and positive constant curvature. Studying the conformal mapping of surfaces in isothermal coordinates, as Gauss had done, Liouville exhibited a remarkable class of surfaces (the Liouville surfaces) for which the linear element can be brought into the form

$$ds^2 = \big(\varphi(u) + \psi(v)\big)(du^2 + dv^2).$$

The geodesic lines on these surfaces can be found by quadratures, and this class contains both the second-order surfaces and the surfaces of revolution (appendix to Monge's book, 1850). Two French geometers, Edmond Bour (1831–1866) and Bonnet, and the Italian geometer Delfino Codazzi (1824–1873), took part in a competition announced for 1860 by the Paris Academy of Sciences on the problem of deformation of surfaces.

The prize went to Bour's paper "Théorie de la déformation des surfaces," (*J. Éc. Polyt.*, 1862), and the others received honorable mention. These papers gave a derivation close to that of Minding of conditions for developability of two surfaces. Bour used semigeodesic coordinates and found a number of new cases of deformation. Bonnet, in his paper "Mémoire sur la théorie des surfaces applicables sur une surface donnée," (*J. Éc. Polyt.*, 1865), sharpened the result of Minding,

G. Lamé

pointing out the gap we mentioned above. He then referred one of the surfaces to asymptotic coordinates, in which the linear element has the simple form

$$ds^2 = \lambda \, du \, dv,$$

but did not manage to obtain any new cases of deformation.

Codazzi's approach, in a paper published only after his death entitled "Mémoire relatif à l'application des surfaces les unes sur les autres," (*Mém. div. sav.* 1883), is similar to that of Bour, and the conditions he obtained involved the geodesic and normal curvatures of parametric lines, which turned out to be useful in later research.

Among the interesting results of Bonnet in the memoir just discussed we must mention a generalization of Gauss' theorem on the sum of the angles of a geodesic triangle. This "Gauss-Bonnet theorem" states that for a simple smooth curve bounding a domain Σ of a regular surface

$$\int\limits_{\Sigma} K \, d\sigma + \oint\limits_{C} \frac{ds}{\rho_g} = 2\pi,$$

i.e., the sum of the "total curvature" of the domain Σ and the integral of the geodesic curvature $1/\rho_g$ over the boundary curve is 2π (Fig. 3).

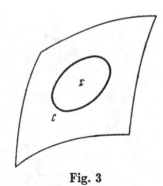

Fig. 3

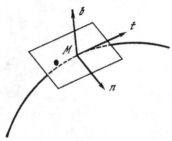

Fig. 4

The crowning achievement in the general theory of space curves was attained soon after the paper of Saint-Venant just mentioned. In his 1847 dissertation, the main portion of which was published as an article "Sur les courbes à double courbure," (*J. math. pures et appl.*, 1852) Frenet, and independently of him J. Serret in his paper "Sur quelques formules relatives à la théorie des courbes à double courbure" (*J. math. pures et appl.*, 1851) obtained the famous "Frenet-Serret formulas" (also called "Frenet's formulas"). The Frenet-Serret formulas, which are analogues of Gauss' formulas for surfaces, connect the direction cosines of the tangent, principal normal, and binormal to a curve and their derivatives with respect to arc length. Nowadays, when the direction cosines of lines are replaced by unit vectors directed along these lines — the vectors $\mathbf{t}$, $\mathbf{n}$, and $\mathbf{b}$ directed along the tangent, principal normal, and binormal of the curve respectively (see Fig. 4 — the components of these vectors in rectangular coordinates coincide with the direction cosines), the Frenet formulas can be written as

$$\frac{d\mathbf{t}}{ds} = k\mathbf{n}, \quad \frac{d\mathbf{n}}{ds} = -k\mathbf{t} + \varkappa\mathbf{b}, \quad \frac{d\mathbf{b}}{ds} = -\varkappa\mathbf{n},$$

where k and $\varkappa$ are the curvature and torsion of the curve; their absolute values are equal to the absolute values of the vectors $d\mathbf{t}/ds$ and $d\mathbf{b}/ds$, i.e., the limits of the ratios of the "angle of intersection" $\Delta\alpha$ between the tangents at two nearby points and the analogous angle $\Delta\beta$ between the osculating planes at these points to the length of the arc Δs between the two points as the arc is contracted to a point. In a plane the curvature is considered positive when the rotation from the vector **t** to the vector **n** is counterclockwise and negative when it is clockwise. In space the curvature is considered to be always positive, and the torsion is considered positive when the osculating helix is right-handed and negative when it is left-handed. The Frenet-Serret formulas can be regarded as differential equations that define the vector **t**, which is the derivative $d\mathbf{x}/ds$ of the radius-vector of a point of the curve with respect to arc length on the curve; consequently the curve itself can be determined from the functions $k = k(s)$ and $\varkappa = \varkappa(s)$, which form the so-called natural equations of the curve. Since the vectors **t**, **n**, and **b** are connected by the additional conditions $\mathbf{t}^2 = \mathbf{n}^2 = \mathbf{b}^2 = 1$, and $\mathbf{tn} = \mathbf{tb} = \mathbf{nb} = 0$; the vector **t** is determined by the choice of initial conditions $\mathbf{t} = \mathbf{t}_0$, $\mathbf{n} = \mathbf{n}_0$, and $\mathbf{b} = \mathbf{b}_0$ at $s = s_0$, i.e., the vector **t** is determined up to a rotation in space and, consequently, the vector **x** is determined up to a rigid motion in space.

The application of these formulas opened a natural route to the study of various questions. Thus, for example, interpreting and finding the curves that Bertrand had found in his "Mémoire sur la théorie des courbes à double courbure," (*J. math. pures et appl.*, 1850) could now be regarded as an exercise for students. However it was a long time before the Frenet-Serret formulas were included in textbooks. Thus they did not appear in the detailed monograph *Théorie nouvelle géométrique et mécanique des lignes à double courbure* (Paris 1860) written by Serret's namesake Paul Serret (1827–1898), a professor at the Catholic University of Paris, nor in the book *Anwendung der Differential- und Integralrechnung auf die allgemeine Theorie der Flächen und der Linien doppelter Krümmung* (Berlin 1872) by Ferdinand Joachimsthal (1818–1861) of the University of Breslau (now Wrocław, Poland), compiled from his lectures during the years 1856–1857 and republished in 1890.

Differential Geometry at Midcentury

Research in differential geometry that was still carried on using classical methods of mathematical analysis began to achieve some success after the mid-nineteenth century not only in France but also in other European countries.

In Britain there were no great creative achievements in this area, and geometers studied more the problems of lines and surfaces of order two, the only exception being the creator of the theory of quaternions William Rowan Hamilton (cf. Bk. 1) who obtained important results in the theory of line congruences in his work on rays. On the other hand it was in Britain that the beautiful geometry textbooks of G. Salmon, which we mentioned above, were created.

In Germany E. Kummer created a general theory of linear congruences (1860), which we shall discuss below. Julius Weingarten (1836–1910), a professor at the Technische Hochschule in Berlin, developed the theory of W-surfaces (for which there exists a functional connection between the two principal radii of curvature) and studied problems of developability of surfaces based on this theory. While doing this, in his paper "Über eine Klasse der Flächen die aufeinander abwickelbar sind," (*J. für Math.*, 1861), Weingarten found expressions for the derivatives $\partial\mathbf{n}/\partial p$ and $\partial\mathbf{n}/\partial q$ in terms of $\partial\mathbf{r}/\partial p$ and $\partial\mathbf{r}/\partial q$, for which the coefficients of these expressions are also functions of the coefficients E, F, G, D, D', and D''. The formulas of Gauss, together with those of Weingarten, constitute a system of differential equations whose unknowns are $\partial\mathbf{r}/\partial p$, $\partial\mathbf{n}/\partial q$, and $\mathbf{n}$, connected in addition by the relations

$$\left(\frac{\partial\mathbf{r}}{\partial p}\right)^2 = E, \quad \frac{\partial\mathbf{r}}{\partial p}\frac{\partial\mathbf{r}}{\partial q} = F, \quad \left(\frac{\partial\mathbf{r}}{\partial q}\right)^2 = G,$$

$$\frac{\partial\mathbf{r}}{\partial p}\mathbf{n} = \frac{\partial\mathbf{r}}{\partial q}\mathbf{n} = 0, \quad \mathbf{n}^2 = 1.$$

If the conditions for integrability of this system are fulfilled, the vectors $\partial\mathbf{r}/\partial p$ and $\partial\mathbf{r}/\partial q$ are determined by the choice of initial conditions, i.e., these vectors are determined up to a rotation in space by giving the coefficients E, F, G, D, D', and D'' as functions of p and q; consequently the vector $\mathbf{r}$ is determined up to a rigid motion in space. Thus the two quadratic forms determine the surface up to a rigid motion.

The theorem that a surface is determined up to a rigid motion by its two quadratic forms was first published by O. Bonnet in his paper "Mémoire sur la théorie des surfaces applicables sur une surface donnée" (1867) already mentioned. This theorem had been proved even earlier by Karl Mikhailovich Peterson (1828–1881), the son of a Latvian peasant and a graduate of the University of Dorpat, where F. Minding was one of his teachers.

This theorem was proved by Peterson in his dissertation "Über die Biegung der Flächen," which was written in 1853 but published (in Russian translation) only in 1952. A brief and very incomplete discussion of this dissertation was given in 1901 by the historian of mathematics P. Stäckel on the basis of a study of the manuscript of the dissertation in the journal *Bibliotheca Mathematica* by A. Kneser, who worked as professor of the University of Dorpat in the years 1889–1900.

Peterson also supplemented the equation of Gauss that connects the coefficients of the first and second fundamental forms of a surface by two more independent equations, which later came to be known as the Codazzi-Mainardi equations, from the names of the Italian geometers who obtained them independently in 1857 and 1868. In his dissertation, based on the three equations just mentioned, Peterson proved a theorem asserting that if the coefficients of two quadratic differential forms (the first must be positive-definite) are connected by such relations, then there exists a surface for which these forms are the first and second fundamental forms, and they determine the surface up to a rigid motion in space. Peterson's subsequent work was carried out while he lived in Moscow, where he carried on active research while teaching mathematics in the German Gymnasium and was one of the founders of the Moscow Mathematical Society. We shall discuss this work of Peterson below.

We note also the work of the German and Swedish geometers on the theory of minimal surfaces, i.e., surfaces of minimal area bounded by a given contour.

Heinrich Friedrich Scherk (1798–1885), a professor of the Hauptschule in Bremen and a student of Jacobi in Königsberg, had found the equations of five minimal surfaces as early as 1835 in his paper "Bemerkungen über die kleinsten Flächen innerhalb gegebener Grenzen," (*J. für Math.*).

In Sweden minimal surfaces were studied by Professor Emmanuel Gabriel Björling (1808–1872), a professor at the University of Uppsala. In his paper "In integrationem aequationis derivatarum partialium superficiei, cujus in puncto uno-quoque principales ambo radii curvedinis aequales sunt signoque contrario," (*Arch. Math. Phys.*, 1844) Björling solved the problem of finding all the minimal surfaces passing through a given strip, i.e., through a family of smoothly varying planes tangent to a given curve. One of the characteristic criteria for a minimal surface is that the "mean curvature" be zero, i.e., the sum of the principal curvatures, and consequently the sum of the principal radii of curvature also, must vanish.

In the course of the mathematical research that began in the context of a general national upsurge and struggle for unification in Italy there were great achievements in the area of differential geometry, some of which were mentioned above. Antonio Maria Bordoni (1789–1860) began to study this subject as early as the 1820's. For many years he was a professor at the University in Pavia. When he became acquainted with the work of Liouville and the ideas of Gauss, he encouraged his colleagues and students to develop them. Among the latter we note Angelo Mainardi (1800–1879) and D. Codazzi, who were mentioned above, as well as Francesco Brioschi (1824–1897), Luigi Cremona (see below) and Eugenio Beltrami (see below). All these mathematicians later became professors of various Italian universities and participated actively in the social and political life of the country. We have mentioned the results of Mainardi and Codazzi in connection with Peterson's dissertation and the competition of the Paris Academy of Sciences. Cremona, whom we shall discuss below, was a student of Chasles and worked mainly in the area of projective and algebraic geometry. Brioschi was the director of the Milan Polytechnic School and conducted research in the area of intrinsic geometry and the theory of determinants. Beltrami was a particularly brilliant mathematician. His professorial career began in 1862 and was spent in Bologna, Pisa, Rome, Pavia, and again in Rome, where he was a member and later president of the National Academy. He worked intensively and productively on the development of differential geometry in the spirit of Gauss, studying problems of developability of surfaces and extending the ideas of Minding mentioned above. In his paper "Ricerche di analisi applicata alla geometria," (*G. mat. Napoli*, 1864), seeking differential invariants for quadratic forms, he introduced two kinds of such invariants generalizing the Lamé parameters and later known as the Beltrami differential parameters. From the modern point of view these invariants are the square-norm of the gradient vector of a scalar field and the divergence of the gradient (i.e., the Laplacian) of a scalar field expressed in the metric of the intrinsic geometry of the surface. We shall return to his interpretation of hyperbolic geometry (1868) below. He also developed the theory of multi-dimensional spaces of constant curvature in the spirit of Riemann (1869), which we shall also discuss below.

K. M. Peterson

Differential Geometry in Russia

Research in the area of differential geometry was also appearing in Russia at this time. The role of K. M. Peterson, who founded Moscow school of geometry, is especially noteworthy.

Peterson's most important paper was "On the ratios and relationships between curved surfaces" (*Mam. C6.*, 1866), devoted to deformation of surfaces, which laid the foundation for a series of papers on the problem of bending on a principal basis, i.e., preserving the conjugacy of a certain net on the surface, the first example of which for deformation of surfaces of revolution on a surface of revolution was found by Minding in the paper already mentioned "Über die Biegung krummer Flächen," (1838). Peterson's paper "On curves on surfaces," (*Mam. C6.*, 1867) and the book *Über Curven und Flächen*, (Moscow/Leipzig 1868) were devoted to differential geometry. Some of the results of these papers of Peterson were later duplicated by G. Darboux and other foreign geometers, but after E. Cosserat's translations of Peterson's main works from 1866–1867 were published in Toulouse in 1905, his work received general recognition. Under Peterson's influence problems of deformation of surfaces were studied by Boleslav Kornelievich Mlodzeevskiĭ (1858–1923), a graduate of Moscow University, a student of the brilliant geometer

O. I. Somov

Vasiliĭ Yakovlevich Tsinger (1836–1907), and later professor of Moscow University and organizer of the Moscow geometric school. These problems were also studied by Dmitriĭ Fëdorovich Egorov (1869–1931), another student of Tsinger. In the early 1920's, together with his student N. N. Luzin, Egorov founded the famous Moscow school of set theory and theory of functions, whose methods of research were to a large degree geometric.

In Petersburg Osip Ivanovich Somov (1815–1876), another graduate of the University of Moscow, developed vector analysis, applying it to geometry and mechanics. In his paper "On higher-order accelerations," (*Зап. Акад. Наук*, 1864) Somov constructed the machinery for differentiation of vector-valued functions and applied it to the study of space curves, using a moving trihedral. In his paper "A direct method of expressing the differential parameters of first and second orders and the curvature of a surface in arbitrary coordinates, rectangular or skew," (*Зап. Акад. Наук*, 1865) certain problems of surface theory are solved by vector methods. The paper of P. L. Chebyshev "Sur la coupure des vêtements" (1878), which is devoted to the problem of "putting clothes on a surface," is famous. This paper introduces the nets known known by Chebyshev's name. We shall discuss below the work of the Kazan' geometer F. M. Suvorov (1845–1911), who studied three-dimensional Riemannian spaces.

The Theory of Linear Congruences

In HM, Vol. 3, we mentioned (cf. p. 193) that the concept of a linear congruence, i.e., a family of straight lines depending on two parameters, was introduced in Monge's paper "Mémoire sur la théorie des déblais et des remblais," (1784). Monge considered only "normal congruences," i.e., congruences normal to a surface. Further stimulus to the development of this theory came from optics. Here we must first of all mention the paper "Optique," (*J. Éc. Polyt.*, 1808) by Étienne-Louis Malus (1775–1812) and Hamilton's paper "Supplement to an essay on the theory of systems of rays," (*Trans. Roy. Irish Acad.*, 1830). Malus' "Optics" contains a great deal of purely optical material (Malus is widely known for the discovery of the phenomenon of polarization of light and other discoveries in optics), but a significant portion of this paper is devoted to the theory of linear congruences. Malus showed that not only normal linear congruences, but also linear congruences of general form have the property that each line of the congruence is the intersection of two orthogonal rotating planes consisting of lines of the congruence. He studied the reflection and refraction of congruences and was the first to study linear complexes, i.e., families of lines depending on three parameters. As to normal congruences, Malus proved that this property is preserved in the reflection of the rays of a congruence from a surface (an error made by Malus in this proof was corrected by Dupin).

Among Hamilton's results on the theory of linear congruences we note "Hamilton's formula," which shows how the "pinch point" of a pair of infinitely near rays of a congruence (the limiting position of the base of the common perpendicular of two rays of the congruence as they approach each other) depends on the direction in the congruence and gives a criterion for the congruence to be normal.

The theory of congruences and the more general "ruled geometry" (the geometry of a manifold of lines and families of lines depending on different numbers of parameters) were founded by J. Plücker in his "System der Geometrie des Raumes in neuer analytischer Behandlungsweise," (*J. für Math.*, 1846). Plücker (who will be discussed below) proposed that the elements of ordinary space be taken as lines rather than points and planes, both of which depend on the same number of parameters (points are determined by three coordinates and planes by the ratios of three coefficients in their equations to the fourth coefficient). Since lines are defined by pairs of linear equations

$$x = rz + \rho, \quad y = sz + \sigma,$$

they depend on four parameters. Plücker invented the terms *complex* and *congruence*. This last term is motivated by the fact that a congruence is formed by the coincident lines of two complexes and is derived from the word *congruens*, meaning *coincident*, reflecting the use of the term *congruence* in Euclidean geometry.

The differential geometry of linear congruences was constructed in analogy with Gauss' differential geometry of surfaces in the paper "Allgemeine Theorie der geradlinigen Strahlensysteme" (*J. für Math.*, 1874) by E. Kummer, whom we mentioned in the chapter on algebra (Bk. 1, pp. 99–106). Kummer considered a linear congruence and a surface intersecting all its lines and defined two quadratic forms, one of which differs from the second fundamental form of the surface in the replacement of the direction cosines of the normal to the surface by the direction cosines of a ray of the congruence, while the second is the metric form for a "spherical mapping" of the congruence analogous to the Gaussian spherical mapping of the surface. It turns out that the two forms of Kummer, like those of Gauss, determine the congruence up to a rigid motion in space.

2. PROJECTIVE GEOMETRY

The Rise of Projective Geometry

Projective geometry acquired a definitive form as an independent discipline in the first half of the nineteenth century, although its prehistory dates back to ancient times, as can be seen from the surviving work of Pappus of Alexandria from the third century (HM, Vol. 1, p. 151). Later the elements of projective concepts gradually came together in books on perspective by artists and architects during the Renaissance (HM, Vol. 1, pp. 321–323, Vol. 2, p. 121). Certain projective concepts occur in the writings of Kepler (HM, Vol. 2, pp. 117–121), Newton (HM, Vol. 2, pp. 127–128), and Leibniz (HM, Vol. 2, p. 126). In the seventeenth century G. Desargues (1639) and after him Blaise Pascal (1640) established a number of important theorems relating to the projective properties of figures, although the subject of projective geometry itself was not yet clearly defined in their writings (HM, Vol. 2, pp. 121–128). After Desargues and Pascal there was no significant progress in the study of projective properties for more than a century and a half, although individual concepts were being clarified. Thus in the works on perspective by Brooke Taylor (1719) and Johann Lambert (1759) it was shown that all the infinitely distant points of a plane lie on a single infinitely distant line (HM, Vol. 3, p. 196). A paper of E. Waring (1762) introduced an analytic notation for a collineation in a plane (HM, Vol. 3, p. 173).

The new progress in this topic in the early nineteenth century was closely connected with the career of Gaspard Monge. His descriptive geometry revived interest in synthetic methods and attracted attention to the method of projection.

Lazare Carnot, Charles Brianchon, Jean Poncelet, Joseph Gergonne, and Michel Chasles, all of whom made important contributions to the development of projective geometry, attended the lectures of Monge, the first in Mézières, the others at the École Polytechnique in Paris.

Lazare Carnot (1753–1823), a prominent activist in the French bourgeois revolution, nicknamed "the organizer of victory," was a soldier and mathematician who published three works on geometry: *De la corrélation des figures en géométrie* (Paris 1801), *Géométrie de position* (Paris 1803), and *Essai sur les transversales* (Paris 1806) (HM, Vol. 3, pp. 198–200), in which he introduced some ideas and concepts that were essential for the rise of projective geometry. Using the concept of a continuous transformation of a figure (a "correlation" in his terminology), he

L. Carnot

stated the so-called principle of continuity ("principle of correlation") according to which certain properties of the transformed figure can be discovered and studied from those of the original (even when the "correlation" leads to imaginary quantities). It is essential to note that he was the first to introduce the cross (anharmonic) ratio of four points of a line taking account of its sign, thereby sharpening Pappus' concept. He then proved that this ratio is invariant for the four points obtained by cutting four lines of a pencil of lines with different secants. In this way he established the harmonic properties of a complete quadrilateral.

Carnot always emphasized the superiority of the synthetic method over the analytic, which was being widely applied at that time, since the former makes it possible to encompass all possible special cases at once.

Charles Julien Brianchon (1783–1864), an artillery captain and later professor in the artillery school, discovered and proved in his memoir "Sur les surfaces courbes du second ordre," (*J. Éc. Polyt.*, 1806) the theorem on the hexagon circumscribed about a conic section dual to Pascal's theorem. In doing this he relied on the polar correspondence with respect to the conic section, thereby taking the first step toward the general principle of duality, which was as yet unknown.

The theory of the polar correspondence was then developed in 1810 by François Joseph de Servois (1767–1847) and in 1812 by Joseph Diez Gergonne

(1771–1859), a professor of mathematics in Nîmes and Montpellier, the founder of the journal *Annales de mathématiques* (usually called *Gergonne's Annals*); Servois introduced the term *pole* and Gergonne the term *polar*.

Poncelet's *Traité des propriétés projectives des figures*

The definition of projective geometry as the study of the projective properties of figures and a systematic exposition of its basic concepts and theorems, which had tremendous influence on its subsequent development, were first given by Poncelet (HM, Vol. 3, p. 201).

Jean Victor Poncelet (1788–1867), a military engineer, studied at the École Polytechnique, and his interest in problems of projective geometry is undoubtedly due to the influence of Monge, whose student he was, and Carnot. Having worked for a time in the École d'Applications, a military school in Metz, at the beginning of 1812 Poncelet was drafted into Napoleon's army; he took part in the Russian campaign and was taken prisoner in November 1812. Poncelet spent two years in Saratov, where the enforced leisure enabled him to organize his plans for projective geometry. He presented his results to those of his fellow prisoners who were graduates of the École Polytechnique. In 1815 he returned to Metz and expounded his results in the *Traité des propriétés projectives des figures* (Paris 1822). The subtitle of the treatise, "A work of utility for those studying the applications of descriptive geometry and geometric operations on land," shows clearly the significant influence of Monge's methods of descriptive geometry on the development of Poncelet's thought.

The publication of the *Traité* was preceded by individual papers of Poncelet published in *Gergonne's Annals* (*Ann. Math.*, 1817–1818), and a memoir "Sur les propriétés des sections coniques," presented to the Paris Academy in 1820, for which the referee's report by Cauchy was published in the same journal (1820–1821). Later Poncelet worked as professor at the École d'Applications in Metz, published a *Cours de Mécanique* in 1826, and moved to Paris in 1835. In Paris he became professor of mechanics at the Sorbonne and the rector of the École Polytechnique. In 1865–1866 he published the second edition of the *Traité des propriétés projectives des figures*.

Poncelet's *Traité* produced a very strong impression on geometers, and its title led to the use of the term "projective geometry."

Poncelet's works contain in explicit the concept of projective properties of plane figures in explicit form, i.e., properties that are preserved under projections and sections. The subject of projective geometry was thereby characterized by the synthetic method. Poncelet called two figures "projectives" if one can be mapped into the other by a chain of projections. Poncelet defines a projective figure as "a figure whose parts have only graphical interrelations...," i.e., relations that are not annihilated by projection, and he defines projective relations or properties to be

J. V. Poncelet

"any relations or properties that hold simultaneously for a figure and a projection of it."[3]

Poncelet showed that a conic section (conic) is a projective figure and that to solve a difficult problem in conics, one should project the conic to circle, solve the problem for the circle, and then carry out the inverse projection. Since the "points of convergence" of parallel lines on the "mapped plane" do not correspond to real points of the projective plane, Poncelet added "ideal" or "infinitely distant" points to all planes, points that project to "points of convergence." Poncelet introduced infinitely distant points using Carnot's principle of correlation, which he called the "principle of continuity." Developing an idea of Carnot on "complex correlation," Poncelet introduced imaginary points of the plane, and, in particular, imaginary infinitely distant points, such as, for example, "cyclic points" — points belonging to all circles in the plane. Two conics can intersect in four real or imaginary points, and two circles in two real and two cyclic points. Poncelet showed that the foci of a conic are the points of intersection of the tangents drawn to it

---

3) Poncelet, J. V. *Traité des propriétés projectives des figures, ouvrage utile à ceux qui s'occupent des applications de la géométrie descriptive et d'opérations géométriques sur la terrain.* Paris 1865, pp. 4–5.

from the cyclic points. Poncelet's book also studied the "projective properties" of three-dimensional figures. Along with the projective properties of figures Poncelet introduced "projective-metric" concepts, whose definition involves the concept of length (such as, for example, the cross ratio). On the basis of these concepts he developed the subject of projective properties of rectilinear figures (on a plane) and conic sections (including Pascal's and Brianchon's theorems).

The central projection of one plane onto another was used to introduce the concept of a projective correspondence (or transformation) of two planes, which Poncelet called a *homography*. This concept was introduced using central projection of a plane onto a plane along with projective transformations of space, in particular homologies. Here the concept of an infinitely distant plane in space was used. Particular attention was paid to the polar transformation, by means of which the duality principle, expressing the equivalent roles of points and lines on the plane, operates. The term *principe de dualité* itself is due to Gergonne, who quarreled with Poncelet over the priority for the establishment of this general law, later justifying it in his paper "Sur quelques lois générales qui régissent les lignes et les surfaces de tous les ordres" (*Ann. Math.*, 1826–1827), no longer resorting to the polar correspondence, but relying on the elementary propositions of projective geometry.

Thus Poncelet's treatise led to the culmination of the initial process of formation of projective geometry. The subject matter of this topic was defined and its basic concepts, laws, and most important theorems, obtained by the application of the synthetic method, were established. However, the definition of the cross ratio contained the metric concept of length of a line segment.

The Analytic Projective Geometry of Möbius and Plücker

Starting in the late 1820's and continuing for nearly half a century the problems of projective geometry became the central occupation of a number of German geometers: A. F. Möbius, J. Steiner, J. Plücker, O. Hesse, Ch. Staudt, and others, who were united by their collaboration in Crelle's *Journal für die reine und angewandte Mathematik*. This area was represented in France by J. V. Poncelet, J. D. Gergonne, M. Chasles, and others, and in Britain by A. Cayley, G. Salmon, and others.

In the papers of Poncelet, Steiner, and Chasles the synthetic method was so closely connected with the subject of projective geometry itself that it was frequently called synthetic geometry. However it was not long until the application of analytic methods began in projective geometry, beginning with the papers of Möbius and Plücker. These methods came to be used more and more widely, although geometers frequently had doubts about the validity of introducing homogeneous coordinates, since their use involved nonprojective concepts.

Homogeneous coordinates, which make it possible to characterize the infinitely distant points of the plane, were introduced in 1827 by Möbius, moreover by a very peculiar method based on the concepts of geometric statics. Independently of Möbius and almost simultaneously with him Karl Wilhelm Feuerbach

(1800–1834) in Germany, the brother of the philosopher Ludwig Feuerbach, and E. Bobillier (1797–1832) in France, were publishing papers using homogeneous coordinates. We shall take the time to discuss first the development of projective geometry by the "analysts" Möbius and Plücker, and then pass to the ideas and results of the "synthetists" Steiner and Chasles.

As a student August Ferdinand Möbius (1790–1868), the son of a dancing teacher in Schulpforta, attended the lectures of Gauss on astronomy at the University of Göttingen (1813–1814), and from 1816 on worked first as an astronomer-observer and later director of the Pleissenburg astronomical observatory in Leipzig. Later on he was simultaneously a professor of mathematics at the University.

Möbius published a large number of individual papers and notes containing valuable and beautiful mathematical results, a two-volume *Lehrbuch der Statik*, (Leipzig 1837) and the book *Der barycentrische Calcül* (Leipzig 1827), which is outstanding in its wealth of profound mathematical ideas. The name of Möbius' book is connected with the "barycentric coordinates" of points introduced in it: if masses m_1, m_2, m_3 are located at the vertices of a fixed triangle $E_1E_2E_3$, then the center of gravity (the barycenter) of these masses can be characterized by the numbers m_1, m_2, m_3 up to a common factor. These numbers, called "barycentric coordinates," are a special case of homogeneous coordinates of points. If the masses m_1, m_2, and m_3 are considered positive, these coordinates are defined only for the interior points of the triangle. The vanishing of one coordinate corresponds to a point of the sides of the triangle, while for points of the plane outside the triangle at least one of the three masses m_1, m_2, m_3 must be assigned a negative value.

Nowadays homogeneous coordinates of points are introduced by choosing a point S outside the plane $E_1E_2E_3$ and directing vectors $\mathbf{e}_1$, $\mathbf{e}_2$, and $\mathbf{e}_3$ along the rays SE_1, SE_2, SE_3. An arbitrary vector $\mathbf{m}$ of space can be represented as a sum $m_1\mathbf{e}_1 + m_2\mathbf{e}_2 + m_3\mathbf{e}_3$. The homogeneous coordinates of a point of the plane are the coefficients m_1, m_2, and m_3 of the decomposition of an arbitrary vector $\mathbf{m}$ directed along a ray SM, in terms of the vectors $\mathbf{e}_1$, $\mathbf{e}_2$, $\mathbf{e}_3$. Möbius' barycentric coordinates can be regarded as the limiting case of such coordinates as the point S recedes to infinity; in this case the vectors $\mathbf{e}_1$, $\mathbf{e}_2$, and $\mathbf{e}_3$, which are directed along parallel lines passing through the points E_1, E_2, E_3 should be regarded as free vectors, and the coordinates m_1, m_2, and m_3 belong to the point of intersection of the plane with the line of the free vector $m_1\mathbf{e}_1 + m_2\mathbf{e}_2 + m_3\mathbf{e}_3$, equivalent to a system of vectors differing from the vectors $\mathbf{e}_1$, $\mathbf{e}_2$, $\mathbf{e}_3$ by the coefficients m_1, m_2, and m_3. Möbius defined analogous barycentric coordinates in space as well by using a tetrahedron $E_1E_2E_3E_4$. Using homogeneous coordinates one can characterize not only ordinary points of the plane, but also the infinitely distant points of the plane when it is completed to the projective plane. In the case of barycentric coordinates the coordinates of an infinitely distant point of the line E_1E_2 are the numbers m and $-m$. To obtain an infinitely distant point of an arbitrary line passing through the point E_1 one must find its point of intersection E_1' with the line E_2E_3, and placing an arbitrary mass at the point E_1, one must place two masses at E_2 and E_3 whose barycenter is the point E_1', while the sum of the masses at the points E_1, E_2, and E_3 is zero.

A. F. Möbius

Barycentric coordinates enabled Möbius to state a whole series of affine and projective properties of two- and three-dimensional figures. In this connection a significant portion of *Der barycentrische Calcül* is devoted to affine and projective transformations, which he was the first to write analytically in homogeneous coordinates, and to similarity transformations and isometries. What we now call a geometric transformation, i.e., a one-to-one correspondence between two figures or regions of space, Möbius called a "relationship" (*Verwandschaft*), possibly under the influence of Euler's term "affinitas" (HM, Vol. 3, p. 168). Möbius called an affine transformation an "affine relationship" or an "affinity." A projective transformation, the most general transformation of the plane under which points lying on the same line (collinear points) map to points having the same property, he called a "collinear relationship" or "collineation." Following an ancient tradition, Möbius called congruent systems of points equal or similar, and he considered not only systems that can be made to coincide by a continuous transformation but also those obtained from each other by reflection from a plane. Möbius showed without using any metric concepts that collineations on the plane are determined by arbitrarily prescribing two corresponding quadruples of points, no three of which are collinear (in three-dimensions two analogous quintuples of points).

The theory of free vectors in the form of mechanical forces, which is closely connected with the barycentric calculus, is the subject of the *Lehrbuch der Statik* mentioned above, which is also heavily permeated with geometric material. In particular, null-systems were considered, special cases of correlative transformations of space that map points into planes in such a way that collinear points map into planes passing through a single line. Null-systems — transformations under which points map into planes passing through these points — are next in simplicity to polar transformations, i.e., transitions from points to their polar planes with respect to second-order surfaces. Before Möbius null-systems had been considered by the Italian geometer Gaetano Giorgini (1795–1874) in his paper "Sopra alcune proprietà de' piani de' momenti," (*Mem. soc. Ital. sci. Modena*, 1828).

In varying lengths, surfaces, and volumes Möbius made systematic use of the principle of signs, the sign being determined according as the direction of circuit of the quantity being measured does or does not coincide with a preassigned "positive" direction. As a result he was the first to give a complete theory of the cross-ratio of a quadruple of points on a line and to prove that it is invariant under projective transformations.

Möbius was the first to introduce unicursal curves (whose coordinates are defined by rational functions of a parameter). The name given to these curves is motivated by the fact that on the projective plane these curves can be drawn without lifting the pen. In particular Möbius found rational parametric representations of the conic sections. Following this route, he was the first to consider third-order curves in three-dimensional space and study their properties.

Julius Plücker (1801–1868), a native of Elbersfeld, studied at the universities of Bonn and Paris and defended his doctoral dissertation in Bonn in 1825. From 1828 to 1831 he was extraordinarius of this university and from 1832 to 1834 a professor at the University of Berlin, which he left after being subjected to attacks from the disciples of Steiner, who were defending the purity of the synthetic method. In 1836 he obtained the position of ordinarius in Bonn in two departments, mathematics and physics. From that time on he conducted active research in these two fields. Thus in the field of physics he also made such remarkable discoveries as the phenomenon of crystallogmagnetism (1847) and the spectra of electric charges (1857). While studying the influence of a magnetic field on charges he came very close to discovering cathode rays, a discovery completed by his pupil Hittorf. However, his most remarkable achievements were in geometry. Although the research in his early papers is based on the synthetic method, by 1828 he was developing his "analytic-geometric method."

Plücker's main ideas and results were expounded in the five following books:

1. *Analytisch-geometrische Entwicklungen*, (Essen 1828–1831).
2. *System der analytischen Geometrie*, (Berlin 1835).
3. *Theorie der algebraischen Curven*, (Bonn 1839)
4. *System der Geometrie des Raumes in neuer analytischen Behandlungsweise*, (Bonn 1846).
5. *Neue Geometrie des Raumes gegründet auf Betrachtung der geraden Linien als Raumelement*, (Leipzig 1868–1869).

In his *Analytisch-geometrische Entwicklungen* Plücker, unlike Möbius, introduced completely general homogeneous projective coordinates on the plane.

J. Plücker

Moreover the route he followed was different. From the analytic point of view these coordinates are simply three independent linear homogeneous functions of the three homogeneous Cartesian coordinates of the points, and are determined up to a common factor. From the geometric point of view, as Plücker showed, the three numbers x_1, x_2, x_3 are proportional (with an arbitrary coefficient of proportionality) to the distances to the sides of an arbitrarily given triangle, each measured with its own unit of length. These coordinates came to be known as "triangular" or "trilinear" coordinates. Similarly in the *System der analytischen Geometrie* Plücker introduced homogeneous projective coordinates in three-dimensional space.

The analytic method made it easier to operate with infinitely distant and imaginary elements and naturally led Plücker to introduce line coordinates, i.e., coordinates of a straight line. If the equation of the line has the form $u_1x_1 + u_2x_2 + u_3x_3 = 0$, then the triple of numbers (u_1, u_2, u_3) will be the line (or tangential) coordinates of this line. Regarding this equation as a condition for the variable point (x_1, x_2, x_3) to lie on a fixed line (u_1, u_2, u_3) or as a condition for the variable line (u_1, u_2, u_3) to pass through the fixed point (x_1, x_2, x_3), Plücker obtained immediately from the symmetry of the expression $u_1x_1 + u_2x_2 + u_3x_3$ a justification for the Poncelet-Gergonne duality principle.

In a similar manner Plücker introduced planar coordinates in three-dimensional space and justified the duality principle in space, in which a point corresponds to a plane.

In his geometric research he applied with great skill the method of abbreviated notation that we have discussed above.

An essentially new idea introduced by Plücker in geometry following the analytic route was the possibility, first pointed out and realized by him, of taking geometric figures other than points as the elements of space, for example, lines, circles, etc. When this is done, manifolds of higher dimension arise naturally (for example, the manifold of circles in a plane is three-dimensional, the manifold of spheres is four-dimensional, the manifold of lines in space also has four dimensions, etc.). The six homogeneous coordinates $p_{ij} = x_i y_j - x_j y_i$ of a straight line in space passing through the points (x_i) and (y_i) (these coordinates are connected by a single quadratic relation) later came to be known as the "Plücker" coordinates, even though they had been introduced earlier by Grassmann in his *Lineale Ausdehnungslehre* (1844).

Plücker made an important contribution to the general theory of plane algebraic curves of higher orders and to the theory of geometric transformations, which will be discussed below.

The Synthetic Projective Geometry of Steiner and Chasles

The largest contribution to the development of projective geometry by synthetic methods after Poncelet was made by Steiner and Chasles. The work of Steiner appeared first, after which the results of the two geometers frequently overlapped.

Jacob Steiner (1796–1863) was born into the family of a Swiss pastor and graduated from primary school in 1815. Inspired by the reforming pedagogical ideas of his teacher Pestalozzi, he worked for a time as a teacher in his pedagogical institute, then moved to Heidelberg (1818–1821), where he studied mostly the work of the French geometers, giving private lessons to support himself. Steiner believed that geometry was best constructed through intense concentration, by contemplating imagined figures. He rejected the application of not only algebra and analysis, but also drawing. The interest in Pestalozzi's pedagogy that arose in Berlin, motivated Steiner to move to Berlin, where he obtained a position as a teacher, taught with great success, and engaged in intensive research, publishing his work in Crelle's *Journal* starting in the first year of its publication (1826). His fundamental work was published in 1832. In 1834 Steiner was elected to the Berlin Academy of Sciences, and a year later he was called to a professorship at the University of Berlin. In his later life, when researching the algebraic geometry of higher-order figures, he frequently published only the statements of his theorems without proof and without any citation of the sources he used.

According to Steiner's program the system of projective geometry should be developed using successive passages from simpler elementary linear geometric shapes to more complicated shapes, and then by use of projective relations, to figures of higher orders. He expounded these ideas in his fundamental work, *Systema-*

J. Steiner

tische Entwicklung der Abhängigkeit geometrischer Gestalten voneinander (Berlin 1832). Figures of the first level are a rectilinear series of points (a row), a pencil of lines, and a pencil of planes; figures of the second level are points and lines of a single plane or a pencil of lines and planes; those of the third level are the points and planes of three-dimensional space. He then established that the basic figures of the first level are projective. Projectivity is characterized by the equality of the cross ratio of corresponding elements, which is introduced using metric quantities. The principle of duality was widely applied. The constructive realization of the projective relations using perspective, i.e., projections and sections, was studied.

Using two projective pencils of lines on a plane a figure of higher order was constructed — a line (or series) of second order — as the set of points of intersection of the corresponding rays. A circle was considered first, and the result was then carried over to the general case of a conic section. Figures of second class (a second-order pencil) were introduced by the duality principle. As a result series and pencils of first order generated different series and pencils of second order (for example, a hyperboloid of one sheet is generated by two projective pencils of planes) and a new route was opened to the study of their various properties.

An important defect in the work of Steiner was the absence of a distinct application of the principle of signs and the use of imaginary elements.

In Steiner's projective geometry the relative weight of constructive problems solvable using only a straightedge or a straightedge and a fixed circle was very high. Problems of these types are equivalent to the construction of the solutions of linear and quadratic equations respectively. Problems of the second type include constructing the fixed points of a projective correspondence in a row. Since the problems of constructing a finite number of points that can be solved using a compass and straightedge are problems of these two types, it follows that every such problem can be solved using only a straightedge given that one circle has been drawn. Such constructions, as we have seen, had been studied by Lambert and Poncelet and are a special case of the constructions of al-Farabi, Abu'l Wefa, and Leonardo da Vinci (HM, Vol. 1, pp. 230, 323). Steiner devoted a special book *Die geometrischen Constructionen ausgeführt mittels der geraden Linie und eines festen Kreises* (Berlin 1833) to these constructions.

Other works of Steiner that deserve to be mentioned include his paper on geometric maxima and minima (*J. für Math.*, 1842), in which the isoperimetric problem in particular was beautifully studied by elementary methods; however, no proof of the existence of a solution is given, since the necessity of such a proof had not yet been noticed.

Michel Chasles (1793–1880) became a student at the École Polytechnique in time to catch the last years of Monge's teaching, and while there he wrote a paper on the hyperboloid of one sheet (1813). However he then abandoned the scholarly life, settled in his provincial homeland, took up banking, and accumulated a vast fortune. Only some 25 years later, in 1837, did his remarkable work *Aperçu historique sur l'origine et le développement des méthodes de la géométrie* (Paris 1837) appear. It aroused wide interest and contained a number of original results.

In 1841 Chasles became a professor of mechanical engineering at the École Polytechnique, and in 1846 a chair of higher geometry was established at the Sorbonne for him. In 1839 Chasles was elected a corresponding member of the Paris Academy of Sciences, and in 1851 he became an Academician. His most intensive research and teaching activity began at the Sorbonne and continued for more than thirty years. After publishing a number of papers and his *Traité de la géométrie supérieure* (Paris 1852), he became one of the leading French geometers. For that reason what happened to him in the 1860's was all the more regrettable. Taking an interest in the history of science, Chasles acquired from a certain person, among other old manuscripts, the original of an unknown letter of Pascal, in which the latter supposedly anticipated Newton's theory of gravitation. Chasles began publishing sensational excerpts from these letters in the *Comptes Rendus* of the Paris Academy of Sciences. These excerpts evoked ever-increasing doubts and objections in the scholarly world. A few years later it turned out that Chasles had been the victim of a clever trickster: expert testimony proved that all the documents this man had sold were forgeries.

In his scholarly research Chasles followed the traditions of French geometers, applying synthetic methods. He may have arrived at his system of projective geometry on his own, independently of Steiner, but the works of these two have much in common. Thus the foundation for both was the invariance of the cross-ratio of four points of a line and four lines of a pencil, called by Chasles the *anharmonic ratio*, and this relation itself was introduced using metric concepts. Chasles, however, like

M. Chasles

Möbius, in studying the lengths of directed line segments, made clear and distinct use of the principle of signs, so that his theory received a more complete exposition than did that of Steiner. Like Steiner, Chasles studied the conic sections, regarding them as formed using two projective pencils or series; unlike Steiner, however, he frequently spoke of geometric relations in an imaginary domain and regarded the real parts of figures as having been accidentally separated from the imaginary parts. Chasles introduced into projective geometry the general concept of a *correlation* — a correlative transformation of space, special cases of which are a polar transformation and a null-system. He also gave a detailed exposition of the theory of collineations, which, following Poncelet, he called "homographies."

An especially important role in arousing interest in projective geometry was played by Chasles' textbook of higher geometry and his *Aperçu historique sur l'origine et le développement des méthodes de la géométrie*, far more than half of which consisted of his original remarks. The book was very soon translated into a number of European languages, including Russian. Chasles was the first to exhibit the role of the many predecessors of modern scholars in the foundation of projective geometry — Euclid, Pappus, Desargues, and others. He analyzed the rise of the study of polar correspondence and the principles of continuity and duality. He exhibited the profound connection between particular individual research papers

Ch. von Staudt

and emphasized the natural idea that new discoveries were prepared for by the works of earlier scholars. However Poncelet and several other French geometers were hostile to Chasles' book, believing that he was only trying to disparage the value of their work.

Staudt and the Foundation of Projective Geometry

In the mid-nineteenth century there was an acrimonious controversy between the proponents of the synthetic and analytic methods in projective geometry, the two sides accusing each other of mixing projective and metric concepts. Indeed the basic concept that is applied in the synthetic presentation of projective geometry, the cross-ratio of four points of a line, was introduced through consideration of the lengths of intervals. On the other hand the introduction of projective coordinates was also based either on concepts of the distances to the sides of a triangle or on formulas applied in statics, or in a slightly different manner, but always invoked rectangular or skew Cartesian coordinates, whose definition involved metric concepts. Thus the problem arose of freeing projective geometry from the admixture of metric concepts contained in its foundations.

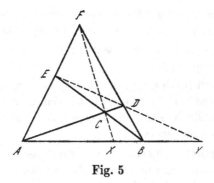

Fig. 5

This problem attracted the attention of Staudt, and he studied it for many years, as a result of which he achieved significant progress in founding a purely projective geometry independent of any metric concepts.

Christian von Staudt (1798–1867), a native of Rothenburg and the scion of a noble family, was, like Möbius, a student of Gauss, who guided him into astronomy and number theory. Staudt first worked at the Gymnasium, first in Würzburg, and then in Nürnberg; from 1835 until the end of his life he was a professor at the University of Erlangen. In Erlangen he concentrated on geometric problems. Here he wrote the book *Geometrie der Lage* (Nürnberg, 1847), whose title underscores its descent from Carnot's book; he also wrote three *Beiträge zur Geometrie der Lage* (Nürnberg 1856, 1857, 1860). Where the projective geometry of Poncelet, Möbius, Plücker, Chasles, and Steiner was closely connected with elementary geometry, and a projective invariant of four points of a line or four lines of a pencil, such as the cross ratio of these four points or lines, was defined in terms of the ratio of lengths of intervals on the line or the sines of the angles between the lines of a pencil, Staudt succeeded in freeing projective geometry from elementary geometry. In his *Geometrie der Lage* Staudt introduced a harmonic quadruple of elements independently of the concept of the cross ratio following a purely projective route, using a complete quadrangle or quadrilateral (Fig. 5). The projective nature of the two forms of the first step is characterized by the preservation of harmonicity. To deduce the fundamental theorem that a projective correspondence of two figures of the first level can be defined by arbitrarily giving two points of the corresponding elements he proved a preliminary lemma on the double points of two projective series lying on the same line, namely that the number of double points in this case cannot be more than two if the correspondence is not an identity; it should be noted that the proof was carried out with insufficient rigor, since the axiomatization of continuity was not yet established at that time. From this lemma and theorem it follows that the new definition of a projective correspondence is equivalent to the definition given by Steiner, which relied on the preservation of the cross ratio.

Staudt went on to define projective correspondences of figures of the second and third levels — collineations and correlations. For example a collineation of two planes was introduced as a pointwise one-to-one correspondence that preserves the rectilinearity of a row. Correlations were defined similarly.

A polarity was distinguished as a special case of a correlation of two plane fields with a common support.

A conic section (taking account of imaginary figures) was introduced as the set of points belonging to their polars in a given polarity. Surfaces of second order were introduced similarly. The projective properties of nondegenerate second-order figures were studied on this basis.

Following a purely projective route Staudt associated with each point of the projective line either a real number or the symbol ∞. The construction of such a projective scale is based on the fact that arithmetic operations $x \mapsto x + a$ and $x \mapsto ax$ on the usual number line can be regarded as affine transformations, and consequently as special cases of projective transformations. Therefore, assigning to any three distinct points of the projective line the symbols 0, 1, and ∞, for any two points corresponding to numbers a and b one can construct points corresponding to the numbers $a + b$, $a - b$, ab, and a/b, and consequently points corresponding to all integers and rational numbers, after which points corresponding to irrational numbers can be constructed by passage to the limit. In this way Staudt was able to give a purely projective definition of homogeneous coordinates on the plane and in three-dimensional space by means of purely projective operations. To define such coordinates on a plane, as Staudt showed, it suffices to give the vertices of a triangle $E_1 E_2 E_3$ and some point E not lying on the lines $E_i E_j$. In this case, assigning the symbols 0 and ∞ to the points E_i and E_j, and the symbol 1 to the intersection of the line $E_i E_j$ and EE_k, we define a projective scale on the line $E_i E_j$. Similarly to define such coordinates in three-dimensional space it suffices to define the vertices of a tetrahedron $E_1 E_2 E_3 E_4$ and a point E not lying in the planes of $E_i E_j E_k$. The geometric calculus defined by Staudt came to be known as *Wurfsrechnung*, since Staudt had given the name *Wurf*, meaning *throw*, to numbers that correspond to a quadruple of points on a line in a projectively invariant way. It can be shown that a Wurf is the cross ratio of the given quadruple of points, but a rigorous proof of this fact eluded Staudt for the reasons given above.

Relying on the construction of a rational net and introducing projective coordinates using Wurfs without any metric concepts, Staudt was still unable to define the irrational values of the coordinates with sufficient rigor. This defect was noted and subsequently corrected by F. Klein in his paper "Über die sogenannte Nicht-Euklidische Geometrie," (*Math. Ann.*, 1871–1872).

In his *Beiträge* Staudt showed that imaginary points, which had previously been introduced in a purely analytic manner, could also be defined synthetically using elliptic involutions on a line, i.e., involutive projective transformations of the line having no real fixed points. We remark that an elliptic involution is defined by a real figure, namely by two arbitrary pairs of real points that separate each other (the double points of an involution separate both of these pairs harmonically). To distinguish one of a pair of complex-conjugate points from the other Staudt proposed defining a preferred direction on the line of support of the involution.

Cayley's Projective Geometry

In Britain the development of projective geometry came to rely on analytic methods.

The textbooks of G. Salmon on analytic geometry, which we mentioned above, contained a great deal of material on projective geometry; Salmon is also the author of very popular textbooks on the algebraic geometry of higher-order curves and the theory of invariants (*Higher Plane Curves*, Dublin 1852; *Modern Higher Algebra*, Dublin 1859). A. Cayley and J. J. Sylvester (who were primarily algebraists) and Salmon formed the "invariant triplets," so named because of their work on invariant theory.

Arthur Cayley obtained exceptionally important results in the theory of algebraic forms that he created and in the coordinate exposition of projective geometry. After his geometric research it became especially clear that the justification of purely projective geometry was of fundamental importance.

Cayley introduced the so-called projective metric in his paper "A sixth memoir upon quantics," (*Philos. Trans.*, 1859). After certain supplements and clarifications of F. Klein (1871) it followed from this paper that not only Euclidean geometry, but also other metric geometries called non-Euclidean (the geometries of Lobachevskiĭ and Riemann) could be regarded as special forms of general projective geometry, when the space is supplemented by a certain fixed second-order figure, the "absolute" in Cayley's terminology (for more details see below). It should be noted that even before Cayley a student of Chasles, Edmond Laguerre (1834–1886), had found a projective form for the size of an angle in a Euclidean plane in his "Note sur la théorie des foyers," (*Nouv. ann. math.*, 1853): According to "Laguerre's formula" the angle φ between two straight lines can be expressed in terms of the cross-ratio W of these lines and two conjugate-imaginary "isotropic lines" (lines joining the point of intersection of the given lines with the cyclic points of the plane of these lines) by the relation

$$\varphi = \frac{l}{2} \ln W.$$

In this sense projective geometry contains metric geometry, and hence the justification of it should be constructed independently of metric concepts in order to avoid a logical vicious circle.

Cayley himself avoided this error, developing projective geometry in a purely analytic manner. For example, he defined a point in the plane to be a triple of numbers (x_1, x_2, x_3), not all zero and determined up to a common nonzero factor; a line (u_1, u_2, u_3) was defined similarly. The condition for the point to lie on the line was given by the equality $u_1 x_1 + u_2 x_2 + u_3 x_3 = 0$, etc.

The subsequent development of mathematics showed the superiority of analytic methods in geometry. Despite the beauty of the proofs of certain theorems by synthetic methods; despite the intuitiveness and great pleasure that this approach offers in the contemplation of geometric figures; finally, despite the great assistance of spatial intuition in the solution of particular problems, on the whole these methods are cumbersome and inaccessible and frequently preclude the formulation and solution of more general problems.

Thus, for example, profound general principles of classification of individual geometric disciplines according to groups and subgroups of transformations, as we shall see, were obtained by F. Klein using analytic methods in his "Erlanger program" of 1872.

3. ALGEBRAIC GEOMETRY AND GEOMETRIC ALGEBRA

Algebraic Curves

Analytic methods were applied with 'great success in the algebraic geometry of curves and surfaces of higher order and in the development of the theory of invariants, where other methods happen to be inapplicable. The mathematicians Jacobi, Hesse, Grassmann, Aronhold, Cayley, Sylvester, Salmon, Clebsch, Gordan, Beltrami, Cremona, and others worked in this area.

In the eighteenth century third- and fourth-order curves were discussed in a number of books on analytic geometry after the theory of second-order curves (HM, Vol. 3, p. 164). Courses of analytic geometry in the nineteenth century included only the topic of lines and conics, while the theory of algebraic curves was combined with projective geometry. Nowadays, as a rule, algebraic curves are studied on the projective plane rather than the affine or Euclidean plane, and for simplicity of computation this plane is assumed complex. A classical exposition of such a theory is Plücker's *Theorie der algebraischen Curven* (1839). Introducing linear (tangential) coordinates of lines u_1, u_2, u_3, Plücker considered the equation

$$\varphi(u_1, u_2, u_3) = 0,$$

which defines a one-parameter family of lines, as the equation of the envelope of the family. This equation is called *tangential* since the curve here is defined not by its points but by its tangents. The degree of the point equation of this curve is called its *order*, and the degree of its tangential equation is called its *class*. Plücker showed that on the complex projective plane the order (n) of a curve, its class (m), the numbers of its nodes (δ), cusps $(\varkappa)$, bitangents (ι), and points of inflection (τ), in the absence of any other singularities, are connected by the formulas

$$m = n(n - 1) - 2\delta - 3\varkappa, \qquad n = m(m - 1) - 2\iota - 3\tau,$$

$$\tau = 3n(n - 2) - 6\delta - 8\varkappa, \qquad \varkappa = 3m(m - 2) - 6\iota - 8\tau,$$

which now bear his name.

Comparing the number of parameters on which the curve depends with the number of constants in the equation, Plücker discovered a number of errors made by Euler in his classification of fourth-order curves.

In his paper "Theorie der Abel'schen Functionen" (*J. für Math.*, 1857) Riemann introduced an important characteristic of plane algebraic curves; he denoted this characteristic by the letter p. Alfred Clebsch (1833–1872), a professor at the University of Göttingen and one of the founders of the journal *Mathematische Annalen*, called the quantity p the *genus* (*Geschlecht*) of the curve in his paper "Über diejenigen ebenen Curven, deren Coordinaten rationale Functionen eines Parameters sind" (*J. für Math.*, 1864). Riemann showed that when $p = 0$, the coordinates of a curve can be expressed as rational functions of one parameter; when $p = 1$ they can be expressed as elliptic integrals, and for $p > 1$ by hyperelliptic Abelian integrals. The curves of order zero are the unicursal curves studied by Möbius; curves of genus 1 are called *elliptic* or *bicursal*.

In his paper "Über die Singularitäten algebraischer Curven," (*J. für Math.*, 1865) Clebsch showed that for curves for which Plücker's formulas hold the genus is related to the numbers n, m, δ, $\varkappa$, ι, and τ by the formulas

$$p = \frac{(n-1)(n-2)}{2} - \delta - \varkappa = \frac{(m-1)(m-2)}{2} - \tau - \iota.$$

Among other papers on the theory of plane curves we note the paper of Gustave Roch (1839–1866), a professor at the University of Halle, "Über die Anzahl der willkürlichen Constanten in algebraischen Functionen," (*J. für Math.*, 1865), in which the author, while developing the ideas of Riemann's "Theorie der Abel'schen Functionen," proved the theorem now known as the Riemann-Roch Theorem.

Algebraic Surfaces

In 1849 Cayley and Salmon each published a paper on triple tangent planes to third-order surfaces (*Cambridge and Dublin Math. J.*). In his paper Cayley established that any smooth cubic surface contains a certain number of lines, and Salmon showed that the number of these lines is 27. A paper of J. Steiner "Über die Flächen dritten Grades" (*J. für Math.*, 1856) is devoted to the synthetic theory of cubic surfaces. The 27 lines on a cubic surface soon became the subject of special research by Steiner (*Monatsber. Preuss. Akad. Wiss.*, 1857), L. Schläfli, whom we shall discuss below in connection with his work on multi-dimensional geometry (*Quart. J.*, 1858), J. de Jonquières (*Nouv. ann. math.*, 1859), and J. J. Sylvester (*Comptes Rendus*, 1861), whom we mentioned in the chapter on algebra (Bk. 1, 82–84). While cubic surfaces had at first been considered in complex space, Schläfli also considered such surfaces in real space, in which he distinguished (according to the real or imaginary nature of the lines and planes in which coplanar triples of these lines lie) five classes of real surfaces. The 27 lines on a smooth cubic surface in complex space can be described as follows: the equation of such a surface can be brought to the canonical form $ace - bdf = 0$, where a, b, c, d, e, f are linear polynomials. If we denote the intersection of the planes $a = 0$ and $b = 0$ by the symbol ab, then nine of the 27 lines on the surface in question are the lines ab, ad, af, cb, cd, cf, eb, ed, and ef. If we denote the three lines belonging to the surface and intersecting the lines ab, cd, and ef by $(ab, cd, ef)_i$, where $i = 1, 2, 3$, the other 18 of the 27 lines are the lines $(ab, cd, ef)_i$, $(ad, cf, eb)_i$, $(af, cb, ed)_i$, $(ab, cf, ed)_i$, $(ad, cb, ef)_i$, and $(af, cd, eb)_i$.

In the papers just mentioned and their successors it was proved that these 27 lines constitute 45 coplanar triples and 36 so-called double sextuples, also discovered by Schläfli. These "double sextuples" have the property that each line of each of the sextuples fails to intersect one of the lines of each of the other sextuples, but does intersect the other five.

Triple tangent planes to these surfaces had been had been studied previously in the early papers of Cayley and Salmon. In the course of several decades of the nineteenth century a large number of properties of these surfaces were found. The five forms of smooth real cubic surfaces discovered by Schläfli have respectively 27, 15, 7, 3, and 3 real lines and 45, 15, 5, 7, and 13 real planes of coplanar lines.

The significance attributed to these surfaces in the nineteenth century can be seen from Sylvester's words to the effect that the "cubic icosaheptagram" ought to be engraved on the tombstones of Cayley and Salmon, just as the cylinder, cone, and sphere were engraved on Archimedes' tomb. After 1869, when Christian Wiener (1826–1896), a professor of descriptive geometry at the Technische Hochschule in Karlsruhe, succeeded in making a model of a cubic surface with all 27 lines on it, Sylvester announced that this was one of those discoveries "which must forever make 1869 stand out in the Fasti of Science."[4] The Soviet mathematician Yu. I. Manin, who studied the problem of the 27 lines on a cubic surface from a modern point of view in his monograph *Cubic Forms* (1972), has written: "The configuration of the twenty-seven lines on a smooth cubic surface has been the subject of entire books... Their elegant symmetry is at once awe-inspiring and annoying."[5]

The general theory of algebraic surfaces in complex space was constructed by Max Noether (1844–1921), a student of Clebsch, a professor of the University of Erlangen, and father of the famous algebraist Emmy Noether (1882–1935), in his papers "Zur Theorie des eindeutigen Entsprechens algebraischer Gebilde" (*Math. Ann.*, 1870–1875) and "Extension du théorème de Riemann-Roch aux surfaces algébriques," (*Comptes Rendus*, 1886) and others. It was also developed in the papers of Federigo Enriques (1871–1946), professor of the Universities of Bologna and Rome, in his *Introduzione alla geometria sopre le superficie algebraiche* (Rome 1896), and by several other scholars. After the appearance of multi-dimensional geometry the theory of algebraic manifolds of spaces arose, which is rather more algebraic than geometric in nature.

---

4) Henderson, A. *The Twenty-seven Lines upon the Cubic Surface*. Cambridge 1911, pp. 2–3. See also Manin, Yu. I. *Cubic Forms* [in Russian]. Moscow 1972, pp. 195–196.
5) Manin, Yu. I. *Op. cit.*, p. 121.

Geometric Computations Connected with Algebraic Geometry

A special topic in algebraic geometry, called "enumerative geometry," grew out of Plücker's method of computing the numbers of parameters of algebraic curves and their equations. This topic was founded by Professor Hermann Schubert (1848–1911) of the "Johanneum" in Hamburg, who wrote a book entitled *Kalkül der abzählenden Geometrie* (Leipzig 1879).

In his paper "Die n-dimensionalen Verallgemeinerungen der fundamentalen Abzahlen unseren Raumes," (*Math. Ann.*, 1866) Schubert extended these methods to multi-dimensional geometry and in particular established that the dimension of the manifold of all m-dimensional planes of n-dimensional space — the so-called Grassmann manifold (see below) is $(m+1)(n-m)$. He also found the dimensions of the "Schubert manifolds" discovered by him; these are manifolds of planes having intersections of given dimensions with a nested system of fixed planes (such a system of planes is now called a *flag manifold*). Schubert's research was continued by Hieronymus Georg Zeuthen (1839–1920), a professor at the University of Copenhagen and better known as an historian of mathematics, in his *Lehrbuch der abzählenden Methoden der Geometrie* (Leipzig 1914). The principles of computing parameters, which were insufficiently justified in the work of Schubert, were justified using topological methods in 1929 by the famous algebraist (also a well-known historian of mathematics) B. L. van der Waerden.

In the book *Ausdehnungslehre* (1862) by H. Grassmann, which we shall discuss in more detail below, an original geometric calculus was proposed, which in its essence belongs to projective geometry and is an outgrowth of Maclaurin's "organic geometry" (HM, Vol. 3, p. 156). If x is an arbitrary point of a conic and a, b, and c are three fixed points of the same conic, while A and B are two fixed lines in the plane of the same conic, then by finding the intersection of the line xa with the line A, the line joining the resulting point of intersection to b, the point of intersection of the resulting line with the line B, and the line joining the resulting point of intersection to c, Grassmann found that this last line again passed through x, so that any point x of the conic could be constructed in this way. Grassmann expressed this fact as the vanishing of the "planimetric product" $xaAbBcx = 0$ (Fig. 6). This method can be applied to define and construct any algebraic curve. For example, a third-order curve is characterized by the equality $xaAbxCcDdx = 0$.

Grassmann's *Lineale Ausdehnungslehre*

We have already mentioned that a significant portion of Möbius' "barycentric calculus" is taken up by the study of "affine relationship." We note that Möbius made systematic use not only of the oriented lengths of line segments, but also of oriented areas of plane figures and oriented volumes of solids, where the "sign" of the area of a triangle is determined by the order in which its vertices or the sides emanating from a single vertex are listed, while the "sign" of a the volume of a tetrahedron is determined by the order in which the edges emanating from

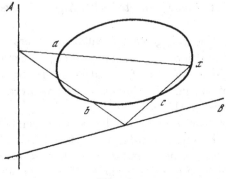

Fig. 6

a vertex are listed. Möbius showed that both the ratios of the oriented lengths of line segments and the ratios of oriented areas and volumes are preserved under affine transformations.

An important role in the establishment of affine geometry as a separate branch of geometry was played by Grassmann's *Lineale Ausdehnungslehre* (Leipzig 1844).

Hermann Grassmann (1809–1877) was born in Stettin (now the city of Szczecin in Poland) to the family of a pastor. There was a family tradition of interest in science and the arts. He first studied theology and philology in Berlin, and began to study mathematics independently in 1832. Also in Berlin Grassmann began to work as a mathematics teacher. In 1836 he returned to Stettin, where he worked as a mathematics teacher in the Gymnasium for the rest of his life. Grassmann devoted his leisure time to various sciences. His mathematical research was the most significant, but he also studied physics: the subject of electric current, the theory of color and the theory of vocal sounds, these last papers being quite highly esteemed by the famous physicist and physiologist Helmholtz, who, as we shall see below, also studied geometry. Grassmann was also a famous linguist — an expert in Sanskrit and the author of a glossary to the *Rig Veda*. He compiled a collection of German folk songs; at the same time he was the editor of a newspaper in Stettin, an active Mason, and a religious activist. Grassmann made a significant mark in all areas of his multi-faceted activity; the only area in which Grassmann was not at the summit of his field was teaching in the Gymnasium. He was happy when only a few students came to hear him, and he encouraged the others to amuse themselves by every means available to them.

The main concepts introduced in the *Lineale Ausdehnungslehre* were "extended figures," "extended magnitudes," and "systems" of various levels — we would now say of various dimensions. An "extended figure" of some level is what is called in modern terms a manifold, and a "system" is a vector space. Grassmann first defined an extended figure of first level as "the collection of elements through which a generating element passes under continuous motion," and, in particular, he defined a "simple extended figure" as one "obtained under continuous extension

H. Grassmann

of the same basic variation," i.e., an oriented arc of a continuous line, a special case of which is a segment of a straight line. Line segments are considered equal if they are generated by "the same variation," and to each class of equal oriented line segments an "extended magnitude or extension of first level" is assigned, i.e., a free vector. Grassmann also defines a "system of first level" as the "collection of elements that can be obtained by continuing the same or opposite variation," i.e., an abstract line. Grassmann went on to define a "system of second level," i.e., an abstract two-dimensional plane:

> If I now start with two basic variations of different kinds and subject an element of the first basic variation (or the one opposite to it) to an indefinite continuation, and subject the element thus changed to a second kind of variation, also continued indefinitely, I call the collection of elements so formed a *system of second level.*[6]

Then in a completely analogous manner Grassmann defines "systems" of third and higher levels, i.e., three-dimensional and multi-dimensional spaces:

---

6) Grassmann, H. *Gesammelte mathematische und physikalische Werke.* Leipzig 1894, Bd. 1, Erster Teil, S. 47.

If I then take a third basic variation that does not map the same initial element into an element of this system of second level and which I shall therefore call independent of the first two and subject an arbitrary element of this system of second level to this third variation (or the one opposite to it), continued indefinitely, then the collection of elements thus formed is a system of third level.[7]

The plane of ordinary space is a system of second level, and "the whole infinite space" is a system of third level.

To any two "elements" α and β Grassmann assigned the "interval" $[\alpha\beta]$ and proved the following theorem: *If $[\alpha\beta]$ and $[\beta\gamma]$ are arbitrary variations, then $[\alpha\gamma] = [\alpha\beta] + [\beta\gamma]$.*[8] It is obvious that Grassmann's "intervals" are bound vectors, and his "variations" are free vectors. Grassmann applied these concepts to "geometry" (three-dimensional space) by assigning to each pair of points X, Y of this space the "interval" $[XY]$, and to mechanics, using "intervals" to represent velocities, accelerations, and forces.

An important peculiarity of Grassmann's multi-dimensional geometry is the "outer product" of intervals: "the outer product of n intervals is the extended quantity of level n obtained when each element of the first interval generates a second, each element so generated generates a third, etc."[9] In other words the outer product of two intervals is a parallelogram, the outer product of three intervals is a parallelepiped, while the outer product of n intervals is what is nowadays called an n-dimensional parallelepiped. Grassmann applied the outer product of two and three intervals to the determination of the area of a parallelogram and the volume of a parallelepiped, and also to the determination of the static moment of force and the condition for equilibrium of forces in mechanics.

Subsequently Grassmann issued a new book under the shorter title *Die Ausdehnungslehre* (Leipzig 1862), which was a significantly revised version of the original *Lineale Ausdehnungslehre*. In this book he introduced the concept of linear dependence of quantities

$$\mathbf{a} = \beta\mathbf{b} + \gamma\mathbf{c} + \cdots$$

($\mathbf{a}$ is produced numerically from the quantities $\mathbf{b}$, $\mathbf{c}$,... using the numbers $\beta, \gamma, \ldots$), "units" (linearly independent basis elements), "extensive quantities" (expressions produced numerically from a system of units that Grassmann wrote in the form $\alpha_1\mathbf{e}_1 + \alpha_2\mathbf{e}_2 + \cdots$ or in abbreviated form $\sum \alpha e$), the sum and difference of extensive quantities $\sum \alpha e \pm \sum \beta e = \sum(\alpha \pm \beta)e$, the product of an extensive quantity by a number $\sum \alpha e \beta = \sum(\alpha\beta)e$; the "inner product" $\mathbf{a}|\mathbf{b}$ of two extensive quantities, and the "outer products" $[\mathbf{ab}]$, $[\mathbf{abc}]$... of two or several extensive quantities. Grassmann's "extensive quantities" are our vectors of an abstract vector space. Specific pictures of a directed line segment, which Grassmann called a *Stab* (*staff*), were connected with such vectors. The "inner product" of two extensive quantities

---

7) *Ibid.*, S. 52.

8) *Ibid.*, S. 56.

9) *Ibid.*, S. 89–90.

is essentially the dot product, while the outer products of two or three extensive quantities are respectively the cross product and the vector triple product. Grassmann denoted vectors by boldface Latin letters, a convention that remains to this day.

Besides the multiplication of "extensive quantities" Grassmann defined in this book the particular product of points and lines that we have discussed above. Grassmann's books of 1844 and 1862, despite their nearly identical titles, are quite different from each other. Where verbal statements predominate in the first book, the second book is written in the language of formulas; where the first book expounds (multi-dimensional) affine geometry, the second contains multi-dimensional Euclidean geometry.

Grassmann regarded his vector calculus as a realization of Leibniz' ideas on "analysis situs" (HM, Vol. 2, pp. 126–127), which was mentioned in connection with projective geometry (HM, Vol. 3, p. 200) and will be discussed further in connection with topology; Grassmann's calculus seems to be closest to Leibniz' program. Grassmann devoted a special book to an idea of Leibniz, his *Geometrische Analyse geknüpft an die von Leibniz erfundene Geometrische Charakteristik* (Leipzig 1847).

Hamilton's Vectors

W. R. Hamilton arrived at the concept of a vector independently of Grassmann (Bk. 1, pp. 74–76). The term *vector* itself is due to Hamilton. For Hamilton the concept of a vector was closely connected with the quaternions he had introduced, and he expounded the vector calculus systematically in his *Lectures on Quaternions* (1853). Hamilton regarded quaternions as the formal sums of real numbers, which he called "scalars" (since such numbers could be arranged like the steps of a ladder, a scale), and vectors $ai + bj + ck$ (from the Latin word *vector*, meaning *carrier*, since such expressions determined the displacement from the point (x, y, z), which he called the *vehend* (*thing to be carried*) to the point $(x + a, y + b, z + c)$, called the *vectum* (*thing carried*). These three terms were analogous to the terms *divisor*, *dividend*, and *quotient*, but only the term *vector* caught on.) Hamilton regarded a quaternion α as the sum of a scalar $S\alpha$ and a vector $V\alpha$, and pictured vectors by directed line segments going from the "vehend" to the "vectum." In contrast to Grassmann, for whom the applications of vector analysis to affine geometry were distinct from its applications to metric geometry, Hamilton's vector calculus had a metric character from the very beginning. Multiplying two vectors $\alpha = a_1 i + a_2 j + a_3 k$ and $\beta = b_1 i + b_2 j + b_3 k$, Hamilton obtained the quaternion $\alpha\beta = S\alpha\beta + V\alpha\beta$. The scalar part

$$-S\alpha\beta = a_1 b_1 + a_2 b_2 + a_3 b_3$$

was later called the "scalar product" of the vectors by J. W. Gibbs, and the vector part

$$V\alpha\beta = (a_2 b_3 - a_3 b_2)i + (a_3 b_1 - a_1 b_3)j + (a_1 b_2 - a_2 b_1)k$$

the "vector product." Hamilton made wide application of vector algebra in mechanics and physics. He was the first to write the condition for equilibrium of systems of forces defined by the vectors β_i applied at the points with radius vectors α_i as the two vector equalities $\sum \beta_i = 0$, $\sum V\alpha_i\beta_i = 0$.

Besides vector algebra Hamilton also founded vector analysis. By introducing the differential operator $\nabla = i\dfrac{\partial}{\partial x} + j\dfrac{\partial}{\partial y} + k\dfrac{\partial}{\partial z}$, which he called *nabla* from the biblical *nebela*, a musical instrument resembling a harp, of triangular shape. Hamilton defined the gradient of a scalar field a and the divergence and curl of a vector field α as the formal products ∇a, $S\nabla\alpha$, and $V\nabla\alpha$.

Hamilton's vector analysis was applied to the theory of an electromagnetic field by the British physicist James Clerk Maxwell (1831–1879) in his *Treatise on Electricity and Magnetism* (London 1873), in which he predicted the existence of electromagnetic waves, which were subsequently discovered by Heinrich Hertz (1857–1894) and made the basis of radioelectronics. Maxwell's treatise attracted the attention of physicists to vector calculus; and the vector calculi of Hamilton and Grassmann were combined and vector algebra was given its modern form in the books *Elements of Vector Analysis* (New Haven 1881–1884) by Josiah Willard Gibbs (1839–1903), a professor of physics at Yale University, and *Electromagnetic Theory* (London 1903) by the British engineer and physicist Oliver Heaviside (1850–1925), who lived nearly his whole life as a layman, but became a member of the British Royal Society in 1891.

Hamilton also introduced linear vector functionals, which he denoted

$$\Phi\xi = \alpha \cdot S\xi\beta + \alpha' \cdot S\xi\beta' + \alpha'' \cdot S\xi\beta'' + \cdots,$$

and showed that a sum of this type can always be reduced to the sum of three terms. At present functionals of this type are written in the form

$$\Phi\mathbf{x} = \mathbf{a} \cdot \mathbf{b}\mathbf{x} + \mathbf{a}' \cdot \mathbf{b}'\mathbf{x} + \mathbf{a}'' \cdot \mathbf{b}''\mathbf{x}.$$

and the operator $\Phi = \mathbf{a}\cdot\mathbf{b} + \mathbf{a}'\cdot\mathbf{b}' + \mathbf{a}''\cdot\mathbf{b}''$ is called a linear operator. It is also called an *affinor* and a *homography*, and an expression of the form $\Phi = \mathbf{a} \cdot \mathbf{b}$ is called the *dyadic, operator,* or *tensor* product of the vectors $\mathbf{a}$ and $\mathbf{b}$ (the expression $\mathbf{a} \cdot \mathbf{b}$ is called a *dyad*). The linear operators form a ring isomorphic to the ring of matrices of order equal to the number of dimensions of the space. Hamilton proved a number of theorems on linear operators equivalent to Cayley's theorems on matrices. In particular, among them was the Cayley-Hamilton equation

$$\Phi^3 - (\mathrm{Tr}\,\Phi)\Phi^2 + (\mathrm{Inv}\,\Phi)\Phi - (\mathrm{Det}\,\Phi)I = 0,$$

a relation satisfied by linear operators on three-dimensional space and 3×3-matrices.

Linear operators and operator products arose again in Gibbs' *Elements of Vector Analysis*.

4. NON-EUCLIDEAN GEOMETRY

Nikolaĭ Ivanovich Lobachevskiĭ and the Discovery of Non-Euclidean Geometry

The attempts over many centuries to prove Euclid's fifth postulate led to the appearance of a new geometry in the early nineteenth century, distinguished from Euclidean geometry by the fact that the fifth postulate does not hold. This geometry is now known as *hyperbolic* or *Lobachevskiĭ* geometry, after the name of the first man to publish an exposition of it.

Nikolaĭ Ivanovich Lobachevskiĭ (1792–1856) was born in Nizhniĭ Novgorod in the family of a minor functionary. Having lost her husband early, Lobachevskiĭ's mother secured her son's acceptance into the Kazan' Gymnasium. After graduating in 1807 Lobachevskiĭ entered the newly opened University of Kazan', with which he remained in association to the end of his life. A great influence on Lobachevskiĭ was a friend of Gauss, Professor M. F. Bartels (1769–1836), who was invited to Kazan' in 1808. Bartels later worked at the University of Dorpat (now Tartu). The brilliant young student Lobachevskiĭ annoyed the reactionary university administration with his "unrealistically high opinion of himself, stubbornness, and insubordination," as well as his "disturbing escapades," in which the author of one report on him discerned "signs of atheism."[10] However the professors, especially Bartels, defended the refractory student, and in 1811 Lobachevskiĭ successfully completed his university studies and obtained the master's degree. Having become an instructor at the university, Lobachevskiĭ continued to work under the supervision of Bartels for a while. In 1816 he was appointed extraordinary professor and in 1822 he became ordinary professor. In 1820 he became dean of the faculty of physics and mathematics, and in 1827 rector of the university. In this post, which he occupied until 1845, Lobachevskiĭ manifested great organizational talent. He saved the university after a fire and an epidemic of cholera. The majority of University buildings were built under his leadership, and the library that now bears his name was completed. Lobachevskiĭ also exerted great influence on teaching in nearly all departments. In 1845 he ceased to work at the University, but he remained one of the leaders of the extensive educational district of Kazan' to the end of his life.

One of the prerequisites to Lobachevskiĭ's geometric discoveries was his materialistic approach to problems of knowledge. Lobachevskiĭ was firmly convinced that a material world exists objectively, independently of human consciousness and of the possibility of knowing it. In his speech "On the most important subjects of education" (Kazan', 1828) Lobachevskiĭ quoted with sympathy the words of Francis Bacon:

> Give up striving in vain to extract all of wisdom from the mind alone; ask nature, who contains all truth and will answer without fail and satisfactorily all your questions.

---

10) Kagan, V. F. *Lobachevskiĭ* [in Russian]. Moscow/Leningrad 1948, p. 58.

N. I. Lobachevskiĭ

Lobachevskiĭ went on to point out that the very laws of logical inference are reflections of real regularities of the world:

Reason means the known laws of reasoning in which the first active causes of the Universe are, as it were, imprinted and which thus reconcile all our conclusions with natural phenomena.[11]

In his essay "On the foundations of geometry," which was the first publication of the geometry he had discovered, Lobachevskiĭ wrote,

The first concepts with which any science begins must be clear and reduced to the smallest possible number. Only then can they serve as a secure and sufficient basis for knowledge. Such concepts are acquired from the senses; one should not trust innate ideas.[12]

Lobachevskiĭ thereby rejected the idea of the a priori nature of geometric concepts advocated by Immanuel Kant, from which it had been deduced that the only conceivable geometry was that of Euclid.

11) Modzalevskiĭ, L. B. *Material for a Biography of N. I. Lobachevskiĭ* [in Russian]. Moscow/Leningrad 1948, p. 323.

12) Lobachevskiĭ, N. I. *Collected Works* [in Russian]. Moscow 1946, Vol. 1, p. 186.

Lobachevskiĭ's first geometric work, bearing the title "Geometry" and written in 1823, was printed only after his death. This original instructional aid reflects Lobachevskiĭ's thoughts on the foundations of geometry. One of his attempts to prove the fifth postulate also occurred about this time.

By 1826 Lobachevskiĭ had arrived at the conviction that the fifth postulate is independent of the other axioms of Euclidean geometry and on 11(23) February 1826 he presented a paper at a faculty meeting under the title "An abbreviated exposition of the foundations of geometry with a rigorous proof of the theorem on parallels," in which he discussed the beginnings of the "imaginary geometry," he had discovered, as he called the system later known as Lobachevskiĭ (hyperbolic) geometry. The 1826 paper became part of Lobachevskiĭ's first publication on non-Euclidean geometry, which bore the title "On the foundations of geometry" and was printed in the journal of Kazan' University, the *Kazan' Messenger*, in 1829–1830. His memoirs "Imaginary geometry," "An application of imaginary geometry to certain integrals," and "New foundations of geometry with a complete theory of parallels," all published in the *Research Notes of Kazan University* in 1835, 1836, and 1835–1838 respectively, were devoted to the further development and applications of the geometry he had discovered. The revised text of "Imaginary geometry" appeared in French translation in Crelle's *Journal für Mathematik* in Berlin. Also in Berlin a separate book by Lobachevskiĭ in German, entitled *Geometrische Untersuchungen zur Theorie der Parallellinien*, appeared in 1840. Finally, in 1855 and 1856 he published *Pangeometry* in Kazan' in both Russian and French. Lobachevskiĭ's geometry received general recognition from mathematicians only after his death. Lobachevskiĭ's colleague at the University of Kazan', Pëtr Ivanovich Kotel'nikov (1809–1879) proclaimed openly in his *Antrittsrede* in 1842:

> I cannot pass over in silence the fact that millennia of diligent attempts to prove one of the basic theorems of geometry with complete mathematical rigor, that the sum of the angles of a rectilinear triangle is two right angles, has inspired a venerable distinguished professor of our university to undertake a marvelous labor — to construct the entire subject of geometry on a new assumption: the sum of the angles of a rectilinear triangle is less than two right angles — a labor that will sooner or later be appreciated.[13]

Gauss valued highly the "Geometrische Untersuchungen," which earned Lobachevskiĭ corresponding membership in the Göttingen Academy (formerly the Academy of Sciences of the Duchy of Hannover) in 1842. However, Gauss did not express his opinion on the new geometry in print.

---

13) *A Survey of the Lectures Given at Kazan' University during the 1842–1843 Academic Year. A speech of Professor Ordinarius Kotel'nikov and report for the year 1841–1842* [in Russian]. Kazan' 1842, pp. 17–18.

Gauss' Research in Non-Euclidean Geometry

Gauss' high opinion of Lobachevskiǐ's discovery was due to the fact that Gauss had been interested in the theory of parallels since the 1790's and had arrived at the same conclusions as Lobachevskiǐ. Gauss never published his views on this question; they have been preserved only in his notebooks and in a few letters to friends. In 1799 Gauss wrote to his fellow student at the University of Göttingen Farkas (Wolfgang) Bólyai (1775–1856) of his studies of the theory of parallel lines.

> To be sure I have achieved a great deal, which would pass for a proof of the fifth postulate for most people, but in my eyes this proves exactly nothing: for example, if one could prove that a rectilinear triangle exists having area larger than any prescribed area, I would be able to prove all of geometry with complete rigor. Most people consider this an axiom, but I do not. Thus, it might be that the area is always less than a certain bound, no matter how far into space the vertices of the triangle are extended. I have many such propositions, but I do not find any of them satisfactory.[14]

In 1804 Gauss wrote to F. Bólyai of the latter's attempted proof of the fifth postulate in his *Theoria parallelarum* (Marosvásárhely 1804): "Your method does not satisfy me... However I still hope that some time before I die these underwater rocks will allow us to cross on them."[15] As can be seen, at that time Gauss had not yet abandoned his attempts to prove the fifth postulate. In 1816, in a letter to the astronomer C. L. Gerling (1788–1864), after establishing that if the fifth postulate is rejected there must exist an absolute unit of length, Gauss announced: "I find nothing contradictory in this. It would even be desirable if Euclidean geometry were not true, since we would then have a common measure a priori."[16] These words show that in 1816 Gauss still considered Euclidean geometry "true" in the sense of physical reality. But by 1817, in a letter to the astronomer W. Olbers (1758–1840), Gauss wrote,

> I am becoming more and more convinced that the necessity of our geometry cannot be proved, at least by human reason and for human reason. It may be that in the next life we shall arrive at views on the nature of space that are now inaccessible to us. Until then geometry cannot be placed on the same level with arithmetic, which exists purely *a priori*, but rather with mechanics.[17]

From this one can see the source of Gauss' doubts: he had originally been a proponent of Kant's opinion on the a priori nature of mathematical concepts; but by reflecting on the theory of parallels he arrived at the conclusion that in any case there is no such a priori nature in geometry. Possibly this is the reason Gauss did not publish his paradoxical discoveries. In a letter to Gerling in 1818 he wrote:

---

14) *On the Foundations of Geometry* [in Russian]. pp. 101–102.

15) *Ibid.*, p. 102

16) *Ibid.*

17) *Ibid.*, p. 103.

"I am glad that you have the courage to speak as if you recognized the falseness of our theory of parallels and along with it all of our geometry. But the wasps whose nest you are stirring up will fly straight for your head."[18] Gauss' reference to "stirred-up wasps" apparently referred to the proponents of traditional views of geometry and of the a priori nature of mathematical concepts.

János Bólyai

Non-Euclidean geometry was discovered independently of Lobachevskiĭ and Gauss by the remarkable Hungarian mathematician János Bólyai (1802–1860), son of Farkas Bólyai. János Bólyai was born in the Transylvanian city of Marosvásárhely (now Tirgu-Mures in Romania). After graduating from the military/engineering academy in Vienna he served in the fortress of Temesvár (now Timişoara). J. Bólyai became interested in the problem of parallels under the influence of his father and as early as 1823 wrote to him, "To be sure I have not yet reached my goal, but I have obtained rather remarkable results — I have created an entire world out of nothing."[19] His father, who had despaired of his own attempts to prove the fifth postulate, implored his son to abandon these studies:

> You must not try to master the theory of parallel lines by this route: I know this route, I have gone over it to the end, I have lived through this dark night and buried every ray of light, every joy of my life in it. I implore you, leave the subject of parallel lines in peace; you should avoid it as you would avoid sensuality; it will deprive you of your health, your leisure, and your peace. It will destroy all the joy of your life. This dense fog is capable of swallowing up a thousand towering Newtons and will never lift from the earth. The miserable human race will never attain perfect truth, even in geometry![20]

After J. Bólyai had arrived at the same ideas as Lobachevskiĭ and Gauss, his father did not understand him. Nevertheless he proposed publishing a brief exposition of the discovery in an appendix to his textbook of mathematics, which appeared in 1832. The full name of J. Bólyai's work was "Appendix scientiam spatii absolute veram exhibens: a veritate aut falsitate Axiomatis XI Euclidis (a priori haud unquam decidenda) independentem." It is usually called *The Appendix* for short. In 1833 J. Bólyai retired; his discovery was not recognized during his lifetime. Gauss, to whom F. Bólyai sent the Appendix, understood it, but did nothing to promote the recognition of J. Bólyai's discovery. In 1837 J. Bólyai entered a competition for a prize of the Jablonow Society of Leipzig on the problem of "perfecting the geometric theory of imaginary numbers." In this work J. Bólyai rediscovered Hamilton's theory of pairs, which had been published in 1833–1835. Written in extremely condensed form and with references to the *Appendix*, which was inaccessible to the jury, Bólyai's paper was not properly appreciated. All this drove J. Bólyai into a deep depression, from which he never really recovered.

---

18) *Ibid.*

19) Bólyai, J. *Appendix.* Moscow/Leningrad 1950, p. 18.

20) *Ibid.*, pp. 18–19.

J. Bólyai's *Appendix* was also written in extremely condensed form, involving a great deal of ad hoc notation; this was because F. Bólyai allowed his son too little space to discuss his discovery. Therefore it was not easy to glean the essence of J. Bólyai's discovery from his exposition. Perhaps the only one who understood this essay during the author's lifetime was Gauss. In a letter from Gauss to Gerling, written shortly after the publication of the Appendix, we find, "I consider this young geometer von Bolyai a genius of the first magnitude."[21] However to F. Bólyai himself Gauss wrote,

> Now a few words about your son. If I begin by saying that I should not praise this work, you will be astonished at first, but I cannot do otherwise. To praise it would be to praise myself: the entire content of the essay, the route followed by your son, and the results he obtained, coincide with my own discoveries, some of which date back 30 or 35 years.[22]

Subsequently, after becoming acquainted with the *Geometrische Untersuchungen* of Lobachevskiĭ, Gauss advised the Bólyais, father and son, to read that work. After reading Lobachevskiĭ's paper, J. Bólyai made the silly conjecture that "Gauss — a colossus, who was in possession of great treasures even without this — could not accept the fact that someone had anticipated him in this question. Since he was no longer in a position to block its appearance, he reworked the theory and published it under the name of Lobachevskiĭ."[23]

Hyperbolic Geometry

In his memoir "On the foundations of geometry" (1829) Lobachevskiĭ first of all reproduced his paper of 1826. At the beginning of this part he wrote: "Who would not agree that no mathematical science should begin with concepts as obscure as those with which we, following Euclid, begin geometry, and that no lapse of rigor should be tolerated anywhere in Mathematics such as we are forced to admit in the theory of parallel lines."[24] Then follow the words quoted above on "first concepts" with which any science begins.

Having next defined the basic concepts of geometry that are independent of Euclid's fifth postulate and remarking that the sum of the angles of a rectilinear triangle cannot be larger than π, as is the case for spherical triangles, Lobachevskiĭ announced,

> We have seen that the sum of the angles of a rectilinear triangle cannot be larger than π. It remains only to assume that this sum is equal to π or less than π. Either one may be assumed without any subsequent contradiction, whence two geometries result: one, useful up to the present because of its simplicity, is in accord with all actual measurements; the other, imaginary

---

21) *On the Foundations of Geometry* [in Russian]. p. 112

22) *Ibid.*, p. 113.

23) Bólyai, J. *Appendix.* p. 31.

24) Lobachevskiĭ, N. I. *Collected Works* [in Russian]. Vol. 1, p. 185.

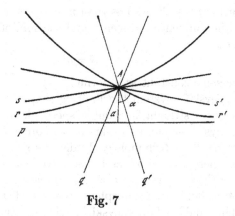

Fig. 7

geometry, which is more general and hence more laborious in its computations, admits the possibility that lines may depend on angles.[25]

Lobachevskiĭ showed that in "imaginary geometry" the sum of the angles of a triangle is always less than π and that two lines may fail to intersect when they form angles with a transversal adding up to less than π. Parallel lines are defined as those that do not intersect, but can be be obtained by passage to the limit from intersecting lines. Two lines pass through each point of the plane parallel to a given line lying in the plane; these lines divide the pencil of lines passing through the given point into four regions, through two of which pass lines intersecting the given line, while through the other two pass lines that do not intersect this line and cannot be obtained by limiting passage from those that do intersect; such lines are called *divergent* lines. The parallel lines separate the intersecting lines from the divergent lines (cf. Fig. 7, where the lines r and r' are drawn through the point A parallel to the line p, the lines q and q' pass through the point A and intersect the line p, and the lines s and s' diverge from the line p). Lobachevskiĭ called the angle α between the line drawn through A parallel to the line p and the perpendicular from A to p the *angle of parallelism* and showed that the function $\alpha = \Pi(a)$ that expresses the dependence of this angle on the length a of the perpendicular can be written (in modern notation) in the form

$$\Pi(a) = 2 \arctan e^{-a/q}, \tag{1}$$

where q is a certain constant. When $a \neq 0$, the angle of parallelism is always acute. It tends to $\pi/2$ as $a \to 0$, while the constant q can serve as an absolute unit of length in the hyperbolic plane analogous to the absolute unit of angle in Euclidean space. Lobachevskiĭ also established that divergent lines have a common perpendicular and recede from each other on both sides of it, while two parallel lines approach each other and the distances of points of one of them from the other tends to 0 as these points recede to infinity. The sum of the angles of a triangle in

25) *Ibid.*, pp. 194–195.

hyperbolic geometry is always less than π, and if δ is the "angular defect" of the triangle, i.e., the difference between π and the sum of its angles, then the area of the triangle S is

$$S = q^2\delta, \tag{2}$$

where q is the same constant as in formula (1).

In Lobachevskiĭ's system the limiting shape of a circle as its radius tends to infinity is not a straight line but a special kind of curve called a "limiting circle." Nowadays these curves are known as *horocycles*, from Greek words equivalent to Lobachevskiĭ's terminology. Under the same circumstances the limiting position of a sphere is not a plane but a curved surface called by Lobachevskiĭ a "limiting sphere" and now known as a *horosphere*. Lobachevskiĭ noted that Euclidean geometry holds on a horosphere, the role of straight lines being played by horocycles. This enabled Lobachevskiĭ, relying on Euclidean trigonometry on the horosphere, to deduce the plane trigonometry of his system of geometry. The name "imaginary geometry" emphasizes that this geometry bears the same relation to Euclidean "practical" geometry that imaginary numbers bear to real numbers. His words, "Ideas are acquired through the senses, one should not believe in innate ideas," show that Lobachevskiĭ found no contradiction in the consequences of the assumption that the fifth postulate did not hold, and he came to the conclusion that Euclidean geometry is not the only conceivable one. Lobachevskiĭ immediately posed the question of experimental testing as to which geometry holds in the real world — "practical" geometry or "imaginary," and to this end he decided to measure the sum of the angles of a triangle with very large sides. However, having computed from the latest astronomical calendar the sum of the angles of the triangle formed by two diametrically opposite positions of the Earth in its orbit and the star Sirius, assuming one of these angles a right angle and the other is the angle of parallelism, he found that this sum differs from π by an amount less than the error of angle measurement in his day. "After which," he wrote, "one can imagine the extent to which this difference, on which our theory of parallels is based, justifies the precision of the computations of ordinary Geometry and makes it possible to regard the accepted principles of the latter as if they were rigorously demonstrated."[26]

This shows that Lobachevskiĭ's phrase "a rigorous demonstration of the theorem on parallel lines," in his 1826 paper was intended to mean that it was impossible to establish experimentally which of the two geometries holds in the real world, whence it follows that in practice one can use "practical geometry" without any danger of falling into error.

The most complete exposition of Lobachevskiĭ's system is his *New Foundations of Geometry with a Complete Theory of Parallels* (1835–1838). This work begins with the words: "Contact constitutes a distinct property of bodies and makes them geometric, when we retain this property, not taking account of any other, whether essential or accidental..."[27]. He further defines sections of bodies,

---

26) *Ibid.*, p. 200.
27) *Ibid.*, Vol. 2, p. 168.

spaces, congruence of bodies ("identity") and equal measure ("equality"), "the three principal sections" that divide a body into eight parts, and by use of them, surfaces, lines, and points; distances, and then spheres; planes (the geometric loci of points equidistant from two points), and lines. Thus Lobachevskiĭ's exposition of geometry was based on the purely topological properties of contact and section; congruence of bodies and equality of intervals were essentially defined using motion. Lobachevskiĭ went on to give a detailed exposition of the geometry he had discovered. At the end he wrote,

> In nature we know only motion, without which sense impressions are impossible. Thus all other concepts, for example, geometric concepts, are produced by our mind artificially, being contained in the properties of motion; for that reason space in and of itself does not exist for us. Hence there can be no contradiction in our mind if we assume that certain forces in nature follow one geometry and others a different one of their own.

The idea of the possibility of different geometric properties in different parts of space and their dependence on "forces," i.e., on matter, is a distant anticipation of the ideas of Einstein's general theory of relativity. Lobachevskiĭ went on to state the conjecture that his geometry might hold "either outside the visible world or in a dense sphere of molecular attractions."[28]

In his later work Lobachevskiĭ introduced coordinates and computed a whole series of new definite integrals from geometric considerations. To this topic he devoted his paper "Application of imaginary geometry to certain integrals," (*Research Notes of Kazan' University*, 1836). Many of them were included in the "Tables d'intégrales définies" (1858) published by the Dutch mathematician David Bierens de Haan (1822–1895), and came to be included in later handbooks.

J. Bólyai's "Absolute Geometry"

We have already pointed out that J. Bólyai's *Appendix* was written very concisely, using a number of abbreviations: Bólyai denoted the infinite line passing through a and b by $\widetilde{ab}$, the ray with vertex a passing through the point b by $\widetilde{ab}$, the angle with sides $\widetilde{ab}$ and $\widetilde{ac}$ by abc, and a right angle by R. The beginning of the *Appendix* says:

> If the line $a\widetilde{m}$ does not intersect a line $b\widetilde{n}$ in the same plane, but intersects every line $b\widetilde{p}$ (in abn), we shall denote this fact by writing $b\widetilde{n}|||a\widetilde{m}$. It is clear that such a $b\widetilde{n}$ exists and for any point b (outside $a\widetilde{m}$) there is only one such, and $bam + abn$ is not larger than R, for if bc is rotated about b until $bam + abc = 2R$, there will be a *first* instant when $b\widetilde{c}$ does not intersect $a\widetilde{m}$, and then $bc|||am$. It is also clear that $bn|||em$ wherever e lies on $\widetilde{am}$.[29]

Here the parallel lines bc and am are intended in the sense of Euclidean geometry when the fifth postulate holds (the "eleventh axiom" in Bólyai's terminology)

---

28) *Ibid.*, pp. 159–160.
29) *On the Foundations of Geometry* [in Russian]. p. 72.

and in the sense of hyperbolic geometry when this postulate fails. Here already we encounter a formulation applicable to either geometry, which is characteristic of Bólyai. Bólyai was striving to state as many facts as possible from both geometries in such a formulation. Thus, for example, he stated the law of sines for both geometries in the form

$$\frac{\sin A}{\bigcirc a} = \frac{\sin B}{\bigcirc b} = \frac{\sin C}{\bigcirc c},$$

where $\bigcirc r$ is the length of the circle of radius r. Bólyai called Euclidean geometry "system Σ" and the new geometry "system S." Whatever was true in both systems he called "absolute"; it was "absolute" geometry, as the title of his essay shows, that constituted the "science of space, absolutely true, independently of the truth or falsity of Euclid's eleventh axiom." The modern term "absolute geometry," which is derived from Bólyai's term, is not completely equivalent to it: we take absolute geometry to consist of the facts that are independent of the validity of the fifth postulate, whereas Bólyai included in his "absolute" statements certain facts that are different in the two geometries.

The Consistency of Hyperbolic Geometry

Having deduced the trigonometric formulas for his new system in his first paper "On the foundations of geometry," Lobachevskiĭ remarked that, "these equations become... [the equations of] spherical Trigonometry if the sides a, b, and c are replaced by $a\sqrt{-1}$, $b\sqrt{-1}$, $c\sqrt{-1}$; but throughout ordinary Geometry and spherical Trigonometry only the contents [i.e., ratios] of lines occur: consequently ordinary geometry, Trigonometry, and this new Geometry, will always be mutually consistent."[30] This means that if we write the law of cosines, the law of sines, and the dual law of cosines for spherical trigonometry on a sphere of radius r as

$$\cos\frac{a}{r} = \cos\frac{b}{r}\cos\frac{c}{r} + \sin\frac{b}{r}\sin\frac{c}{r}\cos A,$$

$$\frac{\sin A}{\sin(a/r)} = \frac{\sin B}{\sin(b/r)} = \frac{\sin C}{\sin(c/r)},$$

$$\cos A = -\cos B\cos C + \sin B\sin C\cos(a/r),$$

the formulas for hyperbolic trigonometry can be written in the same form with the sides a, b, and c of the triangle replaced by the products ai, bi, ci. Since multiplying the sides a, b, and c by i is equivalent in these formulas to multiplying the radius of the sphere by i, setting $r = qi$ and using the known relations

$$\cos(ix) = \cosh x, \quad \sin(ix) = i\sinh x,$$

---

30) Lobachevskiĭ, N. I. *Collected Works* [in Russian]. Vol. 1, p. 206.

we can rewrite the corresponding formulas of hyperbolic trigonometry in the form

$$\cosh\frac{a}{q} = \cosh\frac{b}{q}\cosh\frac{c}{q} - \sinh\frac{b}{q}\sinh\frac{c}{q}\cos A,$$

$$\frac{\sin A}{\sinh(a/q)} = \frac{\sin B}{\sinh(b/q)} = \frac{\sin C}{\sinh(c/q)},$$

$$\cos A = -\cos B\cos C + \sin B\sin C\cosh(a/q).$$

Lobachevskiĭ himself did not use the functions $\cosh x$ and $\sinh x$; instead he used combinations of the function $\Pi(x)$ he had introduced and the trigonometric functions. The constant q in these formulas is the same one as in formulas (1) and (2).

Lobachevskiĭ actually proved the consistency of his system by introducing coordinates on both the plane and in space, thereby constructing an arithmetical model of the hyperbolic plane and hyperbolic three-space. However Lobachevskiĭ himself saw the evidence for the consistency of the geometry he had discovered in the connection just exhibited between the formulas of his trigonometry and those of spherical trigonometry. This inference of Lobachevskiĭ's was not correct. In his memoir he had proved that the formulas of spherical trigonometry follow from his geometry, but in fact, in order to assert that the consistency of the trigonometric formulas implies the consistency of hyperbolic geometry, it would have been necessary to prove that all propositions of the latter can be deduced from its trigonometric formulas and "absolute geometry" — the propositions that are independent of the fifth postulate. Lobachevskiĭ attempted to carry out such a proof in his *Imaginary Geometry*, where he wrote, "Now, departing from geometric constructions and taking a brief detour, I intend to prove that the main equations that I have found [in the paper cited above] for the dependence of the sides and angles of a triangle in imaginary Geometry can be adopted with profit in Analysis and will never lead to false conclusions in any relation."[31]

Lobachevskiĭ went on to connect the trigonometric formulas with the propositions of absolute geometry and deduced from this assertion that the sum of the angles of a triangle is less than π, which, as is known, is equivalent to Lobachevskiĭ's postulate. However, this reasoning still does not constitute a complete proof of consistency, since the formulas of spherical trigonometry themselves, from which it follows that the sum of the angles of a triangle is greater than π, if regarded as formulas of plane trigonometry, would contradict the axioms of absolute geometry. Lobachevskiĭ's arguments actually prove only that his trigonometric formulas are consistent.

However, Lobachevskiĭ's arguments can be made the basis of a complete proof of the consistency of his geometry, though it is necessary to use some methods that were unknown in his time. To do this one must make use of Poncelet's idea of imaginary points in space, introduced in his *Traité des propriétés projectives des figures*. If we supplement real Euclidean space by all of its imaginary points, we

---

31) *Ibid.*, Vol. 3, p. 49.

obtain complex Euclidean space. Every algebraic and analytic line or surface in real space can be regarded as part of the line or surface in complex space defined by the same equations, and the distances between points and the angles between lines in complex space are given by the same formulas as in real space. For that reason the trigonometric relations on the sphere of complex space are expressed by the same formulas as in real space. Thus the geometry of the hyperbolic plane can be realized on a sphere of imaginary radius qi in a subspace of complex space, whose points have rectangular x- and y-coordinates that are real ($x = \bar{x}$, $y = \bar{y}$), and z-coordinates that are purely imaginary ($z = -\bar{z}$). This subspace can be regarded as a real affine space in which the distance d between the points with rectangular coordinates x_1, y_1, z_1 and x_2, y_2, z_2 is defined by the formula

$$d^2 = (x_2 - x_1)^2 + (y_2 - y_1)^2 - (z_2 - z_1)^2.$$

Such a space was defined much later by H. Poincaré (1906) and H. Minkowski (1908) in connection with the interpretation of the special theory of relativity and is now called *pseudo-Euclidean space*. The sphere of radius qi in this space has the form of a hyperboloid of two sheets (the geometry of the hyperbolic plane is realized on each nappe of such an hyperboloid), and in this space there are also spheres of real radius having the shape of one-sheeted hyperboloids and spheres of zero radius having the shape of cones.

This interpretation, which demonstrates vividly the consistency of Lobachevskiĭ's planimetry, explains why the formulas of hyperbolic trigonometry can be obtained from those of spherical trigonometry by replacing the radius of the sphere by qi. This sphere has purely imaginary radius and is the "imaginary sphere," concerning which J. H. Lambert wrote that he "was nearly compelled to draw the conclusion that the third hypothesis holds in some imaginary sphere" (HM, Vol. 3, p. 218).

We note that the interpretation of the hyperbolic plane on a sphere of purely imaginary radius in pseudo-Euclidean space also has certain properties; for example, motions of the hyperbolic plane can be represented by rotations of the sphere, circles can be represented as sections of the sphere by Euclidean planes; curves that are the geometric loci of the points equidistant from lines ("equidistant curves") can be represented as sections of a sphere by pseudo-Euclidean planes; and horocycles can be represented as sections of the sphere by isotropic planes (obtained by limiting passages from both Euclidean and pseudo-Euclidean planes). Instead of studying one nappe of the sphere of imaginary radius it is often more convenient to to interpret the hyperbolic plane as a complete sphere, but with diametrically opposite points identified. Hyperbolic three-dimensional space admits an analogous interpretation in the 4-dimensional pseudo-Euclidean space that is applied to interpret the space-time of special relativity.

The coordinates x, y, and z of a point of a sphere of imaginary radius can be regarded as homogeneous coordinates of a corresponding point of the hyperbolic plane. Such coordinates were used by K. Weierstrass (from different considerations) in his seminar on hyperbolic geometry, conducted in 1870 at the University of Berlin. As a result they are called Weierstrassian coordinates.[32]

Propagation of the Ideas of Hyperbolic Geometry

The new system of geometry did not receive recognition during the lifetime of its creators. Except for the speech of P. I. Kotel'nikov just mentioned, we do not know any official positive reactions to Lobachevskiĭ as the creator of a new geometry. There were no reactions of any kind to Bólyai's *Appendix*. Gauss, as stated, avoided publishing his discoveries, confining himself to passing remarks in letters to a few friends. This situation did not change until the 1860's. During the years 1860–1865, soon after the death of Gauss, his correspondence with the astronomer Schumacher (1780–1850) was published, containing in particular a letter of Gauss on Lobachevskiĭ's *Geometrische Untersuchungen*. Gauss wrote

> This essay contains the foundations of the geometry that would have to hold and would constitute a rigorous, coherent whole if Euclidean geometry were not true... Lobachevskiĭ calls it 'imaginary geometry.' You know that I have shared these views for 54 years (since 1792) with a certain development of them that I do not care to mention here; thus I did not find anything really new to me in Lobachevskiĭ's essay. But in the development of the subject the author has not followed the same route that I followed. It has been carried out masterfully by Lobachevskiĭ in a truly geometric spirit. I consider it my duty to direct your attention to this essay, which will surely afford you a very rare pleasure.[33]

In 1865 there appeared a "Note on Lobatschewsky's imaginary geometry" (*Phil. Mag. London*) by A. Cayley. In this note Cayley compared Lobachevskiĭ's trigonometric formulas with spherical trigonometry, and although, as can be seen from the note, Cayley did not understand the essence of Lobachevskiĭ's discovery, his note promoted recognition of this discovery.

In 1866 in Bordeaux and Paris there appeared a French translation of Lobachevskiĭ's *Geometrische Untersuchungen* together with excerpts from the correspondence of Gauss with Schumacher made by Professor Guillaume Jules Hoüel (1823–1866) of the University of Bordeaux, and in 1867 in Paris there appeared the "Essai critique sur les principes fondamentaux de la géométrie," by Hoüel containing an exposition of Lobachevskiĭ's basic ideas. The foundations of hyperbolic geometry were expounded by Professor Richard Baltzer (1818–1887) of the University of Giessen in the second edition of his book *Die Elemente der Mathematik* (Dresden 1867). In the same year Professor Giuseppe Battaglini (1826–1894) of the University of Naples published an article "Sulla geometria imaginaria de Lobatschewsky" *G. mat. Napoli*, 1867) and an Italian translation of *Pangeometry*;

---

32) Kagan, V. F. *Foundations of Geometry* [in Russian]. Moscow/Leningrad 1949, Pt. 1, p. 341.

33) *Ibid.*, p. 119–120.

A. V. Vasil'ev

in 1868 he published an Italian translation of Bólyai's *Appendix*. In 1868 Professor Alekseĭ Vasil'evich Letnikov (1837–1888) of the Moscow Higher Technical School inserted in the third volume of the *Mathematical Collection* a Russian translation of Lobachevskiĭ's *Geometrische Untersuchungen* with a preface in which Lobachevskiĭ's geometric work was characterized as "quite remarkable, but little known," and Professor Erast Petrovich Yanishevskiĭ published in Kazan' a "Historical note on the life and work of N. I. Lobachevskiĭ." Finally, in the same year of 1868, there appeared an article of E. Beltrami on the interpretation of hyperbolic geometry. As a result of these publications hyperbolic geometry became known in all the countries of Europe by 1870; it was then, as we have mentioned, that this geometry became the subject of a special seminar of K. Weierstrass at the University of Berlin.

A large role in the dissemination of the ideas of hyperbolic geometry was played by Fëdor Matveevich Suvorov, a professor of the University of Kazan' whose master's thesis "On the characterization of three-dimensional systems" (1871) was devoted to Riemannian spaces of three dimensions, which are an immediate generalization of three-dimensional hyperbolic space. An even larger role was played by Aleksandr Vasil'evich Vasil'ev (1853–1929), a graduate of the University of Petersburg. Vasil'ev worked as a docent and later professor of the University of Kazan from 1874 until 1907, when he moved to Petersburg. Although algebra was his main specialty, Vasil'ev was a mathematician of wide interests and subsequently edited, jointly with P. S. Yushkevich (1873–1945), the collections *New Ideas in Mathematics* (Petersburg 1912–1915), which made Russian readers acquainted with the ideas of set theory and the foundations of analysis, group theory, the geometric work of F. Klein, the theory of relativity, and other important mathematical discoveries of the time. Vasil'ev was also an historian of mathematics and is the author of a number of studies of the work of Lobachevskiĭ and the historical essay "Whole Number" (Petrograd 1922). He was the first president of the Kazan' Physico-Mathematical Society, which separated from the Kazan' Society of Natural Scientists in 1890. It was under the leadership of Vasil'ev that the Kazan' Physico-Mathematical Society undertook the publication of the first complete collection of Lobachevskiĭ's geometric works, edited by Vasil'ev (Kazan' 1883–1886), the celebration of the hundredth anniversary of the birth of Lobachevskiĭ (1893), which demonstrated the international recognition of the discovery of non-Euclidean geometry, and the international Lobachevskiĭ competitions, which have greatly enhanced the international reputation of the University of Kazan'. The first Lobachevskiĭ prize was awarded in 1897 to Sophus Lie for the third (geometric) volume of his *Theory of Transformation Groups*; the second prize was awarded to W. Killing for a series of papers on non-Euclidean spatial forms and Lie groups (1883–1896). The third prize (1904) was awarded to D. Hilbert for his *Grundlagen der Geometrie* and other geometric work (1895–1900). Since that time the Lobachevskiĭ prize has been awarded to such prominent scholars as F. Schur, H. Weyl, É. Cartan, and A. D. Aleksandrov.

Beltrami's Interpretation

The most convincing argument in favor of the new geometry was furnished by the interpretations of this geometry in Euclidean space that appeared at this time. The first two such interpretations were proposed by Eugenio Beltrami (1835–1900), a professor of mathematics and mechanics at the Universities of Bologna and Rome, in his "Saggio di interpretazione della geometria non-euclidea" (*G. mat. Napoli*, 1868), in which his point of departure was the work of Minding. In this paper Beltrami computed the element of length (the square of the differential of an arc) of the hyperbolic plane in coordinates u, v equal to the distances from the point to two mutually perpendicular lines divided by r (these coordinates are now called *Beltrami coordinates*), and he found that in this system of coordinates the element of length is

$$ds^2 = r^2 \frac{(a^2 - u^2)du^2 + 2uv\,du\,dv + (a^2 - v^2)dv^2}{a^2 - u^2 - v^2}.$$

Then computing the Gaussian curvature of a surface with such an element of length, Beltrami discovered that the Gaussian curvature of the hyperbolic plane is the same at every point, namely $-1/r^2$, i.e., the hyperbolic plane can be regarded as a surface of constant negative curvature.

E. Beltrami

Since from the point of view of its intrinsic geometry every surface can be regarded as an interpretation of any surface developable on it, and a necessary and sufficient condition for developability of surfaces is that their Gaussian curvatures be equal at corresponding points, Beltrami concluded that the hyperbolic plane can be interpreted as any surface of constant negative curvature.

Beltrami established that the surfaces of revolution of constant negative curvature considered by Minding are isometric to the portions of the hyperbolic plane enclosed between two intersecting lines and a circle perpendicular to them, or between two divergent lines, their common perpendicular, and an equidistant curve orthogonal to both of them, or between two parallel lines and a horocycle orthogonal to them. Later (1900) Hilbert proved that every surface of constant negative curvature in Euclidean space is isometric to only a portion or several portions of the hyperbolic plane, but no such surface is isometric to the whole hyperbolic plane.

On the other hand, by considering the points of the Euclidean plane with coordinates numerically equal to the Beltrami coordinates u, v of the hyperbolic plane, Beltrami obtained a second interpretation. Since the coordinates are connected by the condition

$$u^2 + v^2 < a^2, \tag{3}$$

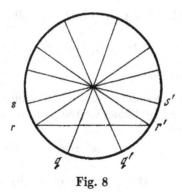

Fig. 8

in this interpretation the whole hyperbolic plane can be depicted as the interior of the disk bounded by the circle

$$u^2 + v^2 = a^2. \tag{4}$$

Beltrami showed that in this situation the straight lines of the hyperbolic plane map to chords of this disk (Fig. 8) and the distance from the point P with coordinates (u, v) to the origin 0 is

$$\rho = \frac{r}{2} \ln \frac{a + \sqrt{u^2 + v^2}}{a - \sqrt{u^2 + v^2}}. \tag{5}$$

Although Beltrami did not give the formula for the distance between two arbitrary points and did not explain how motions of the hyperbolic plane were to be depicted in his interpretation, this interpretation was the first proof, incomplete to be sure, of the consistency of the entire hyperbolic plane.

Cayley's Interpretation

The answer to the questions left open by Beltrami was essentially contained in the paper "Sixth memoir on quantics" (1859) by Arthur Cayley (cf., Bk. 1, pp. 39–40) already mentioned (see above), in which the concept of a projective metric on the plane was introduced.

Cayley wrote the equation of a conic in the projective plane with projective coordinates x, y, and z in the form

$$(a, b, c, f, g, h \ \mathbin{\{} \ x, yz)^2 = 0$$

and associated a distance Dist (P, P') with each pair of points P and P' of this absolute having coordinates x, y, z and x', y', z'. He wrote the distance in the form

$$\cos^{-1} \frac{(a, \ldots \mathbin{\{} x, y, z \mathbin{\{} x', y', z')}{\sqrt{(a, \ldots \mathbin{\{} x, y, z)^2} \sqrt{(a, \ldots \mathbin{\{} x', y', z')^2}}$$

(where $\cos^{-1}$ was the British notation for the arccosine function).

From what Cayley had proved earlier it followed that for points of a single line

$$\text{Dist}\,(P, P') + \text{Dist}\,(P', P'') = \text{Dist}\,(P, P'').$$

Cayley remarked that

the general formulæ suffer no essential modificationation, but they are greatly simplified in form, by taking for the point equation of the Absolute

$$x^2 + y^2 + z^2 = 0.$$

... We then have for the expression of the distance of the points (x, y, z) and (x', y', z')

$$\cos^{-1} \frac{xx' + yy' + zz'}{\sqrt{x^2 + y^2 + z^2}\sqrt{x'^2 + y'^2 + z'^2}}.\quad ^{34}$$

He went on to point out that

if (x, y, z) are ordinary rectangular coordinates in space satisfying the condition

$$x^2 + y^2 + z^2 = 1,$$

the point having (x, y, z) for its coordinates will be a point on the surface of a sphere, and [since the distance defined by Cayley is the spherical distance] we have thus a system of spherical geometry; and it appears that the Absolute in such system is the (spherical) conic, which is the intersection of the sphere with the concentric cone or evanescent sphere.[35]

By a "spherical conic" Cayley had in mind the imaginary disk in which the infinitely distant plane that completes Euclidean space into projective space intersects all the spheres of Euclidean space and the imaginary cone $x^2 + y^2 + z^2 = 0$, which Cayley called the "concentric cone," and the "evanescent sphere." Cayley's metric is thereby realized on the infinitely distant plane that is a projective plane and on the sphere of ordinary space with antipodal points identified. At present the projective plane with the metric so defined is called the *elliptic plane*; for reasons that will become clear below this plane is also called the *Riemannian non-Euclidean plane*, although there are much stronger reasons for calling it the Cayley plane.

Cayley also remarked that

In ordinary plane geometry the Absolute degenerates into a pair of points, viz., the pair of points of intersection of the line infinitely distant with an evanescent circle or, what is the same thing, the Absolute is the two circular points at infinity. The general theory is consequently modified, viz., there is

34) Cayley, A. *Mathematical Papers*. Vol. 2, p. 590.
35) *Ibid.*, p. 591.

not, as regards points, a distance such as the quadrant, and the distance of two lines cannot be in any way compared with the distance of two points.[36]

The "distance between lines" is the angle between the lines in the Euclidean plane or the distance between two parallel lines. In speaking of a conic degenerating to a pair of points Cayley had in mind the conic as a collection of lines, i.e., a second-order pencil, which can degenerate to a pair of ordinary real or imaginary pencils. Cayley did not study the cases when the conic is real or when it breaks into a pair of real pencils. These cases lead to geometries now called hyperbolic (Lobachevskiĭ's geometry) and pseudo-Euclidean. However the great value of the projective metrics he had defined was clear to him, and at the end of his memoir he wrote, "Metric geometry is thus a part of descriptive geometry, and descriptive geometry is all geometry."[37]

Klein's Interpretation

The connection between the Cayley's projective metrics and hyperbolic geometry was established by the German geometer Felix Klein (1849–1925). A native of Düsseldorf, Klein studied at the University of Bonn, where he was a student of Plücker and from 1866 to 1868 his assistant in the physics department. Klein then worked as a professor at the University of Erlangen (1872–1875), at the Technische Hochschule in Munich (1875–1880), at the University of Leipzig (1880–1886), and from 1888 on at the University of Göttingen. In 1871 he established the connection just mentioned between the geometric theories of Lobachevskiĭ and Cayley, which will be discussed in detail below. Having determined that the group of motions of hyperbolic space and the groups of motions of Euclidean space and other projective metrics are subgroups of the group of projective transformations of space, Klein arrived at the general concept of the role of transformation groups in geometry, which he enunciated in his *Antrittsrede* at the University of Erlangen (his "Erlanger program"). Klein played an important role in the assimilation of the ideas of non-Euclidean geometry and group theory by mathematicians, in the foundation of the theory of continuous groups, and in the study of discrete groups of geometric transformations, in particular the so-called Fuchsian groups (whose theory he worked out in a stormy competition with Poincaré), and also the symmetry groups of regular figures (one of his books is devoted to the group of symmetries of the regular icosahedron). For forty years, starting in 1876, Klein was the editor-in-chief of the *Mathematische Annalen*, published in Leipzig. It should also be mentioned that he took an active part in the famous multi-volume *Enzyklopädie der mathematischen Wissenschaften* (Leipzig 1898–1934, Vols. 1–6) and in the reform of mathematics in secondary and higher schools.

In his *Vorlesungen über die Entwicklung der Mathematik im 19. Jahrhundert*, which were given during World War I and published by R. Courant and O. Neugebauer in 1926, Klein described the discovery of his interpretation as follows. He

---

36) *Ibid.*, p. 592.
37) *Ibid.*, p. 592.

F. Klein

became acquainted with Cayley's theory from the book of Salmon, *Conic Sections*, mentioned above, which had appeared in German translation by this time. He first heard of hyperbolic geometry from his friend O. Stolz in the winter of 1869–1870.

> I understood very little of these brief communiqués, but the idea immediately struck me that there was some correlation here. In February of 1870 I gave a paper in Weierstrass' seminar on Cayley's definition of measure and concluded it by asking whether the ideas of Cayley and Lobachevskiĭ were the same. The answer I received was that these are two conceptually very different systems.[38]

Klein wrote that at first he had allowed himself to be dissuaded, and did not return to these ideas until the summer of 1871, again in arguments with Stolz. As a result in that year Klein published an article "Über die sogenannte nichteuklidische Geometrie," in which he showed that in the case when Cayley's "absolute" is a real conic the part of the projective plane inside this conic is isometric to the hyperbolic plane. This paper provoked objections from many quarters, and in particular an accusation that there was a vicious circle here, since projective geometry was usually developed on the basis of Euclidean geometry.

---

38) Klein, F. *Vorlesungen über die Entwicklung der Mathematik im 19. Jahrhundert.* Erster Teil, S. 152.

However Staudt's theory had appeared by that time, providing a foundation for projective geometry independent of Euclidean geometry. Klein devoted the second half of this paper to this problem (1872). Klein made some slight alterations in Cayley's definition, defining the distance between the points A and B as $c \ln W$, where W is the cross-ratio of the points A and B and the points of intersection of the line AB with the absolute. He defined the angle between the lines a and b as $c' \ln W'$, where W' is the cross-ratio of the lines a and b and the tangents to the absolute drawn from their point of intersection. Klein showed that in the case when both of the constants c and c' equal $i/2$, the elliptic plane is obtained, while in the case when $c = 1/2$ and $c' = i/2$, the hyperbolic plane is obtained. Hyperbolic parallels are lines intersecting at a point of the conic. The motions of the elliptic plane and the hyperbolic plane here are represented by collineations mapping imaginary and real conics to themselves. Replacing the conic by an imaginary and an oval surface of second order (quadrics), Klein obtained elliptic space and hyperbolic space. Here the Beltrami interpretation of the hyperbolic plane is the special case of Klein's interpretation when the conic is a circle in the Euclidean plane, thereby solving the problem of representing motions of the hyperbolic plane in the Beltrami interpretation. The case of a real constant c' (an ideal domain of the hyperbolic plane and the pseudo-Euclidean plane) was not considered here by Klein.

Elliptic Geometry

We have already mentioned that elliptic geometry, i.e., geometry with a projective metric whose absolute is an imaginary conic or quadric, was defined for the case of the plane in Cayley's "Sixth memoir on quantics" (1859). Three-dimensional elliptic geometry was defined by Klein in his paper "Über die sogenannte nicht-euklidische Geometrie," (1871). Klein also proposed the terms "elliptic geometry" and "hyperbolic geometry."

The salient facts of the geometry of elliptic space were discovered by William Kingdon Clifford (1845–1879), whom we have mentioned in connection with his work on algebra (Bk. 1, pp. 78–79). Klein wrote, regarding Clifford, "I recall him with particular pleasure, as a man who understood me thoroughly from the first and soon extended my research."[39] In his "Preliminary sketch of biquaternions," (*Proc. Math. Soc. London*, 1873) Clifford first defined poles and polar planes relative to the absolute, and also lines that are mutually polar relative to the absolute, remarking that two points that are polar conjugates relative to the absolute "are distant a quadrant from one another" (i.e., differ by $(\pi/2)r$ or when $r = 1$ by $\pi/2$). For any two lines there are two common perpendiculars on which the smallest and largest distances between the lines are attained. Clifford distinguished the case when it is possible to draw infinitely many common perpendiculars of equal length. In the latter case two given lines and their polars are the rectilinear generators of a quadric, and Clifford called such lines parallel. He went on to prove that "a series of parallel lines meeting a given line constitutes a ruled surface with zero

---

39) *Ibid.*, S. 189.

W. K. Clifford

curvature. The geometry of this surface is the same as that of a finite parallelogram whose opposite sides are regarded as identical."[40] Clifford's "parallels" are nowadays called *paratactic lines*; in contrast to the parallels of Euclidean ("parabolic") geometry and hyperbolic geometry, which intersect in a point of an infinitely distant plane or in a point of the "absolute." Paratactic lines are intersecting lines, but, like Euclidean parallels, equidistant from each other. The surface constructed by Clifford can be defined as the geometric locus of the points equidistant from a line and its polar. This surface is a ruled quadric surface obtained by revolving one of the two paratactic lines about the other, its rectilinear generators of both families being paratactic with its axes. This surface, which is isometric to a rhombus of the Euclidean plane with opposite sides identified, is the simplest example of a space with Euclidean geometry and finite volume. At the same time it is the simplest example of a solution of the so-called Clifford-Klein problem, to find spaces with a Euclidean metric nonisometric to the Euclidean space as a whole.

---

40) Clifford, W. *Mathematical Papers*, 1885, p. 193.

5. MULTI-DIMENSIONAL GEOMETRY

Jacobi's Formulas for Multi-dimensional Geometry

We have seen that the ideas of multi-dimensional geometry were stated many times by various mathematicians in their attempts at a geometric interpretation first of algebraic degrees greater than 3 (HM, Vol. 1, p. 230, 278, 299–301), and then of mechanical problems (HM, Vol. 3, p. 183). However the true geometry of multi-dimensional spaces arose only when it became necessary to give a geometric interpretation of functions of several variables in connection with the development of the theory of algebraic forms of n variables and n-fold integrals. Many mathematicians of the first half of the nineteenth century studied the theory of n-fold integrals, including Ostrogradskiĭ, Jacobi, and others.

Of the papers of this time devoted to n-fold integrals Jacobi's 1834 paper "De binis quibuslibet functionibus homogeneis secundi ordinis per substitutiones lineares in alias binas transformandis, quae solis quadratis variabilium constant; una cum variis theorematis de transformatione et determinatione integralium multiplicium," (Bk. 1, p. 69), is of crucial importance in this connection. Here, besides the problem of computing multiple integrals, an important problem of the theory of algebraic forms is solved: *Find linear substitutions*

$$\begin{aligned}
y_1 &= \alpha'_1 x_1 + \alpha'_2 x_2 + \cdots + \alpha'_n x_n, \\
y_2 &= \alpha''_1 x_1 + \alpha''_2 x_2 + \cdots + \alpha''_n x_n, \\
&\qquad\qquad\cdots\cdots\cdots\cdots\cdots\cdots \\
y_n &= \alpha_1^{(n)} x_1 + \alpha_2^{(n)} x_2 + \cdots + \alpha_n^{(n)} x_n,
\end{aligned}$$

for which the following would hold:

$$y_1 y_1 + y_2 y_2 + \cdots + y_n y_n = x_1 x_1 + x_2 x_2 + \cdots + x_n x_n.\text{"}[41]$$

Such a substitution, as Jacobi showed, satisfies the equations

$$\begin{aligned}
\alpha'_\varkappa \alpha'_\lambda + \alpha''_\varkappa \alpha''_\lambda + \cdots + \alpha_\varkappa^{(n)} \alpha_\lambda^{(n)} &= 0, \\
\alpha'_\varkappa \alpha'_\varkappa + \alpha''_\varkappa \alpha''_\varkappa + \cdots + \alpha_\varkappa^{(n)} \alpha_\varkappa^{(n)} &= 1,
\end{aligned}$$

which are the conditions for orthogonality of the matrix $\alpha_\varkappa^{(\lambda)}$. He then solved the problem mentioned in the title of the paper, i.e., he found a linear substitution $x_\varkappa = \sum_\lambda b_\varkappa^{(\lambda)} y_\lambda$, that simultaneously reduced the two quadratic forms

$$V = \sum_{\varkappa,\lambda} a_{\varkappa,\lambda} x_\varkappa x_\lambda, \quad W = \sum_{\varkappa,\lambda} b_{\varkappa,\lambda} x_\varkappa x_\lambda$$

---

41) Jacobi, C. G. J. *Gesammelte Werke*. Berlin 1884, Vol. 3, S. 199.

to

$$V = G_1 y_1 y_1 + G_2 y_2 y_2 + \cdots + G_n y_n y_n,$$
$$W = H_1 y_1 y_1 + H_2 y_2 y_2 + \cdots + H_n y_n y_n.$$

This linear substitution can be interpreted as a substitution that simultaneously reduces to canonical form the equations of two quadrics $V = 1$ and $W = 1$.

The most remarkable part of Jacobi's paper was the computation of an "$(n-1)$-fold integral extended over all positive values of the real variables $x_1, x_2, \ldots, x_n$ satisfying the equation

$$x_1 x_1 + x_2 x_2 + \cdots + x_n x_n = 1.$$"

Jacobi's solution has the following form: "If n is even, then

$$S = \frac{(\pi/2)^{n/2}}{(n-2)(n-4)\cdots 2},$$

and if n is odd,

$$S = \frac{(\pi/2)^{(n-1)/2}}{(n-2)(n-4)\cdots 3}.\text{"}42.$$

These integrals are expressions for the volumes of the surfaces of segments of a sphere of radius 1 in n-dimensional Euclidean space cut off from the sphere by the solid angle $x_i \geq 0$. To obtain the volume of the surface of the whole sphere these expressions must be multiplied by 2^n, while in the case of spheres of radius r these expressions must be multiplied by r^{n-1}.

We note that in solving what were essentially problems of multi-dimensional geometry Jacobi (like Ostrogradskiĭ) made no use of geometric terminology.

Cayley's Analytic Geometry of n Dimensions

The terminology of multi-dimensional geometry first appeared in a mathematical work in a small memoir of Cayley entitled "Chapters in the analytical geometry of (n) dimensions," (*Phil. Mag. London*, 1843). However, the paper itself is purely algebraic, and the term "geometry of n dimensions" occurs only in the title. The paper is concerned with systems of homogeneous equations in n variables:

$$A_1 x_1 + A_2 x_2 + \cdots + A_n x_n = 0$$
$$\cdots\cdots\cdots\cdots\cdots\cdots\cdots\cdots\cdots$$
$$K_1 x_1 + K_2 x_2 + \cdots K_n x_n = 0$$

and the "reciprocal equations" relative to these systems defined by some homogeneous function of order 2:

$$U = \frac{1}{2}\Big[\sum(\alpha^2)x_\alpha^2 + 2\sum(\alpha\beta)x_\alpha x_\beta\Big];$$

---

42) *Ibid.*, S. 267.

the left-hand sides of the "reciprocal equations" are the determinants of the partial derivatives $\partial U/\partial x_\alpha$ and the coefficients of the equations of the given system. Cayley showed that if a given system has r independent equations, there will be $n - r$ in the reciprocal system. From the point of view of multi-dimensional projective geometry the variables $x_1, x_2, \ldots, x_n$ can be regarded as the coordinates of points of $(n - 1)$-dimensional space, the system of r independent equations defines an $(n - r - 1)$-dimensional plane in this space, the function U defines the quadric $U = 0$, and the reciprocal equations define an $(r - 1)$-dimensional plane that is the polar for the $(n - r - 1)$-dimensional plane relative to the quadric. Cayley was not yet in possession of multi-dimensional terminology, but the fact that he had in mind the picture we have described can be seen from his concluding words: "In the case of four variables the above investigations demonstrate the following properties of surfaces of the second order,"[43] after which he presents examples of theorems on quadrics, lines, and planes of three-dimensional projective space.

It is possible that Cayley arrived at the ideas of multi-dimensional geometry under the influence of Hamilton's quaternions: as early as 1845 Cayley published the paper "On Jacobi's elliptic functions and on quaternions," (*Phil. Mag. London*), mentioned above (Bk. 1, p. 77), in which he not only considers quaternions, but also generalizes them to octonions ("Cayley numbers"), which can be interpreted as vectors in an 8-dimensional space.

Grassmann's Multi-dimensional Geometry

The first fundamental investigation in multi-dimensional geometry was Grassmann's *Lineale Ausdehnungslehre* (1844), in which affine geometry was developed not only for three-dimensional space, but for multi-dimensional space also. That it was Grassmann who succeeded in establishing an articulated theory of multi-dimensional space seems to be due largely to the fact that Grassmann was not under the influence of other great mathematicians of his time, and to his predilection for philosophical generalizations.

After defining "first-level figures," "second-level figures," and "third-level figures," Grassmann remarked "since this method of formation is theoretically applicable without restriction, I can define systems of arbitrarily high level by this method. . . geometry goes no further, but abstract science knows no limits."[44] As can be seen, Grassmann interpreted geometry to be only the geometry of ordinary space, the only thing to which he applied the term "space," while what we now call the geometry of multi-dimensional spaces he called *Ausdehnungslehre*. Multi-dimensional spaces themselves he called "systems" of level higher than 3.

In his *Ausdehnungslehre* (1862) Grassmann defined an n-dimensional vector space and the "outer product" of vectors of this space, which are now called simply n-vectors. Grassmann represented outer products as linear combinations of outer products of basis vectors, which he called "units." The coefficients of these linear

---

43) Cayley, A. *Collected Mathematical Papers*. Cambridge 1889, Vol. 1, p. 62.
44) Grassmann, H. *Gesammelte mathematische und physikalische Werke*. S. 52–53.

combinations are now called the *Grassmann coordinates* of m-dimensional planes of n-dimensional space; these coordiantes are the basic method of studying m-dimensional planes in n-dimensional spaces, which explains the fact that manifolds of such planes are now called Grassmann manifolds. Addition and multiplication of "extensive quantities" led Grassmann to the algebra of "numbers of variable sign," which we have discussed above.

Plücker's *Neue Geometrie des Raumes*

Another approach to the concept of a multi-dimensional space was proposed by Plücker in his *System der Geometrie des Raumes* (1846), which we mentioned above. In this work he proposed that the basic element of ordinary space be taken as lines, which form a four-dimensional manifold, rather than points.

The ideas of Plücker's ruled geometry were further developed in his *Neue Geometrie des Raumes gegründet auf die Betrachtung der geraden Linie als Raumelement* (1868–1869). The second volume of this book, which was published posthumously, was edited by his student F. Klein. In this book Plücker coordinates were introduced, which are special cases of the Grassmann coordinates for $n = 3, m = 1$: if a line passes through a point with projective coordinates x_i and y_i, the Plücker coordinates of the line are $p_{ij} = x_i y_j - x_j y_i$. These coordinates, like x_i and y_i are defined up to a multiple, and when points with coordinates x_i and y_i are replaced by points with coordinates $z_i = \alpha x_i + \beta y_i$ and $t_i = \gamma x_i + \delta y_i$, they are multiplied by $\alpha\delta - \beta\gamma$; the coordinates p_{ij} are also connected by the relation $p_{01}p_{23} + p_{02}p_{31} + p_{03}p_{12} = 0$.

Plücker defined "linear complexes" to be complexes determined by one linear relation between the coordinates p_{ij} and "linear congruences" to be determined by two complexes. Such congruences are subdivided into hyperbolic (sets of lines that intersect two intersecting lines), elliptic (sets of lines joining conjugate imaginary points of two conjugate imaginary lines), and parabolic (tangents to a ruled quadric at points of one of its rectilinear generators); the last of these is a limiting case of the first two.

Schläfli's *Theorie der vielfachen Kontinuität*

In 1851 Ludwig Schläfli (1814–1895), a professor at the University of Berne, carried out wide-ranging research on multi-dimensional Euclidean geometry, which he presented to the Vienna Academy of Sciences. However his work was not published in its entirety until 50 years later. Still, the most important results of this work were published by Schläfli in the articles "Réduction d'un intégrale multiple qui comprend l'arc du circle et l'aire du triangle sphérique comme cas particuliers," (*J. math. pures et appl.*, 1855) and "On the multiple integral $\int^n dx\, dy \ldots dz$ whose limits are $p_1 = a_1 x + b_1 y + \cdots + h_1 z \geq 0$, $p_2 \geq 0$, $p_3 \geq 0, \ldots, p_n \geq 0$ and $x^2 + y^2 + \cdots + z^2 < 1$," (*Quarterly J. Math.*, 1858–1860). Schläfli's book was called *Theorie der vielfachen Kontinuität* (Basel 1901) and attempted "to provide

L. Schläfli

a foundation for and to develop a new branch of analysis, which, as a sort of analytic geometry of n dimensions, contains the analytic geometry of the plane and three-dimensional space as the special cases for $n = 2$ and 3."[45] Schläfli noted that the analog of a point of ordinary space is a "solution" ("Lösung") defined by the values of n variables $x, y, \ldots$. Schläfli called the set of all solutions an "n-fold totality," and if $1, 2, 3, \ldots$ equations are given, he called the set of their solutions an "$(n-1)$-fold, $(n-2)$-fold, $(n-3)$-fold,$\ldots$" continuum. As we see, Schläfli, like Grassmann, makes no use of the terminology of multi-dimensional geometry. Later Schläfli introduced "skew" systems of "variables" into the space he was studying, in which the distance between two "solutions" was

$$\sqrt{(x' - x)^2 + (y' - y)^2 + \cdots + 2k(x' - x)(y' - y) + \cdots},$$

and "rectangular coordinate systems," in which this distance is

$$\sqrt{(x' - x)^2 + (y' - y)^2 + \cdots}.$$

---

45) Schläfli, L. *Gesammelte mathematische Abhandlungen*. Basel 1950, Bd. 1, S. 171.

For two linear polynomials

$$p = ax + by + cz + \cdots + hw \text{ and } p' = a'x + b'y + \cdots + h'w$$

Schläfli determined the number of "solutions" for which $p > 0$ and $p' > 0$ simultaneously. Assuming that this set of solutions is to the whole "infinite totality" as a fraction is to a whole, and that the denominator of the fraction is 2π, he called the numerator the "angle between the polynomials p and p'," denoting it by $\sphericalangle(p, p')$. Schläfli defined this angle by the formula

$$\cos \sphericalangle(p, p') = \frac{aa' + bb' + cc' + \cdots + hh'}{\sqrt{a^2 + b^2 + \cdots + h^2}\sqrt{a'^2 + b'^2 + \cdots + h'^2}},$$

where the two square roots in the denominator are assumed positive.

From our point of view Schläfli's "polynomials" are $(n-1)$-dimensional hyperplanes and the angle he defined is the angle between those hyperplanes. Schläfli called planes of any dimension "linear continua" and curved surfaces "higher continua," and he referred to "single continua" (lines) as "paths" and "linear single continua" (straight lines) as "rays." Schläfli went on to define parallel planes and multi-dimensional parallelograms, which he called "parallel schemes." He proved that "the measure of a parallel scheme is the determinant of the orthogonal projections of its edges."[46] That is, if the "projections" of the edges of a "parallel scheme" are $x_0^i, x_1^i, \ldots, x_n^i$, then its volume is the determinant of the matrix (x_j^i). Schläfli went on to define multi-dimensional polyhedra, which he called "polyschemes." He found the volumes of multi-dimensional pyramids and other polyhedra and proved that "the measure of an arbitrary closed n-fold continuum is the square root of the sum of the squares of its projections."[47] He then proved a generalization of Euler's theorem: for a simply connected polyhedron of n-dimensional space the numbers N_p of p-dimensional faces (vertices when $p = 0$, edges when $p = 1$) are related by the equation

$$1 - N_0 + N_1 - N_2 + \cdots - (-1)^p N_p + \cdots + (-1)^n N_{n-1} - (-1)^n = 0$$

(the special case of this formula when $n = 2$ is the equality $N_0 = N_1$, and the special case for $n = 3$ is Euler's theorem $N_0 - N_1 + N_2 = 2$).

It was proved in the subsequent discussion that when $n \geq 5$ there are only three kinds of regular polyhedra — multi-dimensional generalizations of the tetrahedron, cube, and octahedron, having respectively $n + 1$ vertices and faces, 2^n vertices and $2n$ faces, and $2n$ vertices and 2^n faces, while for $n = 4$ there are six kinds of regular polyhedra: in addition to the three kinds just mentioned, which are also possible when $n = 4$, there are also polyhedra with 24 vertices and 24 faces, with 120 vertices and 300 faces, and with 300 vertices and 120 faces. In the second

---

46) *Ibid.*, S. 182.
47) *Ibid.*, S. 183.

part of this work, devoted to the "theory of spherical continua," Schläfli found in particular the surface area of the multi-dimensional sphere, which reduces to the integrals we have already given, which in turn were computed by Jacobi. In the third part he studied "quadratic continua," i.e., multi-dimensional quadrics, and in particular he found the center and principal axes of a "quadratic continuum."

Schläfli's work did not become well-known in his lifetime, and the regular polyhedra of n-dimensional space that he described in his work of 1858–1860 were rediscovered in the late nineteenth century by Washington Irving Stringham (1847–1909) of the University of California at Berkeley in a work entitled "Regular figures in n-dimensional space," (*Amer. J. Math.*, 1880) and by Reinhold Hoppe (1816–1900), a professor at the University of Berlin, in papers entitled "Regelmässige linear begrenzte Figuren von vier Dimensionen," (*Arch. Math. Phys.*, 1882) and "Die regelmässigen linear begrenzten Figuren jeder Anzahl der Dimensionen," (*Arch. Math. Phys.*, 1882).

The concept of an "n-fold extended manifold" was proposed in much more general form than that of Grassmann and Schläfli by B. Riemann in 1854 in his lecture "Über die Hypothesen, die der Geometrie zu Grunde liegen," which we shall discuss below.

The Multi-dimensional Geometry of Klein and Jordan

F. Klein made an important contribution to the propagation of the ideas of multi-dimensional space when he described, in his "Erlanger Programm" (1872) various groups of transformations of three-dimensional spaces; in the concluding paragraph he wrote, "The way to extend the preceding space to the concept of a pure manifold now seems obvious."[48] In the 1870's Klein wrote a number of papers on multi-dimensional geometry and in particular he published an article in which he argued that a closed curve with knots in three-dimensional space can be untied in four-dimensional space, and that the same is true of linked circles in three-dimensional space (1876). Klein later recalled, "Here again [as in the case of non-Euclidean geometry] the obstacle to progress were the philosophers. The latter lacked understanding of the immanent meaning that is inherent in mathematical theories... But, besides the opposition from philosophers, who asserted that n-dimensional space is meaningless, there arose an unexpected difficulty of quite the opposite character. There appeared philosopher/enthusiasts, who deduced from the existence and fruitfulness of the mathematical theory that some real four-dimensional space must exist in nature, from which, in their opinion, it followed that its existence could be proved experimentally."[49] Klein went on to talk about the astronomer and physicist F. Zöllner (1834–1882) and the American medium Henry Slade, "a remarkably skilled conjuror, who by the way, was debunked some years later." Klein told Zöllner of his work in 1876. "This communication," wrote

---

48) *On the Foundations of Geometry* [in Russian]. pp. 424–425.

49) Klein, F. *Vorlesungen über die Entwicklung der Mathematik im 19. Jahrhundert.* Erster Teil, S. 168–169.

Klein, "was received by Zöllner with an enthusiasm incomprehensible to me. He thought he had obtained a method for the experimental proof of the 'existence of the fourth dimension' and proposed to Slade that the latter should untie a knotted thread. Slade accepted the proposal... and actually succeeded in doing this to Zöllner's satisfaction. I mention only in passing that a sealed thread was used in this experiment, the seal of which Zöllner pressed together with his thumbs while Slade held his hand above Zöllner's."[50]

Zöllner's "experiments" were also described in an article of Friedrich Engels entitled "Natural Science and the Spirit World":

> According to the latest triumphant reports from the spirit world, it is said that Professor Zöllner has addressed himself to one or more mediums in order with their aid to determine more details of the locality of the fourth dimension. His success is said to have been surprising.... . The spirits prove the existence of the fourth dimension, just as the fourth dimension vouches for the existence of spirits.[51]

In 1875 Camille Jordan (Bk. 1, pp. 66–68) published his "Essai sur la géométrie à n dimensions," (*Bull. soc. math. France*, 1875; there was a short communication under the same title in *Comptes Rendus* in 1872). In contrast to the majority of papers on multi-dimensional geometry of the time Jordan made constant use of everyday terminology:

> We shall consider a point in the space of n dimensions as defined by the coordinates x_1, x_2,..., x_n. A linear equation between coordinates defines a *plane*, k simultaneous linear equations defines a *k-plane*, and $n-1$ equations a *line*, the distance between two points will be $\sqrt{(x_1 - x_1')^2 + \cdots}$, etc.[52]

This terminology coincides with modern notation except for the term k-plane, which is now used for a plane of dimension k rather than a plane of dimension $(n-k)$.

After investigating the conditions for parallelism and perpendicularity and transformations of coordinates, Jordan found the metric invariants of a k-plane and an l-plane, the stationary angles and shortest distance:

> A system consisting of a k-plane P_k and an l-plane P_l passing through a point of space has ρ different invariants, where ρ is the smallest of the numbers k, l, $n-k$, $n-l$. These invariants can be regarded as the angles between multi-dimensional planes (multiplans).[53]

The squares of the cosines of these angles are determined as the eigenvalues of a certain $\rho \times p$ matrix. Jordan also determined the shortest distance between nonintersecting P_k and P_l. He went on to find a canonical form for a rotation in n-dimensional space, i.e., the canonical form of an orthogonal matrix.

---

50) *Ibid.*, S. 270.

51) Engels, F. *Dialectics of Nature*. New York 1940, pp. 306–307.

52) Jordan, C. *Œuvres*. Paris 1962, Vol. 3, p. 3.

53) *Ibid.*, p. 5.

Riemannian Geometry

A new branch of multi-dimensional geometry was founded by B. Riemann (HM, Vol. 3) in his famous address "Über die Hypothesen, welche der Geometrie zu Grunde liegen") given on 10 June 1854 as his *Probevorlesung* at Göttingen University. Gauss was present at this lecture, and it was primarily directed at him. Riemann began as follows.

> The formation of the concept of quantity is possible only when a certain general concept is presupposed, connected with the admission of a number of different states. Depending on whether there exists or does not exist a continuous transition from one state to another we are dealing with a continuous or a discrete manifold; the individual states are called points in the first case and elements of the manifold in the second.[54]

Riemann defined the dimension of a continuous manifold as follows:

> Assume that to a certain concept a continuous set of states is assigned in such a way that it is possible to pass from one state to any other in a certain manner; then the totality of these states forms a singly extended manifold whose distinguishing characteristic is the possibility of continuous displacement at any stage in two directions — forward and backward.[55]

Mapping this set of states continuously into another set of states, Riemann defined a "doubly extended manifold," from it he determined analogously a "triply extended manifold," and repeating this operation n times, an "n-fold extended manifold." This concept essentially coincides with Grassmann's "extension," and Schläfli's "continuum." However, where Grassmann and Schläfli, after defining a multi-dimensional space of quite general form, they actually limited themselves to a multi-dimensional affine or Euclidean space, Riemann defined a metric on an "n-fold extended manifold," which did not reduce to the Euclidean metric, and was in essence a generalization of Gauss' intrinsic geometry of a surface. To be specific, after defining an n-fold extended manifold, Riemann posed the problem of the "metric relations that are possible on such manifolds" and the possibility of "expressing the results of computations in geometric form." Noting that "a secure foundation for both the one and the other was laid down in the famous work of Herr Geheimer Hofrath Gauss on curved surfaces,"[56] Riemann first considered very general assumptions and then limited himself to the case when the differential of arc length ds, i.e., the distance between points with coordinates x_i and $x_i + dx_i$, is a homogeneous algebraic function of the differentials with coefficients which are functions of the coordinates x_i only. Then, passing to the simplest possible cases, Riemann found that such a case is

$$ds^2 = \sum_{i,j} g_{ij}\, dx_i\, dx_j,$$

54) Riemann, B. *Mathematische Werke*. Berlin 1990, S. 305.
55) *Ibid.*, S. 307.
56) *Ibid.*, S. 308.

where $g_{ij} = g_{ji}$ are continuous and twice-differentiable functions of the variables x_i and the quadratic form is positive-definite. Riemann noted that "in particular for space, if we define the position of a point by rectangular coordinates, we have $ds = \sqrt{(dx)^2}$".[57] By "space" Riemann, like Grassmann, understood three-dimensional Euclidean space. Riemann's general expression for ds^2 is a generalization of Gauss' first fundamental form for a surface. Due to Riemann's condition, the n-dimensional space he defined can be regarded as n-dimensional Euclidean space in small regions.

Integrating ds along lines, Riemann obtained the lengths of lines in the space he had defined. Among the different lines joining two points a special place belongs to the shortest (geodesic) lines, defined by differential equations analogous to the differential equations of geodesics on surfaces. Using geodesic lines one can define very intuitively the most important concept of the geometry of the spaces considered by Riemann — the curvature of the space at point in a given two-dimensional direction. To do this one must draw two geodesics from a given point tangent to a given two-dimensional direction, join two points of these lines by a third geodesic, and measure the sum of the angles $A+B+C$ and the area of the resulting "geodesic triangle." ABC. The curvature of space at a given point in a given two-dimensional direction is defined as the limit of the ratio of the angular excess of the geodesic triangle $(A+B+C-\pi)$ to its area as the triangle is contracted to the given point. The angular excess, and hence the curvature, can be both positive and negative. Riemann defined the curvature in a given two-dimensional direction by considering infinitesimal geodesic triangles: in a neighborhood of a point at which he wished to define the curvature Riemann introduced coordinates x_i equal to zero at the point and connected with the distance s from points of this neighborhood to the given point by the relation $s = \sum_i x_i^2$ (at present they are called *Riemannian coordinates*). In these coordinates the linear element near the given point assumes the form

$$ds^2 = \sum_i dx_i^2 + \sum_{i,j,k} c_{ij,k} x_k \, dx_i \, dx_j + \sum_{i,j,k,l} c_{ij,kl} x_k x_l \, dx_i \, dx_j + \cdots,$$

where $c_{ij,k}$ and $c_{ij,kl}$ are the values of the partial derivatives $\partial g_{ij}/\partial x_k$ and $\partial^2 g_{ij}/\partial x_k \partial x_l$ at the point. Riemann remarked that because the coordinate lines are geodesics all $c_{ij,k}$ are equal to 0, and the second-order terms in the expansion of ds^2 constitute a quadratic form in the expressions $x_i \, dx_j - x_j \, dx_i$. If for the sake of uniformity we denote the infinitesimals x_i by δx_i, these expressions can be written in the form $\Delta x_{ij} = \delta x_i \, dx_j - \delta x_j \, dx_i$ and regarded as the coordinates of a parallelogram constructed on the vectors $\{dx_i\}$ and $\{\delta x_i\}$. Therefore the second-order terms in the expansion of ds^2 can be written in the form

$$\Delta \sigma^2 = \sum_{i,j,k,l} R_{ij,kl} \Delta x_{ij} \Delta x_{kl}.$$

---

57) *Ibid.*, S. 310.

Riemann went on to consider, as he described it, the quotient that results from division of $\Delta\sigma^2$ by the area of the triangle with vertices $(0, 0, \ldots, 0)$ $(x_1, x_i, \ldots, x_n)$, and $(dx_1, dx_2, \ldots, dx_n)$ — the triangle constructed on the same vectors $\{dx_i\}$ and $\{\delta x_i\}$, i.e., as we would say, the limit of this quotient as the triangle is contracted to a point. It is this quantity that Riemann took as a measure of the curvature. It differs only in its coefficient from the curvature of space that we defined above at a given point in a given two-dimensional direction (which coincides in the case of a surface with Gaussian curvature measure).

The curved multi-dimensional spaces defined by Riemann in his address of 1854 are now called *Riemannian spaces*, and their geometry is called *Riemannian geometry*.

The quantities $R_{ij,kl}$ form the so-called *curvature tensor*, or *Riemann tensor*. Like the coordinates of any tensor, they have the property that under a change of coordinates x_i the expressions for them in the new coordinates are linear combinations of the expressions in the old coordinates, from which it follows that if they vanish in one coordinate system, then they vanish in any coordinate system.

Riemann succeeded in applying the ideas of the geometry he had proposed to solve a practical problem. This application was the subject of a paper he wrote in 1861 bearing the title "Commentatio mathematica respondere tentatis quæstionæ ab Ill. Academiæ Parisiensi propositæ: Trouver quel doit être l'état calorique d'un corps solide homogène indéfini pour qu'un système de courbes isothermes, à un instant donné reste isothermes après un temps quelconque, de telle sorte que la température d'un point puisses s'exprimer en fonction du temps et de deux autres variables indépendantes."

The paper Riemann submitted to this competition, not quite finished and written in an extremely compressed form, was not understood by the jury, and was not awarded the prize. The manuscript was not published until it appeared in the first edition of Riemann's collected works (1876). Here Riemann brought the heat equation

$$\sum_i \frac{\partial}{\partial s_i}\left(\sum_j b_{ij}\frac{\partial u}{\partial s_i}\right) = h\frac{\partial u}{\partial t}$$

into its simplest form. This problem is equivalent to transforming the quadratic form $\sum_{i,j} b_{ij}\, ds_i\, ds_j$ to a sum of squares. A necessary and sufficient condition for such a transformation to be possible is that the following expression vanish:

$$K = \frac{1}{2}\frac{\sum\limits_{i,j,k,l}(ij,kl)(ds_i\delta s_j - ds_j\delta s_i)(ds_k\delta s_l - ds_l\delta s_k)}{\left(\sum\limits_{i,j}b_{ij}ds_i\, ds_j\right)\left(\sum\limits_{i,j}b_{ij}\delta s_i\delta s_j\right) - \left(\sum\limits_{i,j}b_{ij}\, ds_i\delta s_j\right)^2}.$$

This expression is invariant under a change of variables. In regard to this expression, which he denoted (III), Riemann wrote,

The expression $\sqrt{\sum\limits_{i,j} b_{ij}\, ds_i\, ds_j}$ can be regarded as a linear element in a general n-fold extended space lying outside our intuition. If in this space we draw all possible shortest lines from the point $(s_1, s_2, \ldots, s_n)$ whose initial directions are characterized by the relations

$$\alpha\,ds_1 + \beta\,\delta s_1 : \alpha\,ds_2 + \beta\,\delta s_2 : \ldots : \alpha\,ds_n + \beta\,\delta s_n$$

(α and β being arbitrary quantities), these lines form a certain surface that can be thought of as situated in the usual space of our intuition. In that case the expression (III) will be a measure of the curvature of the surface at the point $(s_1, s_2, \ldots, s_n)$.[58]

The quantities (ij, kl), the so-called quadruply-indexed symbols of Riemann, are the same as the ones we denoted $R_{ij,kl}$ above and can be expressed in terms of the coefficients g_{ij} of the Riemann fundamental quadratic form and their derivatives. Nowadays they are written in the form

$$R_{ij,kl} = \sum_h \left(\frac{\partial \Gamma^h_{jk}}{\partial x_i} - \frac{\partial \Gamma^h_{ik}}{\partial x_j} + \sum_r \Gamma^h_{ir}\Gamma^r_{jk} - \sum_r \Gamma^h_{jr}\Gamma^r_{ik} \right) g_{hl},$$

where the quantities Γ^i_{jk} are defined by the relations

$$\sum_l g_{il}\Gamma^l_{jk} = \frac{1}{2}\left(\frac{\partial g_{ik}}{\partial x_j} + \frac{\partial g_{ij}}{\partial x_k} - \frac{\partial g_{jk}}{\partial x_i} \right) \tag{6}$$

(Riemann denoted the quantities Γ^i_{jk} by p_{ijk}); by use of the quantities Γ^i_{jk} the equation of the geodesic lines of Riemannian space can be written in the form

$$\frac{d^2 x_i}{dt^2} + \sum_{j,k} \Gamma^i_{jk} \frac{dx_j}{dt}\frac{dx_k}{dt} = 0. \tag{7}$$

This equation determines the parameter t of the geodesic lines up to a transformation $t \mapsto at + b$.

Riemann conducted a separate study of the curved multi-dimensional spaces of nonzero constant curvature, which have the same degree of mobility as Euclidean space. When the measure of the curvature is α, the expression for ds can be reduced to the form

$$\left(1 + \frac{\alpha}{4}\sum x^2\right)^{-1} \sqrt{\sum dx^2}.$$

When $n = 2$, this becomes the expression for the linear element of a surface of constant curvature obtained by Minding in terms of the Gaussian curvature of the surface.

An example of an n-dimensional Riemannian space of constant curvature is the sphere in $(n + 1)$-dimensional Euclidean space, and an example of an n-dimensional Riemannian space of constant negative curvature is n-dimensional hyperbolic space, first defined by E. Beltrami in his "Teoria fondamentale degli spazî di curvatura costante," (*Ann. Mat.*, 1868).

---

58) *Ibid.*, S. 435.

Riemann's Idea of Complex Parameters of Euclidean Motions

The last part of Riemann's paper is entitled, "An application to space." We have already noted that the multi-dimensional space defined in the paper was called a *manifold* by Riemann, and he reserved the term *space* to mean physical space. After enumerating "conditions that are necessary and sufficient for the definition of metric relations in space," (the vanishing of the curvature at every point for each two-dimensional direction, the constancy of the curvature in space, and the "independence of the lengths of lines" from their locations), he concluded that "displacements, or changes in position are complex quantities expressible in terms of three independent units." [59] These last words of Riemann mean that the number of parameters that define a motion of space is finite, or, as geometers began to say after they had assimilated the concept of a group, that the number of parameters of the group of motions is finite. However the expression of the motions of space using complex quantities expressed in terms of these "three independent units" was not found until the very end of the nineteenth century, in a book by Prof. Aleksandr Petrovich Kotel'nikov (1865–1944) entitled *The Cross-Product Calculus and Certain of its Applications to Geometry and Mechanics* (Kazan' 1895), and soon afterwards in the book *Geometrie der Dynamen* by Prof. Eduard Study (1862–1922) of the University of Greifswald. It was shown in these works that the manifold of oriented lines of three-dimensional Euclidean space can be mapped onto a sphere in three-dimensional Euclidean space, not real Euclidean space, however, but one constructed over a particular kind of complex numbers introduced by W. Clifford in his "Preliminary study of biquaternions." These numbers are now called *dual numbers* (a term due to Study) and can be defined as expressions of the form $a + b\varepsilon$, $\varepsilon^2 = 0$. When this is done, as Kotel'nikov and Study showed, the dual distance between two points of this sphere representing two lines, forming an angle φ, and having shortest distance d, is equal to the dual number $\varphi + \varepsilon d$, while the motions of Euclidean space are represented by rotations of the dual sphere, and consequently depend on three dual parameters. It is possible that Riemann had been struggling to formulate these three parameters, although the nature of the complex numbers in terms of which the motions of Euclidean space can be expressed was not clear to him.

Riemann's Ideas on Physical Space

Next, after touching on "the extension of spatial constructions in the direction of the infinitely large," Riemann wrote,

> In explaining nature questions of the infinitely large are futile. The situation is quite different with questions of the infinitely small. Our knowledge of causal connections is intimately bound up with the precision with which we can study phenomena on an infinitely small scale. The success achieved over

---

59) *Ibid.*, S. 315.

the last few centuries in the understanding of the mechanics of the external world have been due almost exclusively to the precision of construction that became possible as a result of the discovery of infinitesimal analysis and the application of the fundamental elementary concepts that were introduced by Archimedes, Galileo, and Newton, and which are used by modern physics. Meanwhile in those areas of natural science where there are not yet any fundamental concepts to enable the analogous constructions to be performed, phenomena are being studied on the infinitesimal scale of space, to the extent that this can be done with a microscope, with the aim of establishing causal connections. Therefore questions of the metric relations of space on the infinitely small scale are by no means idle.

If we assume that bodies exist independently of their location, so that the curvature is everywhere constant, it follows from astronomical observations that this curvature cannot be nonzero. Or, if it is nonzero, one can at least say that the portion of the Universe accessible to telescopes is insignificant in comparison with a sphere of the same curvature.[60]

These last words of Riemann's show that, like Lobachevskiĭ, he was interested in measuring the sum of the angles of cosmic triangles "in the portion of the Universe accessible to telescopes" and knew that the results of these measurements at that time gave no reason to doubt that the sum of the angles is 180°. Riemann went on, again like Lobachevskiĭ, to surmise that "on the infinitesimal level" geometry could be different from what it is "in the large":

The empirical concepts on which the establishment of metric spatial relations is based — the concept of a rigid body and a light ray, seem to lose all clarity on the infinitesimal level. Therefore it is quite conceivable that the metric relations of space on the infinitesimal level do not correspond to geometric assumptions; indeed, we must adopt this assumption if the observed phenomena are to be made simpler by its use.[61]

As we see, Riemann was not only taking account of the molecular and atomic structure of matter; in an era of nearly unchallenged supremacy of the wave theory of light he was returning to the corpuscular conception of light, which brought him close to the modern conception of light as a stream of photons.

Noting that at the beginning of his address he had remarked that in the case of a discrete manifold the principle of metric relations is preserved in the very concept of a manifold, while in the case of a continuous manifold this principle is adjoined to space from without (i.e., in the case of a discrete manifold one can define distance in it by measuring the "steps" between the discrete points, while in the case of a continuous manifold one must give the coefficients of the metric form as a function of the coordinates of the points), Riemann wrote:

It follows from this that either the reality that gives rise to the idea of space forms a discrete manifold, or one must attempt to explain the appearance of

---

60) *Ibid.*, S. 317.
61) *Ibid.*, S. 317.

metric relations by something external — connecting forces acting on that reality.

One can hope to find answers to these questions only if, starting from the already existing concept confirmed by experiment whose basis was laid by Newton, we gradually perfect it, being guided by the facts that cannot be explained by it. The kind of investigations that are carried out in the present paper, namely those starting from general concepts, are useful only to prevent restricted concepts and ingrained prejudice from hindering progress toward knowledge of the connections in nature.

Here we stand on the threshold of a domain belonging to a different science — physics, and the present day does not give us occasion to cross it.[62]

In these words Riemann brilliantly anticipated both the "appearance of metric relations" resulting from the gravitational properties of matter established by Einstein's general theory of relativity and the ideas of discrete or "quantized" space that are being discussed in modern physics.

The Work of Christoffel, Lipschitz, and Suvorov on Riemannian Geometry

Riemann's research was continued by Elwin Bruno Christoffel (1829–1900), a professor at the Technische Hochschule in Zürich and then at the University of Strasbourg. In his paper "Über die Transformation der homogenen Differentialausdrücke zweiten Grades" (*J. für Math.*, 1869) he considered the conditions under which the Riemannian geometry defined by a form $\sum_{i,j} g_{ij} \, dx_i \, dx_j$ coincides with that defined by a form $\sum_{i,j} h_{ij} \, dx_i \, dx_j$. Where Riemann had posed the problem of generalizing Gauss' intrinsic geometry of surfaces to multi-dimensional spaces, Christoffel's problem was to generalize the problem of development of surfaces. The condition Christoffel found for the identity of the two geometries was that the differential forms $\sum_{i,j,k,l} R_{ij,kl} \, dx_i \, dx_k \, \delta x_j \, \delta x_l$ computed for the two qudratic forms be the same. Christoffel denoted the quantities Γ^i_{jk} by the symbols $\begin{Bmatrix} jk \\ i \end{Bmatrix}$, and the quantities $\Gamma_{jk,i} = \sum_l g_{il}\Gamma^l_{jk}$ by $\begin{bmatrix} jk \\ i \end{bmatrix}$, as a consequence of which these quantities are often called the "Christoffel symbols" of first and second kind respectively.

Simultaneously with Christoffel similar questions were answered by Rudolf Lipschitz (1832–1903), a professor at the University of Bonn, in his "Untersuchungen in Betreff der ganzen homogenen Funktionen von n Differentialen," (*J. für Math.*, 1870).

The same problem for three-dimensional Riemannian spaces was solved independently of Christoffel and Lipschitz by Fëdor Matveevich Suvorov (1845–1911), a native of the Perm' province, a student of Kotel'nikov and a professor at the University of Kazan', who did a great deal to popularize Lobachevskiǐ's discovery. Suvorov's master's thesis "On the characterization of three-dimensional systems,"

---

62) *Ibid.*, S. 318.

E. B. Christoffel

published in Russian in Kazan' in 1871 and in a brief French exposition in Paris in 1873, was devoted to the problem of the equivalence of two quadratic differential forms in three-dimensional Riemannian spaces. (Suvorov's doctoral dissertation, defended in 1884, was devoted to imaginary points and lines on the plane and their application to the construction of conics.)

The Multi-dimensional Theory of Curves

The creation of multi-dimensional Euclidean geometry soon led to the construction of differential geometry in a multi-dimensional space. C. Jordan, in his paper "Sur la théorie des courbes dans l'espace à n dimensions," (*Comptes Rendus*, 1874), constructed a multi-dimensional analogue of the Frenet-Serret formulas. Jordan considered a curve in n-dimensional space defined in rectangular coordinates by functions $x_1 = f_1(s), \ldots, x_n = f_n(s)$, taking the parameter s to be arc length. Considering a variable point of the curve with coordinates $X_1, \ldots, X_n$ "infinitely close to its predecessor" Jordan wrote, "Its coordinates, taking account of infinitesimals of order $n - k + 1$, will satisfy a system of linear equations, and the set of these equations defines the osculating k-plane to the given curve,"[63] We recall

---

63) Jordan, C. *Œuvres.* T. 4, p. 337.

F. M. Suvorov

that Jordan defined a "k-plane" to be the intersection of k "planes," which in modern notation are $(n-1)$-dimensional hyperplanes, and consequently Jordan's "osculating k-plane" is actually the osculating $(n-k)$-dimensional plane of the curve, i.e., the limiting position of the $(n-k)$-dimensional plane passing through $n-k+1$ points of the curve as all these points are contracted to a single point. When $k = n - 1$, this is the tangent line. Jordan went on to state an important property of "osculating k-planes": "Two adjacent osculating k-planes intersect in the osculating $(k+1)$-plane."[64] The expression "two adjacent k-planes" is a colloquial usage, since in fact between any two osculating planes of any dimension one can construct an osculating plane of the same dimension whose point of contact lies between the points of contact of the first two planes. But the statement formulated by Jordan has a precise meaning, namely that if we consider $n-k+2$ points of the curve and draw $(n-k)$-dimensional planes through the first and last of these and an $(n-k+1)$-dimensional plane through all these points, then when all of $n-k+2$ of these points are contracted to a single point, the $(n-k+1)$-dimensional plane, the two $(n-k)$-dimensional planes and the $(n-k-1)$-dimensional plane of intersection of two $(n-k)$-dimensional planes become respectively the $(n-k+1)$-dimensional osculating plane, the $(n-k)$-dimensional osculating plane, and the

___
64) *Ibid.*

$(n - k - 1)$-dimensional osculating plane, while in the general case two $(n - k)$-dimensional planes of n-dimensional space intersect in an $(n - 2k)$-dimensional plane, and when one of the $(n - k)$-dimensional planes approaches the other this intersection becomes an $(n - 2k)$-dimensional plane.

Jordan continued:

The points of space can be referred to a new rectangular coordinate system analogous to the system considered by Serret for ordinary space, namely: 1) the osculating plane at the point $x_1, \ldots, x_n$; 2) the plane perpendicular to the preceding plane containing the osculating 2-plane (biplan); 3) the plane perpendicular to the osculating 2-plane, containing the osculating 3-plane (triplan), etc., as far as the plane perpendicular to the tangent. The new coordinates $\xi_1, \ldots, \xi_n$ will be connected with the old ones by the orthogonal relations

$$X_\rho = a_{\rho 1}\xi_1 + \cdots + a_{\rho n}\xi_n."^{65}$$

The phrase "orthogonal relations" means that the matrix $(a_{\rho\sigma})$ is an orthogonal matrix and, consequently, the columns $a_{\rho 1}, \ldots, a_{\rho n}$ of this matrix can be regarded as the coordinates of the vectors of an orthonormal basis. Jordan continued:

Let us now consider a point of a curve infinitely close to the preceding point, obtained by replacing s by $s+ds$. We draw through this point a new system of rectangular coordinates analogous to the preceding. Let $\xi_1 + d\xi_1, \ldots, \xi_n + d\xi_n$ be the coordinates relative to the new axes. Then the following relations will hold:

$$\frac{d\xi_1}{ds} = c_1\xi_2, \ldots, \frac{d\xi_\rho}{ds} = -c_{\rho-1}\xi_{\rho-1} + c_\rho\xi_{\rho+1},$$

$$\ldots, \frac{d\xi_n}{ds} = -c_{n-1}\xi_{n-1} + 1,$$

where the coefficients $c_1, \ldots, c_n$ by definition are the *curvatures* of the given curve at the point $x_1, \ldots, x_n$.

The formulas

$$\frac{da_{\rho 1}}{ds} = c_1 a_{\rho 2}, \ldots, \frac{da_{\rho\sigma}}{ds} = -c_{\sigma-1}a_{\rho,\sigma-1} + c_\sigma a_{\rho,\sigma+1},$$

$$\ldots, \frac{dx_\rho}{ds} = a_{\rho n},$$

---

65) *Ibid.*

which make it possible to express the successive derivatives of the coordinates x_ρ as functions of the coefficients a, the curvatures, and their derivatives, can be deduced immediately.[66]

If we write the vectors $\{a_{\rho 1}\}, \ldots, \{a_{\rho n}\}$, which as we have seen constitute an orthonormal basis, in the form $\mathbf{a}_1, \ldots, \mathbf{a}_n$, and the vector $\{x_\rho\}$ in the form $\mathbf{x}$, these last formulas of Jordan can be rewritten as

$$\frac{d\mathbf{a}_1}{ds} = c_1\mathbf{a}_2, \ldots, \quad \frac{d\mathbf{a}_\sigma}{ds} = -c_{\sigma-1}\mathbf{a}_{\sigma-1} + c_\sigma\mathbf{a}_{\sigma+1},$$

$$\ldots, \frac{d\mathbf{a}_n}{ds} = -c_{n-1}\mathbf{a}_{n-1}, \quad \frac{d\mathbf{x}}{ds} = \mathbf{a}_n.$$

Nowadays the vectors $\mathbf{a}_\sigma$ are denoted $\mathbf{t}_{n-\sigma+1}$, and Jordan's "curvatures" c_σ are denoted $-k_{n-\sigma}$, the curvature k_i being called the ith curvature of the curve. Thus the formulas found by Jordan are now written as

$$\frac{d\mathbf{x}}{ds} = \mathbf{t}_1, \quad \frac{d\mathbf{t}_1}{ds} = k_1\mathbf{t}_2, \ldots,$$

$$\frac{d\mathbf{t}_i}{ds} = -k_{i-1}\mathbf{t}_{i-1} + k_i\mathbf{t}_{i+1}, \ldots, \frac{d\mathbf{t}_n}{ds} = -k_{n-1}\mathbf{t}_{n-1}.$$

These formulas are the multi-dimensional generalization of the Frenet-Serret formulas. The geometric meaning of the curvature k_1 is obvious — it is the absolute value of the vector $\dfrac{d\mathbf{t}_1}{ds}$, which equals the limit of the ratio of the angle $\Delta\alpha_1$ between the tangents at two nearby points of the curve to the arc length Δs between the points; the geometric meaning of the absolute value of the curvature k_{n-1} is also obvious — it is the absolute value of the vector $d\mathbf{t}_n/ds$, equal to the limit of the ratio of the angle $\Delta\alpha_{n-1}$ between the osculating $(n-1)$-dimensional planes at two nearby points of the curve to the arc length Δs between these points. Jordan also showed the geometric meaning of all the intermediate curvatures k_i: "The angle φ_k between two adjacent osculating k-planes is given by the very simple formula

$$\varphi_k = ds \cdot c_k."^{[67]}$$

The meaning of this concise statement is that if we again consider $n - k + 2$ points of the curve, draw $(n - k)$-dimensional planes through the first and last of them, and denote the angle between them by $\Delta\alpha_k$, then the limit of the ratio of this angle to the arc length Δs between the first two of these points as all the points are contracted to the first point equals the curvature k_{n-k}, i.e., in other words, the curvature k_i is the limit of the ratio of the angle between the i-dimensional osculating planes to the curve at two nearby points to the arc length between these points.

---

66) *Ibid.*

67) *Ibid.*, p. 338.

We have seen above that in 1872 Jordan determined the stationary angles between m-dimensional planes in n-dimensional Euclidean space. The result of Jordan just given shows that if we take two osculating m-dimensional planes at two nearby points and find the stationary angles between these planes, the ratio of all these angles, except one, to the arc length Δs between these points as all the points are contracted to a single point tends to zero, and the ratio of the last of these angles to Δs tends to the curvature k_{n-k}.

We remark that in Jordan's notation all the curvatures c_i have the same sign, but he used both right- and left-handed coordinate systems; nowadays only right-handed systems are used, and the curvatures k_{n-1} are assigned a positive or negative sign according as the osculating helix of constant curvature k_i to the curve forms a right-threaded or a left-threaded screw.

At the end of the note Jordan found the shortest distance between two osculating planes of different dimensions at two infinitesimally near points of the curve, thereby generalizing the classical Bonnet problem of finding the shortest distance between the tangents at two infinitesimally near points of a curve in ordinary space and the distance between a point of this curve and the osculating plane at an infinitely near point.

We note that Jordan himself did not give the derivation of either the generalization of Frenet's formulas and other properties of curves that he found or the geometric meaning of the curvature. Nowadays the multi-dimensional Frenet formulas are derived in all textbooks of multi-dimensional differential geometry; in contrast the geometric meaning Jordan found for the curvatures k_i is usually not mentioned in these textbooks, although once effective algorithms are established for computing the stationary angles of m-dimensional planes in n-dimensional space the verification of this fact causes no difficulty.

Multi-dimensional Surface Theory

In the same year of 1874, in a paper entitled "Généralisation du théorème d'Euler sur la courbure des surfaces" (*Comptes Rendus*), Jordan laid the foundation of the differential geometry of m-dimensional surfaces in n-dimensional space. Considering the tangent m-dimensional planes to a surface at two infinitesimally near points, Jordan determined the stationary angles of these planes and showed that on a surface there are m mutually orthogonal directions in which the limit of the ratio of the sum $\varphi_1^2 + \varphi_2^2 + \cdots + \varphi_m^2$ to the arc length between the points of tangency has a stationary value. The essence of this theorem is that the sum $\varphi_1^2 + \varphi_2^2 + \cdots + \varphi_m^2$ considered by Jordan is the square of the distance between m-dimensional planes of n-dimensional space passing through a single point in the Riemannian metric in the manifold of these planes that is invariant with respect to the group of rotations of space (this Riemannian space is one of the simplest examples of É. Cartan's symmetric spaces); and the metric in the manifold of m-dimensional planes induces the metric of an m-dimensional Riemannian space in the manifold of tangent planes. If we denote the Riemannian metric forms of

this m-dimensional surface and the m-dimensional Riemannian space in the mani-
fold of planes respectively by $\sum_{i,j} g_{ij}\, du_i\, du_j$ and $\sum_{i,j} h_{ij}\, du_i\, du_j$, then, since these
forms are positive definite, at each point of the surface there exists a basis in which
both forms assume simultaneously the canonical form $\sum_i g_{ii}\, du_i^2$ and $\sum_i h_{ii}\, du_i^2$;
the vectors of the basis are orthogonal in both metrics. Jordan himself formulated
this theorem as follows:

> A k-surface located in a space of $m + k$ dimensions possesses m orthogonal
> dimensions at each point such that the sum of the squares of the angles
> formed by two adjacent tangent planes divided by ds^2 will be maximal or
> minimal.[68]

In accordance with his terminology Jordan took a k-surface to be the intersection of k hypersurfaces, i.e., an m-dimensional surface of $(m + k)$-dimensional
space; and "two adjacent tangent k-planes" is a colloquial expression denoting the
tangent m-dimensional planes at two infinitesimally near points of the surface.
By the expression "maximal or minimal" Jordan meant (when $m > 2$) that only
the necessary condition for an extremum was met, i.e., that the ratio in question
assumed a stationary value.

The following year Richard Beez (1827–1902), a Gymnasium teacher in
Plauen wrote a paper "Zur Theorie des Krümmungsmasses von Mannigfaltigkeiten
höherer Ordnung," ($Z.$ $Math.$ $Phys.$, 1875–1876), in which he constructed the theory of an $(n-1)$-dimensional surface in n-dimensional space. Beez defined the first
and second fundamental forms on this surface, the first being the Riemann form.
In modern notation, if the radius-vector $\mathbf{x}$ of a point of the surface is a function
of curvilinear coordinates $u_1, u_2, \ldots, u_{n-1}$, and the partial derivatives $\dfrac{\partial \mathbf{x}}{\partial u_i}$ and
$\dfrac{\partial^2 \mathbf{x}}{\partial u_i \partial u_j}$ and the unit normal vector to the surface are the vectors $\mathbf{r}_i$, $\mathbf{r}_{ij}$, and $\mathbf{n}$
respectively, the coefficients of these forms can be written as the inner products
$g_{ij} = \mathbf{x}_i \mathbf{x}_j$ and $b_{ij} = \mathbf{n}\mathbf{x}_{ij}$. In this case the condition found by Beez

$$b_{ij}b_{kl} - b_{ik}b_{jl} = R_{il,jk}$$

is one of the conditions for integrability of the multi-dimensional analogues of the
derivation formulas of Gauss and Weingarten. Beez discovered that in the general
case when $n > 3$ this condition makes it possible to express the coefficients b_{ij} of the
second form in terms of the coefficients g_{ij} of the first form and their derivatives,
whence it follows that in the general case an $(n-1)$-dimensional surface in n-
dimensional space is undeformable. The undeformability of even an infinitesimally
small piece of the surface seemed ridiculous to Beez, and he drew the conclusion
that the theorem he had proved was evidence that multi-dimensional geometry is
inconsistent!

---

68) $Ibid.$, p. 343.

Multi-dimensional Projective Geometry

In the second half of the nineteenth century, along with multi-dimensional Euclidean and non-Euclidean geometries mathematicians began to develop multi-dimensional projective geometry as well. For the most part it was multi-dimensional complex projective space and algebraic lines and surfaces in it that were studied, making it possible to extend the algebraic geometry of lines of the projective plane and lines and surfaces of projective three-dimensional space to the multi-dimensional case. This work, as already stated, is more algebra than geometry. The paper of H. Schubert already mentioned, "n-dimensional generalizations of the basic numerical characteristics of our space," belongs to multi-dimensional algebraic geometry.

The Terminology of Multi-dimensional Geometry

We have seen that the creators of multi-dimensional geometry, Grassmann, Schläfli, and Riemann, used a peculiar terminology: they used the words "extension," "continuum," "totality," or "manifold" instead of "space," "solution" or "state" instead of "point," etc. In general they avoided applying the term "geometry" to multi-dimensional geometry. By the end of the nineteenth century, however, the usual geometric terminology, first applied to multi-dimensional geometry by C. Jordan, came to be used more and more frequently. The books of Pieter Hendrik Schoute (1846–1913) *Mehrdimensionale Geometrie* (Leipzig 1902–1905), which contain a systematic exposition of both analytic and synthetic geometry of n-dimensional Euclidean space, do not differ at all from the modern terminology.

In his famous 1895 memoir "Analysis situs" H. Poincaré wrote,

> The geometry of n dimensions studies reality; no one doubts that. Bodies in hyperspace are subject to precise definitions, just like bodies in ordinary space; and while we cannot draw pictures of them, we can imagine and study them. If, for example, the mechanics of more than three dimensions must be condemned as meaningless, the position of hypergeometry is quite different. Indeed, geometry does not have as its sole purpose the direct description of bodies perceived by our sense organs: first of all it is an analytic study of a certain group and consequently nothing hinders the study of new groups that are analogous and more general.
>
> However the question immediately arises: must the language of analysis be replaced by the language of geometry, which loses its advantages as soon as it becomes impossible to use the senses? It turns out that this new language is more precise; moreover the analogy with ordinary geometry can create associations of productive ideas and suggest useful generalizations.[69]

We note that here Poincaré is still using the terms "hyperspace" and "hypergeometry," although in the memoir itself he makes systematic use of the term

---

69) Poincaré, H. *Selected Works* [Russian translation]. Moscow 1972, Vol. 2, p. 457.

"space of n dimensions"; here Poincaré is useing the term "reality" in a completely realist spirit, despite his conventionalist tendencies in the theory of knowledge.

At the same time, even in the late nineteenth and early twentieth centuries some prominent scholars were still very antipathetic to both multi-dimensional geometry and the theory of functions, claiming that no applications to individual problems of classical mathematics and science were to be expected from these disciplines.[70] However in the twentieth century multi-dimensional spaces of the most varied kinds and even infinite-dimensional "Hilbert" spaces and other spaces were highly developed and applied in both mathematics and physics.

6. TOPOLOGY

Gauss' Topology

Leibniz' idea of "analysis situs" (HM, Vol. 2, pp. 126–127) was frequently mentioned by mathematicians of the eighteenth and early nineteenth centuries. To be sure, Carnot and Staudt interpreted this idea as that of nonmetric, mainly projective geometry, and Grassmann interpreted it as the idea of vector calculus. But Euler was already giving his paper on the "Königsberg bridge problem" the title "The solution of a problem belonging to *geometria situs*" (HM, Vol. 3, pp. 204–205), interpreting Leibniz' term topologically.

Gauss understood this term in the same sense. Figures of various kinds of knots can be found in papers of Gauss dating to the year 1794, and his notes "Zur geometria situs" and "Zur Geometrie der Lage für zwei Raumdimensionen" are devoted to knots.

In a notebook on electrodynamics Gauss wrote in 1833, "In geometria situs, which Leibniz foresaw and to which only a few geometers (Euler and Vandermonde) have paid some slight attention, after a century and a half we know little more than nothing."[71] He went on to solve a problem intermediate between "geometry of position" and "geometry of magnitude," consisting of the determination of the number of linkings of two infinite or closed lines. In doing this Gauss remarked that the integral

$$\iint \frac{(x'-x)(dy\,dz' - dz\,dy') + (y'-y)(dz\,dx' - dx\,dz')}{(x'-x)^2 + (y'-y)^2 + (z'-z)^2},$$

---

70) See, for example, Lyapunov, A. M. "The life and works of P. L. Chebyshev," in: *Selected Mathematical Works of P. L. Chebyshev* [in Russian]. Moscow-Leningrad 1946 pp. 19–20. Here, noting the negative attitude of Chebyshev and his students toward the "pseudogeometric refinements in spaces of four and more dimensions," Lyapunov expressed a high opinion of the "profound geometric research of Lobachevskiĭ," who, in his opinion "like Chebyshev, always stayed on the foundation of reality."

71) This note was published by James Clerk Maxwell in his *Treatise on Electricity and Magnetism*. Oxford 1881, Vol. 1, p. 43.

where x, y, and z are the coordinates of points ranging over the first line and x', y', and z' are the coordinates of points ranging over a second line, equals $4\pi m$, where m is the number of linkings of the lines. In the quotation just given Gauss had in mind, besides Euler's paper on the "Königsberg bridge problem," the paper "Remarques sur les problèmes de situation," (*Hist. Acad. Sci. Paris*, 1771 (1774)) by T. Vandermonde, which was devoted to the problem of the "knight's move" in chess, solved by Euler in 1759.

In 1840 Gauss wrote a note in which he introduced the concept of one- and two-dimensional manifolds, which he called a "line" (*Zug*) and a "layer" (*Schicht*), and posed the question of the decomposition of a "layer" bounded by several "lines" into several "layers" bounded by one "line." What is actually involved here is the problem of the connectedness of a surface, which was later solved by Riemann and subsequently became one of the most important problems of combinatorial topology.

Like his notes on non-Euclidean geometry, Gauss' notes on topology were never published, though he sometimes mentioned his ideas on this question in letters to friends.

Generalizations of Euler's Theorem on Polyhedra in the Early Nineteenth Century

Various generalizations of Euler's theorem on polyhedra, according to which the number of vertices N_0, the number of edges N_1, and the number of faces N_2 of a convex polyhedron are connected by the relation

$$N_0 - N_1 + N_2 = 2,$$

played an important role in the foundation of topology.

We have already pointed out (HM, Vol. 3, p. 202) that this theorem holds not only for convex polyhedra but also for all polyedra whose surfaces are in a one-to-one bicontinuous correspondence with a sphere, i.e., are homeomorphic to the sphere, and also for any closed surfaces homeomorphic to the sphere. Here in the case of surfaces the role of faces, edges, and vertices are played by portions of the surface homeomorphic to the interior of a disk, arcs of common boundaries of pairs of such domains, and points belonging to the boundaries of three and more domains.

The first attempt to generalize Euler's theorem to polyhedra of more complicated shape was undertaken by the French mathematician and mechanist Louis Poinsot (1777–1859), a professor at the Ecole Polytechnique, in a paper "Sur les polygones et les polyèdres," (*J. Éc. Polyt.*, 1810). In this paper Poinsot rediscovered the two starlike polyhedra discovered earlier by Kepler and also two new starlike polyhedra that are dual in relation to the first two, and he showed that for the two of these polyhedra for which $N_0 = 12$, $N_1 = 30$, $N_2 = 12$ the Euler relation fails while the following relation holds:

$$2N_0 + N_2 = N_1 + 6.$$

We note that in the preface to this paper Poinsot applied the same term *géométrie de situation* that we encountered in the work of Vandermonde. Nevertheless he connected it only with Euler's problem of the Königsberg bridges and did not see the connection between "geometry of position" and properties of polyhedra.

Cauchy, in his "Recherches sur les polyèdres" (*J. Éc. Polyt.*, 1813), showed that for a net of polygons homeomorphic to the interior of a disk the relation

$$N_0 - N_1 + N_2 = 1$$

holds, and applied this relation to the proof of Euler's theorem. He also proved that for N_3 polyhedra the following relation holds:

$$N_0 - N_1 + N_2 - N_3 = 1.$$

In the same year there appeared the "Mémoire sur la polyédrométrie," (*Ann. Math.*, 1813) by Simon L'Huilier (1750–1840), a professor at the University of Geneva. In this paper conditions were found under which Euler's theorem holds, and it is established that for polyhedra with p handles ("polyhedra of genus p") the more general relation

$$N_0 - N_1 + N_2 = 2 - 2p$$

holds (cf. HM, Vol. 3, p. 203).

Like Euler's theorem, the theorem of L'Huilier holds not only for polyhedra, but also for any surfaces homeomorphic to polyhedra of genus p. We remark that in the case of starlike polyhedra, for which the Poinsot relation given above holds, we have $2N_0 + N_2 - N_1 = 6 - 2p$.

Listing's *Vorstudien zur Topologie*

The term "topology" first appeared in "Vorstudien zur Topologie" (*Göttinger Studien*, 1847), written by Johann Benedikt Listing (1808–1882), a mathematician and physicist at Göttingen and a student of Gauss. This term, now universally accepted but rarely applied until the 1920's, was a neologism coined by Listing himself; Listing proposed using this term, compounded from the Greek words τόπος (place) and λόγος (study), as a replacement for Leibniz' term *geometria situs*, since the word *geometry* suggested measurement, which plays no role in topology. In addition Leibniz' term *géométrie de position* had already been applied to denote projective geometry. Calling the properties of figures that are preserved under continuous transformations "modal," Listing wrote, "We shall use the term *topology* to mean the study of modal relations of three-dimensional figures or the laws of connectivity, mutual position, and order of points, lines, surfaces, solids, and parts of them or the totality of them in space independent of the relations of measure and

J. B. Listing

quantity."[72] The main object of study in the *Vorstudien zur Topologie* is formed by "line complexes," i.e., any lines or curves or collections of lines,"[73] Listing studied knots, chains, and braids and other forms of mutual position of line complexes. Listing's exposition was accompanied by a large number of examples from biology and engineering.

In 1862 Listing published "Der Census räumlicher Complexe, oder Verallge-meinerung des Euler'schen Satzes von der Polyedern," (*Abh. Ges. Wiss.*, Göttin-gen, 1862). Listing took a "three-dimensional complex" to mean "any configuration of points, lines, and surfaces in space. The lines and surfaces may be any desired lines and curves, open or closed, bounded or unbounded."[74] However, Listing con-sidered the "components" of three-dimensional complexes to be not only points, lines, and surfaces, but also solids. Listing used the word *Census* in the title to refer to the classification of complexes. Of the classes of complexes introduced here we mention a "diaphragm" (a surface bounded by a closed curve). Listing defined "simple linkage" of two closed curves to mean that they were so situated relative

---

72) Listing, J. B. *Vorstudien zur Topologie*. S. 35.

73) *Ibid.*, S. 106.

74) *Abhandlungen der königlichen Gesellschaft der Wissenschaften*, Göttingen, **10**, S. 97.

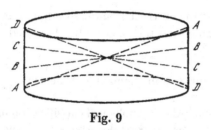

Fig. 9

to each other that each intersected the diaphragm bounded by the other. A "diagram" is a line complex that is the "skeleton" of a three-dimensional complex. In this paper Listing proved a number of theorems on three-dimensional complexes, one of which extends Euler's theorem on polyhedra to three-dimensional complexes homeomorphic to the sphere. Listing also exhibits complexes for which this theorem fails. In the same paper Listing described the so-called Möbius band — a one-sided surface that can be obtained by identifying the points of a finite right circular cylinder that are symmetric with respect to the center of symmetry or by gluing together corresponding points of half of such a cylinder (Fig. 9).

Möbius' "Theorie der elementaren Verwandschaft"

Möbius first described the "Möbius band" in a paper presented to the Paris Academy in 1861 as an entry to a competition on the theme "Improve in some important point the geometric theory of polyhedra." Möbius' paper, written in bad French and containing many new ideas, was not understood by the jury, and like the other papers submitted to the competition, was not awarded the prize. The contents of this paper were published by Möbius in his articles "Theorie der elementaren Verwandschaft," (*Ber. Verh. Sächs.*, Leipzig, 1863) and "Über die Bestimmung des Inhaltes eines Polyeders," *Ber. Verh. Sächs.*, Leipzig, 1865).

We have encountered the term "relationship" when speaking of Möbius' projective and affine geometry; and we have seen that he used this term to mean a geometric transformation. In the paper "Theorie der elementaren Verwandschaft" Möbius understood the word "relationship" in the same sense. Here it is a one-to-one bicontinuous transformation: "Two figures are said to be *in elementary relationship* if every infinitesimal element of one figure corresponds to an infinitesimal element of the other in such a way that two adjacent elements of the first figure correspond to two adjacent elements of the second."[75]

Möbius went on to carry out a classification of homeomorphic surfaces, distinguishing in particular "monions," "binions," "ternions," etc., in "elementary relationship" with surfaces bounded by one, two, three, etc. curves. Then for surfaces of various classes he proved the theorems of Euler and L'Huilier.

In the second article, while sharpening the concept of volume of a polyhedron, he introduced the concept of orientation of polyhedra and surfaces. It was in this connection that he gave the "Möbius band" as an example of a nonorientable

---

75) Möbius. A. F. *Gesammelte Werke*. Leipzig 1886, Bd. 2, S. 517.

surface. He used it to construct examples of "nonorientable polyhedra" for which the theorems of Euler and L'Huilier do not hold and for which volume cannot be defined.

The Topology of Surfaces in Riemann's "Theorie der Abel'schen Functionen"

B. Riemann approached the topological classification of surfaces from a completely different position. In his "Theorie der Abel'schen Functionen" (*J. für Math.*, 1857) Riemann wrote:

> In studying the functions that arise in the integration of total differentials one cannot get by without certain propositions that belong to *analysis situs*. This name, used by Leibniz (although not quite in this sense), should, I believe, be understood as that part of the study of continuously varying quantities in which the variables are not regarded as existing independently of their position and connected by numerical relations, but are studied completely inependently of numerical relations, from the point of view of just the spatial relations of mutual location and connection arising between them.

As a result of Riemann's authority the Leibnizian term "analysis situs" was the most popular name for this area of mathematics for fifty years.

With the algebraic functions $y = f(x)$ defined implicitly by an algebraic equation $F(x, y) = 0$, where $F(x, y)$ is a polynomial in the complex variables x and y, Riemann associated the "multi-sheeted surface" consisting of several "sheets" which pass into one another at "branch points." Among these surfaces, nowadays called Riemann surfaces, Riemann distinguished "simply connected [surfaces], on which any closed curve is the full boundary of some portion of the surface (such, for example, is a disk), and multiply connected [surfaces], which do not posses this property (for example, the annulus formed by two concentric circles),"[77] Riemann pointed out that by using systems of cuts a multiply connected surface can be converted to a simply connected one.

For the sake of better visualization Riemann gave examples of simply connected, doubly connected, and triply connected surfaces.

"*The simply connected surface.* Decomposed into separate parts by any cut. Every closed curve forms the complete boundary of part of the surface (Fig. 10a).

The doubly connected surface. Every curve q that does not separate the surface makes it simply connected. Every closed curve, together with the curve a forms the complete boundary of some part of the surface (Fig. 10b).

The triply connected surface. On this surface every closed curve together with the curves a_1 and a_2 forms the complete boundary of some part of the surface. Every cut that does not separate the surface makes it a doubly connected surface, and two such cuts q_1 and q_2 make it simply connected (Fig. 10c)."[78]

---

76) Riemann, B. *Werke*. S. 123.

77) *Ibid.*, S. 124.

78) *Ibid.*, S. 127–128.

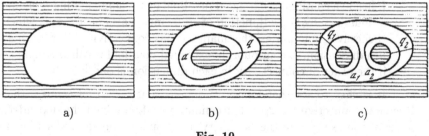

a)   b)   c)

Fig. 10

We have already mentioned that for plane curves Riemann introduced the characteristic p that Clebsch later called the "genus of the curve." This characteristic equals half of the number of cuts needed to make the surface simply connected. (Since Riemann surfaces are closed, this number of cuts is necessarily even.) For the surface of a sphere $p = 0$, while for the torus $p = 1$. For the surface of a sphere with several handles p equals the number of handles. Nowadays the number p is called the *genus of the surface*. In the case of polyhedra this number equals the genus of the surface as defined by L'Huilier.

In his "Theorie der Abel'schen Functionen" Riemann proved that the genus p of a Riemann surface, the number of sheets n, and the number of branch points w are connected by the relation

$$w - 2n = 2p - 2,$$

and observed that "this relation has an essentially nonmetric character belonging to *analysis situs*."[79]

The Multi-dimensional Topology of Riemann and Betti

Riemann extended the topological properties of the surfaces that he discovered to multi-dimensional manifolds. In the table of contents to his address "Über die Hypothesen, welche der Geometrie zu Grunde liegen" Riemann made the following remark on the first section of this address, which bore the title "The concept of an n-fold extended quantity": "Section 1 is simultaneously an introduction to research in *analysis situs*."[80] A collection of his works published in 1876 contained "fragments belonging to *analysis situs*," in which the topological properties of a two-dimensional surface were generalized to an n-dimensional manifold, called by Riemann an "n-chain." Here Riemann defined what are nowadays called called *homologous n-chains*: "an n-chain A is said to be transformable to another n-chain B if A and parts of B jointly form the complete boundary of the interior of an $(n + 1)$-chain." He gave a very important definition:

---

79) *Ibid.*, S. 144–145.

80) *Ibid.*, S. 318.

If inside a continuous extended manifold every nonbounding n-chain becomes a boundary by the adjunction of m definite parts of n-chains which are not themselves boundaries, this manifold has connectivity of order $m + 1$ in the nth dimension. A continuously extended manifold is called *simply connected*, if the order of connectivity in each dimension is one.[81]

It is easy to see that for a two-dimensional surface of genus p the order of Riemann connectivity is $2p + 1$. Riemann also determined the dependence of the order of connectivity of the boundary of a manifold on the order of connectivity of the manifold itself.

Riemann's ideas on the "connectivity" of a multi-dimensional manifold were expounded by his friend Enrico Betti (1823–1892), a professor of the university and director of the Scuola Normale Superiore in Pisa, in the paper "Sopra gli spazî di un numero qualunque di dimensioni," (*Ann. Mat.*, 1871). Betti defined "spaces" (*spazî*) to be manifolds in multi-dimensional Euclidean spaces. Betti denoted the order of connectivity of genus m (in n dimensions) by $p_m + 1$; in the case $p_m = 0$ he called this property *simple connectivity* of genus m.

The ideas of Riemann and Betti formed the foundation for the creation of combinatorial topology by Poincaré in the late nineteenth century.

Jordan's Topological Theorems

C. Jordan made a significant contribution to the solution of various problems of topology. In his paper "Sur la déformation des surfaces" (*J. math. pures et appl.*, 1866), Jordan extended Gauss' problem of finding necessary and sufficient conditions for developability of two flexible and inextensible surfaces on each other without folds and tears and proved that for two flexible and extensible surfaces or parts of them to be developable on each other without folds and tears it is necessary and sufficient that the number of contours bounding parts of the surfaces be the same (if the surfaces are closed, this number is zero) and that the maximal number of closed contours not intersecting any part of themselves or one another that can be drawn on each of these surfaces without separating it be the same for both surfaces. Jordan's proof is based on the decomposition of both surfaces into infinitesimal elements, adjacent elements of one surface corresponding to adjacent elements of the other. The identity of this notation with that of Möbius' 1863 paper on "elementary relationship" makes it very likely that Möbius' paper influenced Jordan. In another memoir the same year, "Des contours tracés sur les surfaces," (*J. math. pures et appl.*, 1866) Jordan posed the question of contours on a surface that can be mapped into one another by a continuous deformation on the surface. This problem, which subsequently came to be called the problem of homotopic contours on a surface, was generalized to multi-dimensional manifolds by Poincaré in his "Analysis situs."

Jordan is also the author of the famous theorem that a closed continuous plane curve separates the plane into two regions, the inside and the outside, and

---

81) *Ibid.*, S. 511.

the inside region is always contained in a disk of finite radius. This theorem was proved by Jordan in an appendix to the third volume of his *Cours d'analyse de l'École Polytechnique*, (Paris 1887) entitled "A remark on certain propositions of the theory of functions." Jordan's proof was carried out as follows. Considering a closed continuous plane curve C without multiple points, for each sufficiently small ε Jordan constructed two polygons, one of which contains the curve C and the other is contained inside it, all the points of these polygons lying at a distance less than ε from the curve C. He then constructed a sequence of such polygons for decreasing values of ε. He defined points outside of the curve C to be those that become exterior to the outside polygons from some index on, points inside the curve being defined analogously. Finally, points of the curve itself are defined as those that are exterior to all the polygons inside C and interior for all the polygons outside C.

The "Klein Bottle"

There is also a section on "analysis situs" in Klein's "Erlanger Programm," devoted to the determination of the role of groups of various geometric transformations in geometry (see below). Klein characterized this branch of geometry as the "search for invariants under transformations made up of infinitesimal deformations."[82]

In his "Bemerkungen über den Zusammenhang der Flächen," (*Math. Ann.*, 1874) Klein proved that the projective plane is homeomorphic to a Möbius band glued together with a disk. If the Möbius band can be obtained by identifying the points of a finite cylinder symmetric with respect to the center of symmetry, the projective plane can be obtained by identifying diametrically opposite points of a sphere, or, what is the same from the topological point of view, by identifying points of a finite cylinder with two bases that are symmetric with respect to its center of symmetry. Under this identification the lateral surface of the cylinder gives the Möbius band and the bases give the disk. From this it can be seen that the projective plane is a one-sided closed surface. Klein himself identified diametrically opposite points of hyperboloids of two sheets and one sheet.

In his lectures *Über Riemanns Theorie der algebraischen Funktionen und ihrer Integrale* (Leipzig 1882), Klein included a separate chapter on "Classification of closed surfaces according to the number p." After beginning with the problem of one-to-one conformal mappings of surfaces Klein remarked that "it is even more important to define an element that is invariant not only with respect to conformal mappings, but also for all one-to-one mappings. This is Riemann's p, the number of cuts that can be made in a surface without disconnecting it."[83] Klein specified that he had in mind transformations defined by continuous functions. Citing Jordan's paper "Sur la déformation des surfaces," Klein went on to introduce the concept of standard surfaces to which all surfaces with various numbers p are homeomorphic: "The sphere and torus will serve as standard surfaces for $p = 0$ and $p = 1$. For

---

82) *On the Foundations of Geometry* [in Russian]. p. 420.

83) Klein, F. *Gesammelte mathematische Abhandlungen.* Berlin 1923, Vol. 3, S. 526.

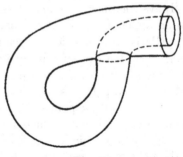

Fig. 11

larger values of p one can consider (as standard surfaces) a sphere endowed with p handles."[84]

In this same work the so-called Klein bottle is defined, a one-sided surface of which, according to Klein, "one can form a conception by turning a piece of rubber tubing inside out and forcing it to intersect itself in such a way that when its ends are joined its outside is connected to its inside."[85] (Fig. 11)

A significant contribution to topology was made by Klein's student Walther von Dyck (1856–1934), who worked mainly in Munich and is the author of the book *On the Analysis Situs of Three-dimensional Spaces* (London 1884) and the papers "Beiträge zur Analysis situs" (*Math. Ann.*, 1885, 1888, 1889). In this work von Dyck proved that the Euler characteristic of the n-dimensional sphere is 2 for even n and 0 for odd n, and the Euler characteristic of n-dimensional projective space is 1 for even n and 0 for odd n. Moreover in the first case projective space is nonorientable, while in the second case it is orientable.

The foundation of combinatorial topology as a science was completed by Poincaré in his frequently cited memoir "Analysis situs" (1895) and the five supplements to it (1899–1902).

7. GEOMETRIC TRANSFORMATIONS

Geometric Transformations in the Work of Möbius

In the work of Möbius considered above an important role is played by the concept of "relationship" in a great variety of forms of "equality" (congruence), "similarity," "affineness," "collineation" (projectivity), and "elementary relationship" (homeomorphism). Each kind of "relationship" is a special case of its successor. For that reason Klein wrote, "Although Möbius had not yet mastered the concept of a group in its modern formulation, the concept of "relationship" is equivalent to it; Möbius is thus a precursor of the 'Erlanger Programm'."[86]

---

84) *Ibid.*, S. 512.

85) *Ibid.*, S. 571.

86) Klein, F. *Vorlesungen über die Entwicklung der Mathematik im 19. Jahrhundert.* Erster Teil, S. 118.

We note also Möbius' paper "Die Theorie der Kreisverwandschaft in rein geometrischer Darstellung," (*Ber. Verh. Sächs. Leipzig,* 1855), in which the theory of circular transformations of the plane was developed, i.e., transformations generated by inversions with respect to circles; in analytic form these transformations can be interpreted as fractional-linear transformations in the complex plane.

The theory of circular Möbius transformations was soon generalized to three dimensions, the essential step being a theorem of J. Liouville, whom we have mentioned many times (Bk. 1, pp. 196–198), published by him in one of the appendices to the 1850 edition of his *Applications d'analyse à la géométrie de G. Monge.* Here Liouville proved that the conformal mappings in three-dimensional space are generated by the inversions with respect to spheres: in the plane conformal transformations are defined by arbitrary analytic functions of a complex variable and comprise a significantly larger class than the circular transformations, so that conformal transformations in three dimensions are the analogs of circular transformations of the plane rather than the conformal mappings.

Helmholtz' Paper "Über die Thatsachen, die der Geometrie zu Grunde liegen"

An important role in the incorporation of the concept of a transformation group into geometry was played by a paper of the famous German physiologist, physicist, and mathematician Hermann Helmholtz (1821–1894) entitled "Über die Thatsachen, die der Geometrie zu Grunde liegen," (*Gött. Nachr.,* 1868). The title itself, which echoes the title of Riemann's 1854 address, shows the connection between the two papers. Helmholtz wrote

> My immediate purpose was, like Riemann, to study which properties of spaces belong to each manifold of a family depending continuously on several variables, when the differences in the manifolds admit quantitative comparison, and which of them, in contrast, not being due to this common character, are peculiar to three-dimensional space.
>
> At the time physiological optics provided me with two examples admitting a three-dimensional representation of variable manifolds, namely the system of colors also mentioned by Riemann, and the system of measurements of the visual field. The two manifolds exhibit certain fundamental differences and induced me to compare them.[87]

Moreover Helmholtz noted that many of the results he obtained coincided with results of Riemann and that, although the publication of Riemann's speech deprived him of the "right of priority" in a number of cases, this coincidence was for him an "important testimony to the correctness" of the route that he chose "in an area of questions that had fallen out of favor due to earlier unsuccessful efforts." Helmholtz went on to point out that his research differed from Riemann's efforts in that he made a more detailed study of the influence of the restrictions he introduced, distinguishing real space from other "multiply extended manifolds,"

---

87) *On the Foundations of Geometry* [in Russian]. p. 367.

H. Helmholtz

on the basis of the fact that ds^2 is a homogeneous function of degree two in the differentials of the coordinates. First among these restrictions Helmholtz placed the requirement of "free mobility of solid figures without change of shape in all parts of space," i.e., the presence of motions that bring congruent figures into coincidence.

Helmholtz based his research on the following "hypotheses":

I. *The space of n dimensions is an n-fold extended manifold...* II. *The existence of movable but invariant (solid) bodies or systems of points is assumed...* III. *Completely free mobility of solid bodies is assumed...* IV. *If a solid body revolves about n − 1 points chosen so that the position of the body depends on only one independent variable, then a revolution without reverse rotation will eventually bring the body back to the initial position from which it departed...* V. *Space has three dimensions.* VI. *Space is infinite in extent.*[88]

In conclusion Helmholtz wrote that

___

88) *Ibid.*, pp. 368–382.

Riemann's research and mine, taken together, thereby show that the postulates given above...constitute a sufficient foundation for developing the study of space.

He made a remark on this last phrase: "They do not distinguish between the geometries of Euclid and Lobachevskiĭ."[89]

The expression "space of n dimensions," which was strange in the 1860's, is explained by the fact that the subsequent hypotheses, as we have seen, reduce this "space" to one of the two three-dimensional spaces claiming to describe physical space — those of Euclid and Lobachevskiĭ. Helmholtz' Hypothesis III requires that space contain a group of geometric transformations not only mapping each point of space to any other point of space, but also mapping each linear element emanating from the first point to any linear element emanating from the second point. If we accept Riemann's requirement that ds^2 be expressible by a quadratic form in the differentials of the coordinates whose coefficients are functions of the coordinates of the point alone, then Helmholtz' Hypotheses I–III single out spaces of constant curvature among the Riemannian spaces thereby defined, i.e., Euclidean space, hyperbolic space, and elliptic space. As can be seen from these hypotheses, Helmholtz does not assume in advance (as Riemann had done) that this quadratic form is positive definite. Without such an assumption Helmholtz' Hypotheses I–III are also satisfied by pseudo-Euclidean space and a sphere of real radius in pseudo-Euclidean space. This possibility is excluded by Helmholtz' Hypothesis IV: in pseudo-Euclidean space, besides the Euclidean planes there are also pseudo-Euclidean planes, in which circles have the shape of hyperbolas. Revolution in such a plane cannot lead back to the initial position "without reverse rotation." The same is true of the spheres just mentioned. Helmholtz' Hypothesis VI excludes elliptic geometry.

Sophus Lie, who founded the theory of continuous groups, wrote in 1886 that this "famous paper of Helmholtz...considers a problem closely related to the new theory of transformation groups,"[90] and went on to propose his own solution to this problem, based on his own theory.

Klein's "Erlanger Programm"

We have already mentioned the "Erlanger Programm" several times, a lecture read by F. Klein in 1872 when he became professor at the University of Erlangen. The full title of this lecture is *Vergleichende Betrachtungen über neuere geometrische Forschungen*, (Erlangen 1872). The discovery of a projective interpretation of hyperbolic geometry in 1871 showed that the group of motions of hyperbolic space is a subgroup of the group of collineations of projective space mapping the "absolute" of hyperbolic space into itself. On the other hand, the group of motions of Euclidean space is a subgroup of the group of similarities (the "principal group" of

---

89) *Ibid.*, p. 382.

90) *Ibid.*, p. 383.

Euclidean space) which in turn is a subgroup of the group of affine transformations — a subgroup of the same group of collineations. At the same time the groups of rigid motions and similarities of Euclidean space are subgroups of the group of conformal transformations of space (in the case of the plane, groups of circular transformations). The group of motions of elliptic space is also a subgroup of the group of collineations. Comparing these facts, Klein arrived at the conclusion that every geometry is really the study of the invariants of some transformation group.

Klein began his "Erlanger Programm" by defining the concept of a transformation group.

> The most important concept needed in the following discussion is that of a group of transformations of space.
> An arbitrary number of transformations of space yields via composition another transformation of space. If a given series of transformations has the property that each transformation obtained by successive application of several transformations belonging to the series itself belongs to the series, we call this series a *transformation group*."[91]

As examples of transformation groups Klein gave "the totality of all motions" of space (he remarked that "each motion is regarded as an operation performed on the entire space") and the "totality of collineations." Klein pointed out that the rotations about a point form a subgroup of the group of motions and that the correlations ("dual transformations") do not themselves form a group, but they form a group when taken together with the collineations.

After pointing out that the geometric properties of a figure are independent of the position occupied by the figure in space, its absolute size, and the orientation in the location of its parts, i.e., that these properties are invariant under motions and similarities of space, as well as under reflections and all transformations composed of them, Klein concluded that "geometric properties are invariant under the transformations of the principal group, and conversely one can say that geometric properties are characterized by invariance under the transformations of the principal group."[92] From "space," i.e., three-dimensional Euclidean space, Klein passed to an arbitrary "manifold."

> In analogy with spatial transformations we speak of transformations of a manifold. They also form groups, but there no longer exists a group distinguished from the others by its importance, as in the case of space: all groups are of equal significance. The following wide-ranging problem thus results as a generalization of geometry:
>
> *Given a manifold and a group of transformations on it, study the properties of figures belonging to the manifold that are invariant under the transformations of the group.*

---

91) *Ibid.*, p. 401.

92) *Ibid.*, p. 402.

In modern terminology, which, however, is usually applied only to a specific group, namely the group of all linear transformations, one can also express this as follows:

Given a manifold and a group of transformations on it, develop the theory of invariants of this group.[93]

Klein then considered "transformation groups, one of which contains another," both in "space" and in "various manifolds": the group of motions, the "principal group," the groups of affine transformations, collineations, and all projective transformations (collineations and correlations), and also "groups of transformations by reciprocal radii" (the groups generated by inversions, i.e., the groups of circular transformations of the plane and conformal transformations of space). Klein also pointed out the extension of the "fundamental group of collinear and dual transformations obtained by introducing suitable imaginary transformations. This step is brought about by a preliminary extension of the class of proper elements of space by the adjunction of imaginary elements."[94]

Transference Principles

In his "Erlanger Programm" Klein considered cases of isomorphism of transformation groups that made it possible to interpret one geometry on another; following L. O. Hesse, Klein called such interpretations "transferences" of one geometry into another. The term "transference" was introduced by Ludwig Otto Hesse (1811–1874), whom we mentioned in the chapter on algebra (Bk. 1, p. 80) in connection with an invariant that he introduced, known as the Hessian in his honor. This was done by Hesse in his paper "Über ein Übertragungsprinzip" (*J. für Math.*, 1866). Hesse's "transference" was based on the fact that if a line of the projective plane is projected from a point of some conic onto that conic (this projection is an analog of the stereographic projection of a line on a circle in the Euclidean plane), then the collineations of the line are represented by collineations of the plane that map the conic into itself, and conversely. This mapping defines an isomorphism of the group of collineations of the line and the group of collineations of the plane mapping the conic into itself.

The group of collineations of the projective line defines a geometry of binary forms on it, i.e., linear forms $\sum_i a_i x^i$ in the two projective coordinates of points of the line. Klein stated Hesse's principle of transference as follows: "The theory of binary forms and the projective geometry of systems of points on a conic section are the same, i.e., to each binary theorem there corresponds a theorem on point systems of a similar kind and conversely."[95] Considering pairs of real and conjugate-imaginary points of a conic and passing from these pairs of points to the lines joining them, then to the poles of these lines with respect to the conic,

---

93) *Ibid.*

94) *Ibid.*, p. 406.

95) *Ibid.*, p. 409.

Klein gave a new form to this principle: "The theory of binary forms and the projective geometry of the plane, which we study using a certain conic section as basis, are equivalent," or, by virtue of the interpretation Klein had given of hyperbolic geometry (which he called *general projective metric geometry* here), "the theory of binary forms and general projective metric geometry on the plane are the same."[96] Klein also pointed out that instead of a conic section in the plane one can take a third-order curve in space, and so forth, having in mind the interpretation of the hyperbolic plane on the so-called norm-curve of three-dimensional and multi-dimensional space, i.e., the curve whose parametric equation in projective coordinates can be brought into the form $x_0 = 1$, $x_1 = t$, $x_2 = t^2,\ldots, x_n = t^n$.

Another example of "transference" considered in the "Erlanger Programm" is "Plücker transference," which is based on the Plücker coordinates $p_{ij} = x_i y_j - x_j y_i$. We have seen that these coordinates are determined up to a common factor and are related by the quadratic equation $p_{01}p_{23} + p_{02}p_{31} + p_{03}p_{12} = 0$. Therefore the Plücker coordinates can be regarded as projective coordinates of a point of a quadric in six-dimensional projective space. Calling the linear forms in four variables x_0, x_1, x_2, x_3 *quaternaries*, Klein deduced that "the theory of quaternary forms coincides with the projective metric definition in a manifold of six homogeneous variables,"[97] having in mind that the quadric in question can be regarded as the absolute of six-dimensional space with a projective metric.

By stereographically projecting an oval quadric on the plane Klein remarked that in this situation the group of collineations of space mapping the oval quadric to itself and leaving the center of projection fixed can be represented as the "principal" group of the plane, and the group of collineations of space mapping the oval quadric into itself can be represented by the group of circular transformations of the plane supplemented by a point at infinity (as the extended plane of a complex variable, not by a line, like the projective plane). Therefore Klein concluded that "elementary plane geometry and the projective study of a surface of second order with one of its points adjoined are the same"[98] and "the geometry of reciprocal radii on the plane and projective geometry on a second-order surface are identical."[99] In a similar way, stereographically projecting the oval quadric in four-dimensional projective space onto a three-dimensional plane in this space, Klein remarked that the group of collineations of four-dimensional space mapping an oval quadric to itself can be represented as a group of conformal transformations of three-dimensional space supplemented by a point at infinity and concluded that "the geometry of reciprocal radii in space reduces to the projective study of the manifold represented by a quadratic equation connecting five homogeneous variables."[100] In these interpretations the points of three- and four-dimensional spaces are represented by circles of the plane or spheres of

---

96) *Ibid.*, p. 410.
97) *Ibid.*, p. 410.
98) *Ibid.*, p. 409.
99) *Ibid.*, p. 413.
100) *Ibid.*, p. 413.

three-dimensional space represented on quadrics by sections of the polar planes of these points. The projective coordinates of these points coincide with the so-called tetracyclic coordinates of the corresponding circles or pentaspherical coordinates of the corresponding spheres introduced by the French geometer Gaston Darboux (1842–1917) in his paper "Sur une classe remarquable de courbes algébriques et sur la théorie des imaginaires," (Bordeaux 1873), as a consequence of which this transference is frequently called the "Darboux transference."

Klein also formulated an interpretation of the oval quadric on a plane in a different way, remarking that the plane supplemented by a point at infinity can be regarded as the extended plane of a complex variable and as the complex projective line: "The theory of binary forms can be represented by the geometry of reciprocal radii on the real plane in such a way that complex values of the variables can also be represented." [101] Regarding the oval quadric as a sphere, Klein formulated the same interpretation in the following form: "The theory of binary forms of a complex variable is represented in projective geometry on a spherical surface." [102]

Together with the conformal transformations of space, which are point transformations of space supplemented by a point at infinity mapping spheres into spheres (regarding planes as special cases of spheres), Klein considered also two groups of transformations of spheres — the so-called higher geometry group of Lie spheres, defined by S. Lie in his paper "Über Komplexe, insbesonders Linien- und Kugelcomplexe, mit Anwendungen auf die Theorie partieller Differentialglei- chungen," (*Math. Ann.*, 1872). The transformations of this group map spheres into spheres preserving tangency, both planes and points being regarded as special cases of spheres. The group of conformal transformations is a subgroup of this group mapping points into points, and these transformations preserve angles between spheres. Another subgroup of this group mapping planes into planes (continuing to regard points as special cases of spheres) is the group of Laguerre transforma- tions defined by Laguerre in his paper "Sur la géométrie de la direction," (*Bull. soc. math. France*, 1880). This group preserves the tangential distance between spheres (the segment of the common tangent of two identically oriented spheres between the points of tangency).

Cremona Transformations

Of great importance in algebraic geometry are the so-called *birational* or *Cremona* transformations, named after the Italian geometer Luigi Cremona (1830–1903), the founder of the Italian school of geometers, a participant in the war for Ital- ian independence and one of the organizers of higher education and research after Italian unification in 1860. A native of Pavia, Cremona was a professor of the Uni- versity of Bologna and the Milan Technical Institute, and after 1873 director of the Engineering College in Rome, which was combined with the department of natural science of the University. Cremona's paper "Sulle trasformazioni geometriche delle

---

101) *Ibid.*, p. 414.
102) *Ibid.*

L. Cremona

figure piane" (*Mem. di Bologna*, 1863) and many papers of later years are devoted to "Cremona transformations." Cremona is also known as the founder of graphical methods of solving the problems of statics, the so-called graphostatics.

Birational transformations are transformations of the projective plane or projective space expressed by rational functions of the projective coordinates and having the property that the inverse transformations are also expressed as rational functions of these coordinates. The simplest birational transformation is inversion with respect to a circle. Another elementary birational transformation of the projective plane is the transformation $x'_i = 1/x_i$, first studied by Plücker in 1830. Other birational transformations are those used by Newton (HM, Vol. 2, pp. 114–117) to obtain third-order curves from the ellipse, parabola, and hyperbola, which he called "hyperbolisms" of the corresponding conic sections. As can be seen from these examples, the order of algebraic curves is not preserved under birational transformations; but, as Cremona showed, the genus p of algebraic curves is preserved.

CONCLUSION

In the development of geometry in the nineteenth century one can trace three main lines:

1) *Increased sophistication of geometric methods and results relating to the geometry of ordinary space, especially in the domain of differential geometry.* In the course of this century mathematicians developed a theory of invariants and forms that determine, up to rigid motion in space, a curve (Serret and Frenet), a surface (Gauss and Weingarten, Peterson and Bonnet), or a rectilinear congruence (Kummer), a significant number of special types of curves, surfaces, and congruences were studied, and the theory of deformation of surfaces received significant development.

2) *Enlargement of conceptions of space.* The main discovery here was non-Euclidean geometry (Lobachevskiĭ, Bólyai, Gauss, Cayley, and Klein). The appearance of the new non-Euclidean geometry led to the transformation of projective geometry, which had originally been a theory of the projective properties of figures in ordinary space (Poncelet, Möbius, Plücker, Steiner, and Chasles) into the geometry of projective space (Staudt and Cayley). In the nineteenth century affine and conformal geometries (Möbius and Liouville) appeared on the same route, and the foundations for symplectic geometry were laid (Möbius), along with multi-dimensional Euclidean and affine geometries (Cayley, Grassmann), followed by multi-dimensional projective and non-Euclidean geometry (Riemann, Beltrami). The theory of intrinsic geometry of a surface (Gauss, Minding, Liouville, Bonnet) led to the appearance of geometry of multi-dimensional Riemannian surfaces. The study of the topological properties of curves and surfaces (Gauss, Listing, Möbius, Riemann) led to the appearance of the topology of multi-dimensional manifolds (Riemann, Betti).

3) *Penetration of algebraic methods into geometry.* This involved first of all the development of the algebraic geometry of curves and surfaces (Plücker, Riemann, Clebsch) and the geometry of transformation groups (Helmholtz, Klein, Lie), which led to Klein's definition of various geometries as theories of the invariants of transformation groups.

While the first line of development of geometry was to a large degree exhausted in the nineteenth century, though interesting studies in the geometry of three-dimensional Euclidean space continued into the twentieth century, especially studies of the deformation of surfaces as well as the axiomatics of this space (which was significantly developed in the late nineteenth century), nevertheless the principal content of the development of geometry in the twentieth century has been along the second and third of these lines. The enlargement of the concept of space in the twentieth century occurred in several directions essentially inherited from the nineteenth-century geometers: at the beginning of the century there appeared the general concept of a topological space (Hausdorff and others) and then the concept of a topological manifold was formulated (L. E. Brouwer), the concepts of Poincaré's combinatorial topology were carried over to topological spaces (P. S. Aleksandrov, A. N. Kolmogorov, L. S. Pontryagin, J. Alexander, H. Hopf, and others). The discovery of Einstein's special theory of relativity (1905) aroused new interest in non-Euclidean geometries. The geometry of pseudo-Euclidean space, which describes the space-time of special relativity, began to develop in the early twentieth century (Poincaré, Minkowski, Klein, Sommerfeld), and then the geometry of Galilean space, which describes the space-time of the classical mechanics of Galileo and Newton (A. P. Kotel'nikov), leading to the appearance of new classes of projective metrics (B. Blaschke, D. M. Y. Sommerville). Of even greater significance for geometry is the discovery of Einstein's general theory of relativity (1916): this discovery aroused new interest in Riemannian geometry, where Levi-Cività discovered parallel transport (1917). The geometry of pseudo-Riemannian space appeared, describing the space-time of the general theory of relativity, and the tensor analysis discovered by G. Ricci in the late nineteenth century was perfected. Attempts to create a unified field theory led to the discovery of an affine connection space (H. Weyl, J. A. Schouten, É. Cartan), and later projective and conformal connection spaces (É. Cartan). These spaces are affine and conformal spaces, both Riemannian and Euclidean. These theories in turn led to the creation of the general concept of a homogeneous space. The theory of differentiable manifolds (O. Veblen, J. Whitehead) arose, on the basis of which the theory of spaces with connections was clarified, and the theory of fiber bundles was created. Merging with topology, the theory of differentiable manifolds generated differential topology. On the basis of the application of topological methods "global differential geometry" arose in both three-dimensional and in multi-dimensional and non-Euclidean spaces (A. D. Aleksandrov). Projective differential geometry received significant development (G. Fubini, E. Čech, S. P. Finikov). One should particularly note the productive application of the methods of geometry of affine connection spaces to this geometry (A. P. Norden). Affine differential geometry (B. Blaschke, P. A. Shirokov), conformal differential geometry, and the geometry of manifolds embedded in other spaces with transformation groups appeared. Algebraic methods, which had begun to penetrate into geometry in the nineteenth century, now became one of the main methods of study. Algebraic geometry received further development and, merging with topology, generated algebraic topology. The application of Lie groups to geometry received significant development. On this route there arose the theory of symmetric spaces (P. A. Shirokov, É. Cartan) and reductive spaces (K. Nomizu, P. K. Rashevskiǐ). While the classical projective, non-Euclidean, and conformal geometries are the geometries of simple Lie groups, the geometry of the other simple Lie groups was constructed in the twentieth century (E. Study, G. Fubini, J. L. Coolidge, É. Cartan, C. Chevalley), followed by the geometry of semi-simple Lie groups (of which only Euclidean and pseudo-Euclidean geometries had been studied previously) and the geometry of more complicated groups. Many of these geometries of simple Lie groups, semi-simple Lie groups, and so forth, are geometries over algebras, which made it necessary to create a general theory of spaces over algebras and more general rings. Another motivation for the creation of this theory came from the papers of D. Hilbert on the foundations of geometry, as a result of which geometries over finite fields appeared in the late nineteenth and early twentieth century (G. Fano, O. Veblen), which received an unexpected application in combinatorial analysis and mathematical statistics.

One should note in particular the theory of infinite-dimensional spaces (Hilbert, S. Banach), which grew out of studies in the theory of functions in the late nineteenth century (S. Pincherle) and the theory of integral equations in the early twentieth century (Hilbert). This theory, on the one hand, is one of the objects of the study of topology in the twentieth century, and on the other hand, the object of a new mathematical discipline, which arose in the early twentieth century at the interface of mathematical analysis, geometry, and algebra — functional analysis.

Chapter 2
Analytic Function Theory

Results Achieved in Analytic Function Theory in the Eighteenth Century

The machinery of power series for representing functions and solving various problems of mathematics and mechanics was used systematically by Newton starting in the 1660's. However it was left to the eighteenth century to perfect the technique of operating with power series, the series used by Newton being supplemented by the series of Taylor and Lagrange. In the eighteenth century, mostly at the initiative of Euler, infinite products, partial fraction expansions, integral representations (gamma function, elliptic integrals), and continued fractions were applied along with power series.

In forming analytic expressions from elementary functions of one or several variables by means of addition, subtraction, multiplication, division, raising to powers, extracting roots, solving (algebraic) equations, and integrating, Euler and his contemporaries were bound to obtain functions analytic everywhere except at isolated singularities, in a neighborhood of which the functions nevertheless admitted power series expansions (containing fractional and negative powers). For that reason, although they interpreted a function in general as an analytic expression, they had sufficient grounds for regarding it as an analytic function. Facts of another sort, in which an analytic expression, for example a convergent series of elementary functions, represented a nonanalytic function, were exceptions in the practice of mathematicians of the time (trigonometric series in the work of D. Bernoulli). The mathematical analysis of the eighteenth century could not accommodate these "exceptions" and for the most part did not consider it necessary to mention them. *Analytically representable functions are analytic functions*: this is one of the basic principles of eighteenth-century mathematical analysis (not stated in exactly these terms, however). This principle was tacitly assumed by almost all of the mathematicians of the first quarter of the nineteenth century. In the late eighteenth century it had been adopted by Lagrange as the starting point for the construction of differential and integral calculus (*Théorie des fonctions analytiques*, 1797).

An entirely new theme in eighteenth-century mathematical analysis was the use of complex values of an independent variable. At the beginning of the first volume of his *Introductio in analysin infinitorum* (1748) Euler emphasized that "even zero and imaginary values are allowed as values of a variable," and went on to give a large number of examples of the fruitfulness of such a broad interpretation of the variable. Thus in this work of Euler a variable was a complex variable. It is curious that in the second volume, in which a variable is given a geometrical interpretation, the interpretation of it is restricted to the domain of a real variable.[1] Some 30 years after the publication of the *Introductio*, in a letter of 20 December 1777 written to the Italian mathematician Lorgna, Lagrange had solid grounds for saying. "I consider it one of the important steps made by analysis in recent years that imaginary quantities no longer present any difficulty and that computations are performed as easily with them as with real quantities."[2]

In the work of Euler the theory of elementary functions of a complex variable received a full development and was perfected by the middle of the eighteenth century. Euler also significantly broadened the usual stock of analytic functions, introducing the gamma function, the zeta function, and the cylindrical functions, establishing the fundamental relations for them and for functions already occurring in the work of other authors (elliptic integrals and elementary functions of theta function type). To be sure, he considered all these nonelementary functions, as a rule, only for real values of the independent variable.

Following Nicolaus Bernoulli, Euler established that the equation $\partial u/\partial x = \partial v/\partial y$ is the condition under which $v\,dx + u\,dy$ is the exact differential of some function (HM, T. 3, p. 342). In connection with problems of fluid mechanics d'Alembert was the first to arrive at the system of equations

$$\frac{\partial u}{\partial x} = \frac{\partial v}{\partial y}, \quad \frac{\partial u}{\partial y} = -\frac{\partial v}{\partial x}. \tag{1}$$

D'Alembert, and after him Euler, obtained pairs of solutions of this system as the real and imaginary parts of an analytic function $f(z) = u + iv$ of a complex variable. It was in the work of Euler in this area that the general character of this result began to appear, since to obtain u and v Euler took $f(z)$ not as particular elementary functions, as d'Alembert had done, but as a general analytic function defined by a power series with arbitrary (real) coefficients.

In his work on integral calculus (communicated to the Petersburg Academy of Sciences in 1777 and published in 1793 and 1797) Euler established the equality

$$\int f(z)\,dz = \int (u\,dx - v\,dy) + i\int (v\,dx + u\,dy),$$

where $f(z) = u + iv$. All that was lacking was the concept of integrals of functions of a complex variable computed along some curve to see here the Cauchy integral theorem that the integral is independent of the path of integration.

---

1) See Markushevich, A. "Fundamental concepts of mathematical analysis and the theory of functions in the work of Euler" [in Russian], in: *Leonhard Euler*. Moscow 1959, p. 119.

2) Lagrange, J.-L. *Œuvres*. Paris 1892, T. 14, p. 261.

Using this last relation Euler showed again that the real and imaginary parts of an analytic function of a complex variable satisfy Eqs. (1). Moreover he was the first to use the device of separation of the real and imaginary parts in the integral of a complex-valued function to compute the integrals of real-valued functions (cf. HM, T. 3, pp. 365–368). It was this device that later served as the point of departure for the investigations of Cauchy.

Finally, in connection with the problem of constructing geographic maps, Euler studied the problem of conformal mapping of surfaces in its general formulation and used a complex variable for the purpose.

In the work of Euler, as well as d'Alembert and later Lagrange connected with various applications of complex numbers to the solution of problems stated directly in terms of real numbers, complex numbers enter the discussion usually as pairs of real numbers having a certain specific meaning, for example as pairs formed from the abscissa and ordinate of a point in a plane, the projections of the velocity of a particle of fluid on the coordinate axes, or pairs of functions. Thus the various particular interpretations of complex numbers (in particular, geometrical interpretations) were incorporated into mathematics starting in the mid-eighteenth century.

However a visual representation of the complex variable as a vector or a point moving over the plane was still not explicitly present, and a visual interpretation of operations on complex numbers was also missing.

To summarize, one can say that in the eighteenth century an extensive amount of factual material was accumulated on which to construct the theory of analytic functions, the fruitfulness of studying them as functions of a complex variable was made clear, and the leading role in this work was played by Leonhard Euler.

Development of the Concept of a Complex Number

The development and propagation of the concepts of complex numbers as vectors or points of the plane played a vital role in the subsequent history of analytic function theory. Euler was quite close to such a concept in the papers where he passed form writing a complex number as $x \pm \sqrt{-1}y$ to the trigonometric form of a complex number $s(\cos \omega \pm \sqrt{-1} \sin \omega)$, for example, in his trigonometric studies conducted around 1755. In his paper "On the representation of a spherical surface on a plane" (1777, 1778, see HM, T. 3, p. 170) he passed from the points (x, y) of the plane (a geographical map) to the complex numbers $x + iy$ and expressed the latter in terms of the longitude and latitude of the points of the sphere, then again returned to the coordinates x and y.

However the first explicit and systematic geometric representation of complex numbers as directed line segments and the corresponding interpretation of operations on them occur in a 1799 paper by the Danish surveyor C. Wessel (1745–1818) "On the analytic representation of directions" ("Om directionens analytiske Betegning, et Forsög anwendt fornemmeling til plane og sphaeriske Polygoners Oplösning," *Danski Vid. Selsk. Skr. N. Samml.*, 1799, 5).

Unfortunately this paper was not noticed at the time by mathematicians, and its contents were not fully appreciated until 100 years later (HM, T. 3, pp. 63–65).

The geometric representation of complex numbers was also studied in the work of the French abbot A. Buée (1748–1826) and the Swiss J. Argand (1768–1822). These papers, dated 1806, and written independently of Wessel, also remained little-known for a long time, or at least, shall we say, insufficiently influential, and this despite the polemic in the pages of the *Annales de Gergonne* (1813–1815, T. 4–5) provoked by the publication of a note by J. F. Français (1775–1833) in that journal leading to similar ideas. Argand held forth twice on this subject in the *Annales de Gergonne*, re-establishing his priority and reproducing with certain improvements a brief and elegant proof of the fundamental theorem of algebra proposed by him in an 1806 paper. In the second of his notes Argand, in particular, introduced the term "modulus [absolute value] of a complex number $a + bi$" in its modern sense (HM, T. 3, pp. 65–67). This term came into general use after Cauchy began to use it starting with his *Analyse algébrique* (1821).

However, in the first quarter of the nineteenth century many mathematicians were very close to the geometric representation of complex numbers. The concept of complex numbers as points of a plane received general recognition after 1831, when Gauss' paper "Theoria residuorum biquadraticorum" was published. This paper included a justification of the complex numbers and a geometric interpretation of them. The term "complex numbers" itself is due to Gauss. In his unpublished papers Gauss exhibited a mastery of the geometric representation of complex numbers as early as his youthful *Tagebuch*.

Gauss' dissertation devoted to the proof of the so-called fundamental theorem of algebra, published in 1799, stands at the juncture of the eighteenth and nineteenth centuries. Here Gauss made extensive use of geometric representations, reducing the problem to the proof that there exist points of intersection of certain algebraic curves in the plane. This proof, although based on a correct idea, is far from being above reproach.

Subsequently Gauss returned to this theorem many times, giving it a statement that has become classical (every algebraic equation of degree at least one has at least one root, real or imaginary), and proposing other proofs of it.[3] The so-called third proof (published in 1816), although its exposition made no direct use of complex numbers and geometric representations, provides an example of the application of the argument principle to the fundamental theorem of algebra.[4]

Cauchy took a long and complicated route to the geometric interpretation of complex numbers. His concept of complex numbers changed throughout almost the entire course of his career. In his *Analyse algébrique* (1821) he classifies "imaginary expressions" (i.e., complex numbers) and "imaginary equations" (i.e.,

---

3) See Book 1, pp. 43–47.

4) See Markushevich, A. "Gauss' papers in mathematical analysis" [in Russian], in: *Karl Friedrich Gauss*, Moscow 1956, pp. 185–187.

equalities containing complex numbers) as symbolic expressions, meaning expressions that "taken literally are imprecise or devoid of meaning, but from which one can deduce precise results by modifying and altering according to definite rules, either the equations themselves or the symbols contained in them."[5] He said explicitly that "an imaginary equation is merely a symbolic representation of two equations between two real variables."

Nearly a quarter of a century later Cauchy again undertook to explain the meaning of the concept of a complex number. The corresponding papers are in collections combined under the general title *Exercices d'analyse et de physique mathématique*, Paris 1840–1847, T. 1–4, occupying four volumes of the second series of his *Œuvres* (XI–XIV). Since we shall have frequent occasion to refer to this work, we shall call it simply the *Exercices* and indicate the volume and year of the first publication. In the paper "Mémoire sur les fonctions de variables imaginaires" in the *Exercices* (1844, T. 3) he repeated his earlier conception, referring to the *Analyse algébrique*. In the interval between these two works Cauchy had proved the theorem that the integral is independent of the path of integration, constructed the theory of residues with its applications, obtained his integral formula, and deduced the power series expansion from it (for all this see below). Nevertheless he continued to emphasize that imaginary numbers "understood literally and interpreted according to the accepted rules, do not refer to anything and have no meaning. The sign $\sqrt{-1}$ is in a way only a device, a tool for computation... ."[6] At the same time he continued to seek a different, more substantive interpretation of complex numbers. Several of his papers in the fourth volume of the *Exercices*, published in installments starting in 1847, are devoted to this effort.

In his "Memoir on the theory of algebraic equivalences substituted for the theory of imaginaries" ("Mémoire sur la théorie des équivalences algébriques, substituée à la théorie des imaginaires") in his *Exercices* (1847, T. 4), Cauchy proposed regarding i as "a real, but indefinite quantity" and an "imaginary equation" as a polynomial equivalence modulo $i^2 + 1$. For example, the rule for multiplication of complex numbers, which he wrote as

$$(\alpha + \beta i)(\gamma + \delta i) \backsimeq (\alpha\gamma - \beta\delta) + (\alpha\delta + \beta\gamma)i,$$

meant only that the difference of the two polynomials on the left- and right-hand sides of the equivalence sign $\backsimeq$ is divisible by $i^2 + 1$.

Thus Cauchy here was constructing the splitting field of a polynomial without assuming in advance the existence of the field of complex numbers. Such a construction was carried out in general form by L. Kronecker in 1882. However, one can discern the same idea as early as Gauss' second proof (1814–1815).[7]

In the end Cauchy still relied on the geometric representation, finding it superior to the earlier algebraic representations. It was in his paper "Mémoire

5) Cauchy, A.- L. *Œuvres complètes*. Paris 1882, Sér. 2, T. 3, p. 153.

6) *Ibid.*, pp. 361–362.

7) See Bashmakova, I. G. "On the proof of the fundamental theorem of algebra" [in Russian], *ИМИ*, **10** (1957), pp. 257–304, and also Book I, p. 47.

A. Cauchy

sur les quantités géométriques" (*Exercices*, 1847, T. 4) that, after recognizing his predecessors, among them Buée and Argand (not mentioning Gauss, however), he proposed "after new and mature reflection" the complete elimination of the symbol $\sqrt{-1}$ and the replacement of the theory of imaginary expressions by a theory of quantities, which Cauchy called "geometric" quantities. Here the geometric quantity r_p is a vector of length r (called its modulus) and polar angle p (the argument or azimuth).

However, in his note "Sur la quantité géométrique $i = 1_{\pi/2}$ et sur la réduction d'une quantité géométrique quelconque à la forme $x + yi$" (*Exercices*, 1847, T. 4) Cauchy resurrected the usual form of a complex number. In doing so he called the geometric quantity $x + yi$ the *affix* of the point $A(x, y)$.

We shall mention one more note in the same fourth volume of the *Exercices*, "Sur les fonctions des quantités géométriques". If $z = x + iy$ is the affix of the point A and $Z = X + iY$ the affix of a moving point B, then "Z must be regarded as a function of z when the value of z determines the value of Z. But for this it suffices that X and Y be certain functions of x and y. Then the location of the moving point A will always determine the location of the moving point B."[8]

---

8) Cauchy, A.-L. *Œuvres complètes*. Sér. 2, T. 4, p. 309.

Thus Cauchy had finally regularized a visual representation of a function of a complex variable, which he had actually been using (without fully recognizing it) in his earlier investigations and which, as we shall see below, had seemed completely natural to Gauss as early as 1811.

Complex Integration

Investigations in which the Eulerian methods of computing definite integrals by use of a complex variable were applied and developed were of the greatest value in the construction of the foundations of analytic function theory.

In a cycle of papers that began with his "Mémoire sur les approximations des formules qui sont fonctions de très-grands nombres" (1782–1783, 1785–1786) and ended with *Théorie analytique des probabilités* (1812, see HM, T. 3, p. 148–150) Laplace developed a method of solving linear difference and differential equations based on the replacement of the unknown function $y(x)$ by integrals of the form $\int \varphi(x) x^s \, dx$ or $\int \varphi(x) e^{-sx} \, dx$, where $\varphi(x)$ is a new unknown function. Here we encounter the famous Laplace transform. These integrals are taken between limits satisfying a certain equation, the limit equation, which is generally nonalgebraic. The roots of this equation are frequently imaginary and sometimes do not exist at all. (This fact did not keep Laplace from referring to them as imaginary, which was the tradition of the eighteenth century!) In the latter case, however, it was a matter of exhibiting ways of letting the variable of integration tend to infinity so that the left-hand side of the "limit equation" tends to zero. Thus, for example, if the limit equation has the form $e^{-x^2} = 0$, one could take the limits of integration as $x = -\infty$ and $x = +\infty$, the integration being taken along the real axis (or along curves asymptotic to it). In other cases Laplace set $x = -\mu + \varpi\sqrt{-1}$ and let ϖ vary from $-\infty$ to $+\infty$, corresponding to integration along a line parallel to the imaginary axis. Having obtained integrals with "imaginary limits," Laplace subjected them to various transformations based on replacing the variable of integration and arrived at integrals of real-valued functions of a real variable. Laplace noted that the transition from real to imaginary variables enabled him to find the values of many definite integrals, and assessed the role of these passages as a sort of induction, recognizing the necessity of an additional verification of results obtained in this way. He noted that Euler had used the passage from real to imaginary in computing integrals simultaneously with him, but that the results obtained by Euler had been published later than his results.

Expressions of the coefficients of a power series as integrals of the sum of the series over a circle in the complex plane occur for the first time in Laplace's *Théorie analytique des probabilités* (Laplace, however, did not use geometric language). To be specific, replacing t^x by $e^{x\varpi\sqrt{-1}}$, i.e., setting $t = e^{\varpi\sqrt{-1}}$, in the equality $u = y_0 + y_1 t + y_2 t^2 + \cdots + y_x t^x + y_{x+1} t^{x+1} + \cdots + y_\infty t^\infty$ (Laplace's way of writing a power series), multiplying both sides of the equality by $e^{-x\varpi\sqrt{-1}} \, d\varpi$, and integrating with respect to ϖ from 0 to 2π, he obtained

$$y_x = \frac{1}{2\pi} \int U \, d\varpi (\cos x\varpi - \sqrt{-1} \sin x\varpi),$$

where U is the result of the substitution $t = e^{\varpi\sqrt{-1}}$ in the expression for the sum of the series. It is obvious that the integral on the right-hand side is

$$\frac{1}{2\pi i} \int\limits_{|t|=1} \frac{U\,dt}{t^{x+1}},$$

i.e., Laplace's formula coincides with the Cauchy integral formulas derived later for the coefficients of a power series.

It is remarkable that Poisson, who himself followed Euler in using the method of "passage from real to imaginary," noted in 1815 that this passage may lead to incorrect results because "an analytic expression of an integral is not always equal to the sum of the differentials" ("Suite du mémoire sur les intégrales définies," *J. Éc. Polyt.*, **17**, 1815). He studied in detail the questions that arose here in his 1820 paper "Sur les intégrales des fonctions qui passent par l'infini entre des limites de l'intégration, et sur l'usage des imaginaires dans la détermination des intégrales définies," *J. Éc. Polyt.*, 1820). Analyzing the formula

$$\int_a^b f'(x)\,dx = f(b) - f(a),$$

Poisson remarked first of all that it may lead to errors if $f'(x)$ becomes infinite. Lagrange had pointed this out earlier, giving the example of the integral $\int_{-1}^{1} \frac{dx}{x^2}$, for which the formula gives the incorrect result -2. Nevertheless in such cases, Poisson continued, one can use the general formula if one "proceeds so that the variable x passes from the limit a to the limit b through a series of imaginary values. Then $f'(x)$ will no longer be infinite for any of the intermediate values, and the definite integral will have its usual meaning," i.e., it can be regarded as the sum of the differentials $f'(x)\,dx$ and hence this formula can be applied. As an example Poisson computed $\int_{-1}^{1} \frac{dx}{x}$, setting $x = e^{iz}$, where the real variable z varies from $(2n+1)\pi$ to 0, and obtained the value $-(2n+1)\pi i$. This same value, according to Poisson, is obtained if the primitive function $\ln x$ is used to compute the integral. We shall then have

$$\int_{-1}^{1} \frac{dx}{x} = [\ln x]_{-1}^{1} = -\ln(-1) = -(2n+1)\pi i.$$

It is obvious that Poisson was on the right path.[9] However the essence of the matter had been formulated in even fuller and more complete form in 1811 by Gauss in his remarkable letter to Bessel of 19 December 1811, which unfortunately was not published until the complete correspondence of the two scholars was published in 1880.

---

9) Yushkevich, A. P. "On the appearance of Cauchy's concept of a definite integral" [in Russian], *Trudy Inst. Ist. Est. i Tekh. Akad. Nauk SSSR*, **1** (1947), pp. 397–401.

... What is meant by $\int \varphi(x)\, dx$ for $x = a + bi$? Obviously if we wish to start from clear concepts, we must assume that x starts from a value for which the integral must equal zero and passes to $x = a + bi$ through infinitesimal increments (each of the form $a + bi$) and then sum all the $\varphi(x)\, dx$. Thus the meaning is completely established. But the passage may take place in infinitely many ways. Just as the set of all real values can be thought of as an infinite line, so the set of all quantities, real and imaginary, can be visualized by means of an infinite plane, each point of which, defined by the abscissa a and ordinate b, will so to speak represent the quantity $a + bi$. The continuous passage from one value of x to another value $a + bi$ is therefore carried out along the line and consequently is possible in many different ways. I now assert that the integral $\int \varphi(x)\, dx$ over two different paths preserves the same value if inside the portion of the plane enclosed between two lines the function $\varphi(x)$ is never equal to ∞. This is a beautiful theorem[10] whose simple proof I shall give on a suitable occasion. It is connected with other beautiful truths involving series expansions. The passage must always be carried out at each point in such a way as never to involve points where $\varphi(x) = \infty$. I insist that such points must be avoided because for them the original fundamental concept of the integral $\int \varphi(x)\, dx$ obviously loses its clarity and easily leads to contradictions.

At the same time, however, it is clear from this how the function generated by the integral $\int \varphi(x)\, dx$ can have many values for the same value of x, specifically depending on whether a single or multiple circuit about the point at which $\varphi(x) = \infty$ is allowed or no such circuit is allowed. If, for example we define[11] $\log x$ by $\int \frac{dx}{x}$ starting from $x = 1$, we can arrive at $\log x$ either by not going around the point $x = 0$ or by encircling it once or several times; each time the constant $2\pi i$ or $-2\pi i$ will be added. From this the different logarithms of every number are completely clear. If, however, $\varphi(x)$ does not become infinite for any finite value of x, then the integral is always a single-valued function x.[12]

Here the complete explanation of the multivaluedness of the logarithm defined as $\displaystyle\int_{1}^{x} \frac{dx}{x}$ is encountered for the first time. Gauss obviously used analogous considerations to explain the multivaluedness of elliptic integrals, without which he could not have understood the double periodicity of the elliptic functions (see below).

---

10) Here, incidentally, it is still assumed that $\varphi(x)$ itself is a single-valued function of x or, at least, that for its values inside this whole region of the plane only one system of values can be assumed without violating continuity. *Gauss' comment.*

11) Gauss' notation for the natural logarithm.

12) *Briefwechsel zwischen C. F. Gauss und W. Bessel*, Leipzig 1880, S. 155–160; see also *C. F. Gauss, Werke*, Göttingen 1866, Bd. 10(1)

The Cauchy Integral Theorem. Residues

Cauchy's first paper of significance in the theory of functions was his "Mémoire sur les intégrales définies" (*Œuvres*, Sér. 1, T. 1), presented to the Academy of Sciences in 1814 and printed in 1825. In this paper Cauchy still lacked the clear understanding of the whole problem of complex integration displayed in the letter of Gauss just quoted. The full picture came together gradually for Cauchy, over a period of nearly three decades. However, his papers were published in a timely manner (except for the delay already noted). For that reason it was these papers rather than Gauss' ideas the formed the foundation for a systematic construction of the general theory. Cauchy's hydrodynamic researches led him to the computation of the definite integrals to which his 1814 paper is devoted. The point of departure for this paper is the relation

$$\int_{x_0}^{X} \int_{y_0}^{Y} f(x,y)\, dy\, dx = \int_{y_0}^{Y} \int_{x_0}^{X} f(x,y)\, dx\, dy. \tag{2}$$

This relation had been pointed out by Euler as early as 1769 (published in 1770). Cauchy used it here as a means of computing single integrals, i.e., as a device for integrating with respect to a parameter. Euler himself had regarded it in this sense (1774). From that time on he, as well as Laplace, gave many examples of the application of this device. Cauchy started from two functions $S(x,y)$ and $V(x,y)$ that were the real and imaginary parts of the expression $F(x,y)$:

$$F(x+iy) = S + iV.$$

Then, as Euler had shown in 1777,

$$\frac{\partial S}{\partial x} = \frac{\partial V}{\partial y}, \quad \frac{\partial S}{\partial y} = -\frac{\partial V}{\partial x}.$$

What interested Cauchy here was only the first of the two relations. Setting

$$f(x,y) = \frac{\partial V}{\partial y} = \frac{\partial S}{\partial x},$$

he obtained

$$\int_{x_0}^{X} \int_{y_0}^{Y} \frac{\partial V}{\partial y}\, dy\, dx = \int_{y_0}^{Y} \int_{x_0}^{X} \frac{\partial S}{\partial x}\, dx\, dy,$$

or

$$\int_{x_0}^{X} [V(x,Y) - V(x,y_0)]\, dx = \int_{y_0}^{Y} [S(X,y) - S(x_0,y)]\, dy. \tag{3}$$

This last relation enabled Cauchy to reduce the computation of one definite integral to the computation of a second integral. From this point of view the analogous formula

$$\int_{x_0}^{X} [S(x,Y) - S(x,y_0)]\, dx = -\int_{y_0}^{Y} [V(X,y) - V(x_0,y)]\, dy, \tag{4}$$

obtained by making the different choice

$$f(x,y) = \frac{\partial S}{\partial y} = -\frac{\partial V}{\partial x},$$

yields nothing new. Only in 1822 did Cauchy arrive at the idea of combining relations (3) and (4) into a single relation in which the complex-valued function $F(x + iy)$ occurs directly. In accordance with this idea, while his 1814 paper was in press in 1825, he added a footnote in which, multiplying (3) by i and adding it to (4), he arrived at the equality

$$\int_{x_0}^{X} F(x + iY)\,dx - \int_{x_0}^{X} F(x + iy_0)\,dx =$$

$$\int_{y_0}^{Y} F(X + iy)\,i\,dy - \int_{y_0}^{Y} F(x_0 + iy)\,i\,dy,$$

which he finally presented as[13]

$$\int_{y_0}^{Y} F(x_0 + iy)\,dy + \int_{x_0}^{X} F(x + iY)\,dx =$$

$$= \int_{x_0}^{X} F(x + iy_0)\,dx + \int_{y_0}^{Y} F(X + iy)\,dy. \qquad (5)$$

This last relation is obviously the Cauchy integral theorem in the elementary special case of integration over a straight-line contour. Cauchy pointed out this geometric meaning in the pamphlet *Mémoire sur les intégrales définies, prises entre des limites imaginaires* (Paris 1825). However, before proceeding to a survey of this pamphlet, let us complete our examination of his 1814 paper.

When Cauchy's predecessors used relation (2), they required only that the limits of integration be constant, i.e., that the double integration extend over a rectangle. Cauchy pointed out that an additional requirement was needed: the function $f(x,y)$ must not become ∞ inside the rectangle or on its boundary. In the cases when this requirement is not met the integrals on the left- and right-hand sides of Eqs. (3) and (4) may differ, and the problem arises of finding new formulas that take account of the difference between these integrals. It was to such formulas, which served as the beginning of the later theory of residues and their applications to the computation of numerous definite integrals, both those already known and new ones, that the major part of the 1814 paper was devoted.

At the very beginning of his pamphlet *Mémoire sur les intégrales définies, prises entre des limites imaginaires*, which we have already mentioned, Cauchy defined the integral

$$\int_{x_0 + y_0\sqrt{-1}}^{X + Y\sqrt{-1}} f(z)\,dz$$

---

13) Cauchy gave this same relation in a note of 1822 and in his 1823 work *Résumé des leçons données à l'École Polytechnique sur le calcul infinitésimal*. In: *Œuvres*. Sér. 2, T. 4, pp. 5–261.

by analogy with the integral of a function of a real variable as the *limit of an integral sum*. To make precise the construction of such a sum one must set $x = \varphi(t)$ and $y = \chi(t)$ $(x + iy = z)$, where $\varphi(t)$ and $\chi(t)$ are functions that are monotonic and continuous for $t_0 \leq t \leq T$ satisfying the conditions

$$\varphi(t_0) = x_0, \quad \chi(t_0) = y_0, \quad \varphi(T) = X, \quad \chi(T) = Y.$$

Choosing such functions is obviously equivalent to defining a certain curve joining the points (x_0, y_0) and (X, Y) in the plane. Cauchy pointed this out in another part of the same paper. Thus the integral of the complex-valued function that he had defined is an integral along a certain curve and through the equations of this curve reduces to an ordinary definite integral, which can be abbreviated as follows:

$$\int_{t_0}^{T} (x' + \sqrt{-1}y') f(x + \sqrt{-1}y) \, dt.$$

Cauchy then states his basic theorem: *If $f(x + y\sqrt{-1})$ is finite and continuous for $x_0 \leq x \leq X$ and $y_0 \leq y \leq Y$, then the value of the integral is independent of the nature of the functions $x = \varphi(t)$ and $y = \chi(t)$.* That is, we might say, it is independent of the shape of the curve of integration inside the rectangle $x_0 \leq x \leq X$, $y_0 \leq y \leq Y$ joining the vertices (x_0, y_0) and (X, Y). This is the Cauchy integral theorem stated for the case of a rectangular domain.

In the proof Cauchy used the derivative of $f(z)$ and, strictly speaking, relied on the continuity of the derivative. However, in the statement of the theorem the existence of the derivative is not mentioned, much less its continuity. (Of course the same is true of the letter from Gauss to Bessel quoted above.) This is due to Cauchy's conviction that a continuous function is always differentiable and that its derivative can be discontinuous only at points where the function itself is discontinuous. This conviction seems to have been based on the fact that in the majority of cases (at least in the first decades of his career) when Cauchy spoke of functions, he had in mind analytic expressions for which the existence of the derivative follows from the rules of differential calculus. The related conviction that the derivative of a continuous function is continuous rested on the tradition going back to Euler of regarding analytic expressions as functions of a complex variable. For that reason the absence of a derivative, for example, for the function $x \sin(1/x)$ at $x = 0$ was explained by the fact that the function was discontinuous at $x = 0$ as a function of a complex variable. It was somewhat more complicated to explain the fact that the derivative of the function $f(x) = \sqrt{x}$ becomes infinite at $x = 0$, and even much later (see, for example, his paper "Sur les fonctions continues de quantités algébriques ou géométriques," *Exercices*, 1847, T. 4) Cauchy continued to insist that his definition of continuity presumes that the function is single-valued and that he regarded as discontinuities the points at which the function becomes multivalued while remaining finite. From this point of view the function $\sqrt{x}$ is not continuous at the point $x = 0$. In explanation of Cauchy's position we remark that in the complex plane there is no neighborhood of the origin in which one can distinguish a continuous and single-valued branch of the function $\sqrt{x}$.

Cauchy carried out the proof of the integral theorem using the techniques of calculus of variations. He replaced the functions $\varphi(t)$ and $\chi(t)$ that define the curve by the nearby curves $\varphi(t) + \varepsilon u(t)$ and $\chi(t) + \varepsilon v(t)$ and by computing the corresponding variation of the integral, verified that it is zero.[14]

In his first papers Cauchy still did not go depart very much from his predecessors, resorting to a complex variable in analysis only as an auxiliary means of a formal character making it possible to solve difficult problems of integral calculus.

A similar position is expressed in the following excerpt from the review of Lacroix and Legendre on Cauchy's paper "Sur les intégrales définies," agreeing with the corresponding opinion of Laplace: "Some investigations in the integral calculus sometimes lead to results in which the passage from real to imaginary can be used as a kind of induction, which, being insufficiently obvious in and of itself, requires confirmation by direct and strict proofs."[15]

Gradually, however, the investigations of Cauchy and others led to such an exceptional wealth of facts and new results that it became clear that it was a matter not of some kind of induction playing a purely auxiliary role and a not very reliable one at that, but of a discipline deserving independent existence — the theory of functions of a complex variable.

Cauchy founded the theory of residues over the period 1826–1829. The name *residue* (literally *that which is left over*) seems to be explained by the fact that Cauchy arrived at this concept while seeking the difference between integrals taken over two paths having a common beginning and end between which there were poles of the function. In this form residues can also be seen in his "Mémoire sur la théorie des intégrales définies" (1814). In the paper "Mémoire sur les intégrales définies, prises entre des limites imaginaires" (1825) he devotes the majority of his attention to the analysis of the cases when $f(z)$ becomes infinite inside the rectangle or on its sides. Here the integrals over different paths have generally unequal values, and Cauchy computes the difference between them under various assumptions. Suppose, for example, $f(z)$ becomes infinite only at one point (a, b) between the two curves of integration, and that the following limit exists and is finite:

$$\mathbf{f} = \lim_{\substack{x \to a \\ y \to b}} [(x - a) + (y - b)\sqrt{-1}[f(x + y\sqrt{-1}).$$

---

14) In his comprehensive book *Mathematical Thought from Ancient to Modern Times* (New York 1972) Morris Kline asserts (pp. 636–637) that this paper of Cauchy was not published until 1874! This is clearly a mistake. The publication in the journal mentioned by Kline (*Bull. sci. math.*, **7** (1874), pp. 265–304; **8** (1875), pp. 43–55, 148–159), was a reprint of the pamphlet that, as mentioned above, had been originally published in Paris in 1825. However, there is indirect evidence that this remarkable paper had not attracted much interest among contemporaries. Thus nearly two decades after its publication the publisher (Bachelier) was still offering it in the list of available works of Cauchy (other works of Cauchy in the same list, already sold out, were noted as "épuisé"). A similar list precedes the list of titles of each of the four volumes of the *Exercices*, 1840–1847.

15) Cauchy, A.-L. *Œuvres complètes*. Sér. 1, T. 1, p. 321.

Then he proves that the difference between the integrals is $\pm 2\pi\sqrt{-1}\mathbf{f}$. The quantity $\mathbf{f}$ is an elementary example of a residue.

The term itself and the formal definition of a residue are first encountered in the paper "Sur un nouveau genre de calcul analogue au calcul infinitésimal" (*Exercices de mathématiques*, Paris 1826, T. 1). Here is how Cauchy introduces and defines this concept:

> If, after the values of x that cause $f(x)$ to become infinite have been found, an infinitely small quantity ε is added to one of these quantities, which we denote x_1, and the function $f(x_1 + \varepsilon)$ is expanded in a series of increasing powers of this quantity, the first terms of the expansion will contain negative powers of ε, and one of them will be the product with a finite coefficient, which we shall call the residue of the function $f(x)$ corresponding to the particular value x_1 of the variable x.[16]

After this paper Cauchy wrote a large number of other papers placed in this and later volumes of his *Exercices de mathématiques (anciens exercices)* (1826–1829), in which he studied the applications of the theory to the computation of integrals, differential equations, the expansion of functions in series and infinite products, theory of equations, and so forth.

Elliptic Functions in the Work of Gauss

On 17 September 1808 Gauss wrote to Schumacher:

> We can now handle the circular and logarithmic functions as easily as one-times-one[17], but the magnificent gold mine holding the precious higher functions remains nearly *terra incognita*. I have worked on this a great deal in the past, and when I have the time I plan to produce my own large work on this subject, which I alluded to in my *Disquisitiones arithmeticae*, p. 593.[18] One is amazed at the extraordinary wealth of new and highly interesting truths and relations provided by these functions (among them are functions connected with the rectification of the ellipse and the hyperbola).... .

Thus the problem was to expand the potential of mathematical analysis by adjoining other important classes of functions to the elementary functions. These new classes were to be mastered just as completely as mathematicians had succeeded in doing for the elementary functions. This process, as we have mentioned, had been going on throughout the eighteenth century. But in the nineteenth century it acquired new characteristics. The power of the general methods of analytic

---

16) *Ibid.*, sér. 2, T. 5, p. 23.

17) Gauss, C. F. *Werke*. Bd. 10 (1), S. 243–245.

18) *Ibid.*, Bd. 1, S. 412–413. There Gauss points out that the principles of the theory of division of a circle can be applied not only to circular functions, but also to other transcendental functions, for example, those depending on the integral $\int \frac{dx}{\sqrt{1-x^2}}$, though the study of such functions was a problem for another extensive work.

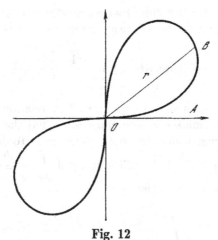

Fig. 12

function theory was tested on the study of special classes. In addition this study made it possible to accumulate new facts and regularities and to pose new questions, which served as an important motivation for the subsequent development of the general theory.

The starting point for Gauss' study of special functions was the construction of the theory of a rather special class of elliptic functions — the lemniscatic functions. Despite all the analogies with the trigonometric functions, which facilitated the first efforts of the young Gauss (starting in 1796), the lemniscatic functions enabled him to discover all the peculiarity of the elliptic functions. Even today their properties can serve as a good propadeutic to the study of the general theory of elliptic functions.[19]

The lemniscatic sine $r = \sin \operatorname{lemn} S$ (Gauss' notation) is defined as the function inverse to the integral that expresses the arc length of the lemniscate $r^2 = \sin 2\theta$: $\widehat{OAB} = S = \int_0^r \dfrac{dt}{\sqrt{1 - t^4}}$ (Fig. 12).

If we set $\dfrac{\varpi}{2} = \int_0^1 \dfrac{dt}{\sqrt{1 - t^4}}$ (one quarter of the length of the entire lemniscate), the lemniscatic cosine is introduced by the relation

$$\cos \operatorname{lemn} S = \sin \operatorname{lemn} (\varpi/2 - S).$$

Applying the method of Euler discussed in chapters V and VI of the first part of Vol. 1 of his *Integral Calculus* (1768) to integrate the equation

$$d\sigma + ds = \frac{d\rho}{\sqrt{1 - \rho^4}} + \frac{dr}{\sqrt{1 - r^4}} = 0$$

19) See Markushevich, A. I. *Remarkable Sines: An Introduction to the Theory of Elliptic Functions* [in Russian]. 2nd ed., Moscow 1975.

(σ and s are two arcs the sum of whose lengths is constant), Gauss obtained the addition theorem for the lemniscatic sine in the form

$$\text{sin lemn}\,(\sigma + s) = \frac{\rho\sqrt{1 - r^4} + r\sqrt{1 - \rho^4}}{1 + r^2\rho^2},$$

where $\rho = \text{sin lemn}\,\sigma$ and $r = \text{sin lemn}\,s$. This relation implies an algebraic relation between the lemniscatic sine and cosine, and the existence of a real period 2ϖ (which is geometrically obvious) is confirmed. To define functions of a purely imaginary argument the limit of the integral $\int_0^r \dfrac{dt}{\sqrt{1 - t^4}}$ is taken as the purely imaginary number $r = iv$. Then

$$\int_0^{iv} \frac{dt}{\sqrt{1 - t^4}} = i \int_0^v \frac{d\tau}{\sqrt{1 - \tau^4}} = iy,$$

whence $v = \text{sin lemn}\,y$ and $iv = \text{sin lemn}\,(iy) = i\,\text{sin lemn}\,y$.

It remains only to apply the addition theorem to obtain the definition of the function of a complex argument $z = x + iy$

$$\text{sin lemn}\,(x + iy) = \frac{u\sqrt{1 - v^4} + iv\sqrt{1 - u^4}}{1 - u^2 v^2}, \tag{6}$$

where $u = \text{sin lemn}\,x$ and $v = \text{sin lemn}\,y$.

Comparison of these formulas with the analogous formula for the trigonometric sine

$$\sin(x + iy) = u\sqrt{1 + v^2} + iv\sqrt{1 - u^2}, \quad u = \sin x, \quad v = \sinh y,$$

immediately reveals two radical differences between the functions $\sin z$ and $\text{sin lemn}\,z$. First, $\sin z$ is a finite-valued function at every point (an entire function), while $\text{sin lemn}\,z$ has an infinite set of poles (at the points where $u = \pm 1$ and $v = \pm 1$), i.e., it is meromorphic in modern terminology. For Gauss this difference between an entire and a meromorphic function was also manifested in the fact that the former can be represented by an everywhere-convergent power series, while the power series of the latter diverges for $|z|$ larger than the pole of the function nearest to the origin. (Gauss said, "The formula for $\text{sin lemn}\,z$ diverges if φ becomes larger than $\varpi/\sqrt{2}$.")[20] Gauss then noted that $\text{sin lemn}\,z$ can be represented as the quotient of two convergent power series $P(z)$ and $Q(z)$ (which will be discussed below). Second, $\sin z$ is a simple periodic function, while it follows from formula (6) that the function $\text{sin lemn}\,z$ has, besides the real period 2ϖ, the purely imaginary period $2i\varpi$ (v is unchanged if y is replaced by $y + 2\varpi$), and consequently is a *doubly periodic function*.

---

20) Gauss, C. F. *Werke*. Bd. 3, S. 406.

Relation (6) reveals that all the zeros of the lemniscatic sine are contained in the formula $\alpha_{m,n} = (m + in)\varpi$ (this is immediately obvious if the right-hand side is represented as $\dfrac{u^2 + v^2}{u\sqrt{1 - v^4} - iv\sqrt{1 - u^4}}$) and all its poles in the formula $\beta_{m,n} = [(2m - 1) + i(2n - 1)](\varpi/2)$, where m and n are arbitrary integers. After this Gauss wrote out formally the simplest entire functions having zeros $\alpha_{m,n}$ and $\beta_{m,n}$ respectively:

$$M(z) = z \prod_{m,n}{}' \left(1 - \frac{z}{\alpha_{m,n}}\right), \quad N(z) = \prod_{m,n}{}' \left(1 - \frac{z}{\beta_{m,n}}\right)$$

(the prime indicates that the values $m = n = 0$ are excluded). He then asserted that

$$\sin \text{lemn } z = M(z)/N(z).$$

It is not difficult to see that the products used by Gauss are not absolutely convergent. They are prefigurations of the σ functions used later by Weierstrass. However, Gauss did not dwell on convergence questions or on the verification of the validity of the last formula at this point. He remarked only that $M(z)$ and $N(z)$ are not periodic and set the goal of replacing them by periodic functions. To do this he introduced the new variable $\zeta = \sin \dfrac{\pi}{\varpi} z$ and regarded $\sin \text{lemn } z$ as a function of ζ. For the latter the zeros have the form $\pm i(e^{n\pi} - e^{-n\pi})/2$ and the poles the form $\pm(e^{(2n-1)\pi/2} + e^{-(2n-1)\pi/2})/2$. He then constructed the functions

$$P(z) = \frac{\varpi \zeta}{\pi} \prod_{1}^{\infty} \left[1 + \frac{4\zeta^2}{(e^{n\pi} - e^{-n\pi})^2}\right],$$

$$Q(z) = \prod_{1}^{\infty} \left[1 - \frac{4\zeta^2}{(e^{(2n-1)\pi/2} + e^{-(2n-1)\pi/2})^2}\right]$$

(this time the products converge and moreover quite rapidly), and asserted that

$$\sin \text{lemn } z = P(z)/Q(z).$$

Gauss did not consider this last transition self-obvious and said that he could prove this formula rigorously.[21] Setting $z = \varpi\psi$, he went on to find the expansions of the periodic functions $P(\varpi\psi)$ and $Q(\varpi\psi)$ in trigonometric series

$$P(\varpi\psi) = 2^{3/4} \sqrt{\frac{\pi}{\varpi}} \left(e^{-\frac{1}{4}\pi} \sin \psi\pi - e^{-\frac{9}{4}\pi} \sin 3\psi\pi + e^{-\frac{25}{4}\pi} \sin 5\psi\pi - \cdots\right),$$

$$Q(\varpi\psi) = 2^{-1/4} \sqrt{\frac{\pi}{\varpi}} \left(1 + 2e^{-\pi} \cos 2\psi\pi + 2e^{-4\pi} \cos 4\psi\pi + \cdots\right),$$

21) *Ibid.*, S. 415.

whence

$$\sin \operatorname{lemn}(\varpi\psi) = 2e^{-\pi/4}\frac{\sin\psi\pi - e^{-2\pi}\sin 3\psi\pi + e^{-6\pi}\sin 5\psi\pi - \cdots}{1 + 2e^{-\pi}\cos 2\psi\pi + 2e^{-4\pi}\cos 4\psi\pi + \cdots}.$$

Gauss also obtained the analogous formulas for $\cos \operatorname{lemn}(\varpi\psi)$. A look at these formulas shows that $P(z)$ and $Q(z)$ (together with the two functions $p(z)$ and $q(z)$ corresponding to $\cos \operatorname{lemn}(\varpi\psi)$) are, up to constant factors, the same as the theta functions later introduced by Jacobi in the special case when (in Jacobi's notation)

$$k = k' = 1/\sqrt{2}, \quad K = K' = \varpi/2, \quad \text{and} \quad q = e^{-\pi K'/K} = e^{-\pi}.$$

Thus, for example, Jacobi's H and Θ_1 can be expressed in terms of Gauss' functions P and Q using the formulas

$$H(\varpi\psi) = 2^{1/4}\sqrt{\frac{\varpi}{\pi}}P(\varpi\psi), \quad \Theta_1(\varpi\psi) = 2^{1/4}\sqrt{\frac{\varpi}{\pi}}Q(\varpi\psi).$$

We have dwelt at such length on the properties of the lemniscatic functions discovered by Gauss in order to show that the derivation of these properties did not require any preliminary development of analytic function theory; in particular it did not require integration in the complex plane. The entire wealth of facts found by Gauss in this area was achieved by the usual methods employed by eighteenth-century analysts and did not go beyond their usual concepts. This holds for the consideration of complex values of variables, for the derivation and use of addition theorems, for the technique of operating with power series (including the inversion of series), where extensive use was made of the method of undetermined coefficients, and for the expansion of functions into infinite products. In addition it becomes manifest from this discussion that Gauss had developed all the fundamental content of the later theory of elliptic functions on the basis of eighteenth-century results (the only things missing were the general properties characteristic of all functions with fixed periods and the general theory of transformation of elliptic functions). Unfortunately Gauss never published these discoveries of his, leaving them in an unfinished, largely draft form, like nearly all the results he obtained in the theory of elliptic functions. For that reason the historical credit for laying the foundation of the theory of elliptic functions remains with Abel and Jacobi, who obtained their results a quarter-century after Gauss.

Although in the construction of the theory of lemniscatic functions Gauss based his reasoning on the analogy with trigonometric functions, later on, when developing the fundamentals of the theory of elliptic functions, he was guided by the analogy with lemniscatic functions.[22] Thus, for example, in November 1799 he

---

22) In connection with this and also with the arithmetic-geometric mean discussed below see Schlesinger, L. " C. F. Gauss. Fragmente zur Theorie des arithmetisch-geometrischen Mittels aus den Jahren 1797–1799," H. 2; "Über Gauss' Arbeiten zur Funktionentheorie," H. 3, in: *Materialen für eine wissenschaftliche Biographie von Gauss*, Gesammelt von F. Klein und M. Brendel. Leipzig 1912.

considered the general elliptic integral of first kind

$$\int_0^x \frac{dx}{\sqrt{1+\mu^2 x^2)(1-x^2)}} = u$$

($\mu^2 = 1$ corresponds to the case of the lemniscate) and wrote an explicit expression for the inverse function $x = S(u)$, which he called the "most general lemniscatic sine." He expressed this function both as a trigonometric series expansion and as the ratio of two entire periodic functions. In doing so he also relied on the arithmetic-geometric mean that he had introduced in the earliest period of his mathematical studies (starting in 1791?). For two numbers a and b $(a > b > 0)$ this mean is defined as the common limit of two sequences

$$a^{(n)} = (a^{(n-1)} + b^{(n-1)})/2, \quad b^{(n)} = \sqrt{a^{(n-1)}b^{(n-1)}} \quad (n \geq 1)$$

with $a^{(0)} = a$ and $b^{(0)} = b$. Gauss denoted it by $M(a, b)$. Around the year 1800 he applied this mean to compute the complete elliptic integral

$$\frac{1}{\pi} \int_0^\pi \frac{d\varphi}{\sqrt{1 - x^2 \cos^2 \varphi}} = \frac{1}{M(1, \sqrt{1-x^2})} = \frac{1}{M(1+x, 1-x)}.$$

This connection is derived following a different route in the later paper "Determinatio attractionis..." (1818).[23] To compute the integral

$$\int_0^{2\pi} \frac{dT}{\sqrt{m^2 \cos^2 T + n^2 \sin^2 T}},$$

which he encountered in determining the secular perturbation exerted by one planet on another, Gauss applied the transformation

$$\sin T = \frac{2m \sin T'}{(m+n) \cos^2 T' + 2m \sin^2 T'},$$

which later came to be called the Gauss transform, and found

$$\int_0^{2\pi} \frac{dT}{2\pi \sqrt{m^2 \cos^2 T + n^2 \sin^2 T}} = \int_0^{2\pi} \frac{dT'}{2\pi \sqrt{m'^2 \cos^2 T' + n'^2 \sin^2 T'}},$$

where $m' = \frac{1}{2}(m+n)$ and $n' = \sqrt{mn}$. Repeating this transform indefinitely, Gauss obtained (in the limit when the coefficients of the squares of the cosine and sine tend to $M(m, n)$)

$$\int_0^{2\pi} \frac{dT}{2\pi (\sqrt{m^2 \cos^2 T + n^2 \sin^2 T})} = \frac{1}{M(m, n)}.$$

---

23) Gauss, C. F. *Werke.* Bd. 3, S. 333.

He used an analogous device to compute the elliptic integral of second kind:

$$\int_0^{2\pi} \frac{\cos^2 T - \sin^2 T}{2\pi\sqrt{m^2\cos^2 T + n^2\sin^2 T}}\, dT.$$

These are the only results of Gauss on elliptic integrals (and functions) published during his lifetime. From the point of view of assuring priority Gauss' choice of material to publish can hardly be considered propitious. In fact he had chosen results that were essentially not new, having been published by Lagrange as early as 1784–1785,[24] while the main wealth of Gauss' discoveries, which anticipated the work of Abel and Jacobi to be discussed below, remained under lock and key.

What did Gauss think afterwards, when his involuntary "rivals" published their first papers on the theory of elliptic functions? After the appearance of the first part of Abel's "Recherches sur les fonctions élliptiques" in Crelle's *Journal* in the fall of 1827, Gauss wrote to Bessel on 30 March 1828,

> It appears that for the time being I won't be able to get back to the work on transcendental functions that I have been conducting for many years (since 1798), since I must first finish up many other things. Herr Abel, I notice, has now preceded me and has relieved me of approximately one-third of these things, the more so as he has carried out all the computations elegantly and concisely. He has chosen the exact same route that I took in 1798; for that reason the large coincidence in our results is not surprising. To my amazement this extends even to the form and partly to the choice of notation, so that many of his formulas seem to be exact copies of my own. To avoid any misunderstanding I note, however, that I do not recall ever having discussed these things with anyone.[25]

If Gauss here regarded elliptic functions as only one-third of his studies in transcendental functions, the other two-thirds must be classified as the theory of arithmetic-geometric means with the rudiments of the theory of modular functions (which will be discussed in the appropriate place) and the theory of hypergeometric functions.

Hypergeometric Functions

The hypergeometric series

$$1 + \frac{\alpha\beta}{1\cdot\gamma}x + \frac{\alpha(\alpha+1)\beta(\beta+1)}{1\cdot 2\gamma(\gamma+1)}x^2 + \frac{\alpha(\alpha+1)(\alpha+2)\beta(\beta+1)(\beta+2)}{1\cdot 2\cdot 3\gamma(\gamma+1)(\gamma+2)}x^3 + \cdots$$

first appeared in a slightly more general form in Chapter XI of the second volume of Euler's *Institutiones calculi integralis* (1769), where it was established in particular that the integral

$$\int_0^1 u^{b-1}(1-u)^{c-b-1}(1-xu)^{-\alpha}\, du$$

---

24) Lagrange, J.-L. *Œuvres.* 1869, T. 2, p. 253ff, especially pp. 271–274.
25) Gauss, C. F. *Werke*, Bd. 10(1), S. 248–249.

can be expanded in a series of the form

$$B(b, c - b)\left[1 + \frac{a \cdot b}{1 \cdot c}x + \frac{a(a+1)b(b+1)}{1 \cdot 2 \cdot c(c+1)}x^2 + \cdots\right]$$

($B(b, c - b)$ is the beta function) and satisfies the differential equation

$$x(1 - x)d^2y + [c - (a + b + 1)x]\, dy\, dx - aby\, dx^2 = 0$$

(the hypergeometric equation).

In Chapter VIII of the same volume Euler considered the still more general equation

$$x^2(a + bx^n)\, d^2y + x(c + ex^n)\, dy\, dx + (f + gx^n)y\, dx^2 = 0,$$

whose solution he sought in the form of a power series. The inhomogeneous case of this last equation was studied by Gauss' friend and teacher J. Pfaff (1765–1825) in his *Disquisitiones analyticae...*, VI (Helmstadt 1797). Pfaff was interested in the cases when the equation could be integrated in closed form. A separate chapter of his book was devoted to the hypergeometric series, where the term "hypergeometric" was applied for the first time in this sense.

Gauss himself had already encountered a special but highly nontrivial case of the hypergeometric series in connection with the arithmetic-geometric mean around the year 1800. At that time he had obtained the expansion

$$\frac{1}{M(1 + x, x - 1)} = 1 + \sum_1^\infty \left[\frac{1 \cdot 3 \cdots (2k - 1)}{2 \cdot 4 \cdots (2k)}\right]^2 x^{2k},$$

which is obviously a special case of the hypergeometric series (with $\alpha = \beta = 1/2$, $\gamma = 1$ and x replaced by x^2). It was the fact that this series is identical to the expansion of the elementary integral $\dfrac{1}{\pi}\displaystyle\int \dfrac{d\varphi}{\sqrt{1 - x^2\cos^2\varphi}}$ that convinced him of the truth of the relation given above. He then remarked that the sum of the series satisfies the differential equation

$$(x^3 - x)\frac{d^2y}{dx^2} + (3x^2 - 1)\frac{dy}{dx} + xy = 0$$

and that the coefficients of the trigonometric expansion of the function $(a^2 + a'^2 - 2aa'\cos\varphi)^{-1/2}$, which plays an important role in the theory of planetary perturbations, could also be expressed by hypergeometric series.

In a letter of 3 September 1805 Gauss[26] informed Bessel that he was in possession of artificial means based partly on seemingly irrelevant researches (probably his study of the arithmetic-geometric mean) enabling him to compute the

---

26) *Ibid.*, S. 237–248.

necessary coefficients quite rapidly. The details he provided show that by 1805, in connection with the needs of astronomical computations, Gauss was studying sums of the hypergeometric series for a fixed value of x and $\alpha = 1/2$ and values of β and γ differing by an integer. In doing this he used continued fractions to compute the ratios of the sums of two hypergeometric series.

All these questions were developed and generalized by Gauss in the following years. They found systematic expression in his "Disquisitiones generales circa serium $1 + \frac{\alpha \cdot \beta}{1 \cdot \gamma} x + \frac{\alpha(\alpha+1)}{1 \cdot 2} \frac{\beta(\beta+1)}{\gamma(\gamma+1)} x^2 + + \frac{\alpha(\alpha+1)(\alpha+2)}{1 \cdot 2 \cdot 3} \frac{\beta(\beta+1)(\beta+2)}{\gamma(\gamma+1)(\gamma+2)} x^3 \cdots,$ " (1813, see above). In the abstract of this paper[27] Gauss noted that there was hardly any transcendental function studied by analysts that could not be reduced to this series, and that in relation to the series his paper was introductory to others, which would contain the results of a study of higher transcendental functions. This paper also has special value in the history of mathematical analysis as the first example of a detailed study of the convergence of a certain class of power series at the endpoints of the interval of convergence.

In the introduction Gauss denoted the sum of the series by $F(\alpha, \beta, \gamma, x)$ (the hypergeometric function, although he did not use this term). He noted further that F is symmetric with respect to α and β, and that for fixed α, β, and γ it is a function of x alone. If $\alpha - 1$ or $\beta - 1$ is a negative integer, it reduces to a rational algebraic function; the element γ cannot be a nonpositive integer, since that would cause infinite terms to appear.

It follows from considering the ratio of each term to its predecessor that the series converges for $|x| < 1$ "from a certain point on, if not from the very beginning [Gauss seems to mean that the absolute values of the terms tend monotonically to zero from some point on] and also leads to a certain finite sum." If $|x| > 1$, the series "diverges if not from the very beginning, then from a certain point on [the absolute values of the terms tend to ∞], so that the sum does not exist."[28] The case $x = 1$ was considered separately (see below). Gauss went on to give expressions of various forms for the elementary functions and the coefficients of the trigonometric series expansion of the function $(a^2 + b^2 - 2ab \cos \varphi)^{-n}$ in terms of hypergeometric functions. His introduction ended here.

The first and briefest section of the paper was devoted to the derivation of linear relations between $F(\alpha, \beta, \gamma, x)$, $F(\alpha', \beta', \gamma', x)$, and $F(\alpha'', \beta'', \gamma'', x)$ when the differences $\alpha - \alpha'$, $\beta - \beta'$, $\gamma - \gamma'$, $\alpha - \alpha''$, $\beta - \beta''$, and $\gamma - \gamma''$ are integers. The relations derived in this section were applied in the second section to find the continued fraction expansions of ratios of the form $F(\alpha', \beta', \gamma', x)/F(\alpha, \beta, \gamma, x)$ when two of the differences $\alpha - \alpha'$, $\beta - \beta'$, $\gamma - \gamma'$ are equal to $+1$ and the third is zero. These are the cases Gauss had encountered in his astronomical computations, as shown by the letter to Bessel quoted above. Gauss did not pose the question of the convergence of these expansions, limiting himself to a formal derivation of them by repeated division of power series. Only later did Riemann prove that these

---

27) *Ibid.*, Bd. 3, S. 196–202.

28) *Ibid.*, Bd. 2.

continued fractions converge in the entire complex plane with a cut from 2 to $+\infty$ along the positive real axis ("Beiträge zur Theorie der durch die Gauss'sche Reihe $F(a, b, c, x)$ darstellbaren Funktionen," 1857).[29]

The third and longest section of Gauss' paper was devoted to the study of the hypergeometric series for α, β, and γ real (as in the entire paper) and $x = 1$. The general case of complex α, β, and γ and the points of the circle $|x| = 1$ was studied completely by Weierstrass in his paper "Über die Theorie der analytischen Fakultäten" (1856).[30]

Noting that all the coefficients from some point on are positive, "in all rigor,...for the benefit of those who are favorably disposed toward the rigorous methods of the ancient geometers" Gauss stated and proved the following propositions: 1) the coefficients increase without bound (if the series does not terminate) when $\alpha + \beta - \gamma - 1 > 0$; 2) the coefficients converge to a finite limit when $\alpha + \beta - \gamma - 1 = 0$; 3) the coefficients decrease to zero when $\alpha + \beta - \gamma - 1 < 0$; 4) in this case [i.e., in case 3)] there are no obstacles to the convergence of the series when $x = 1$ (summam seriei nostrae pro $x = 1$ non obstante convergentia in casu tertio), but if $\alpha + \beta - \gamma \geq 0$, this sum is ∞; 5) the sum is unquestionably finite when $\alpha + \beta - \gamma < 0$.[31]

To establish these assertions Gauss considered the more general case of a series (sequence) M, M', M'', M''',... for which the ratio of each term to its predecessor has the form

$$\frac{t^\lambda + At^{\lambda-1} + Bt^{\lambda-2} + \cdots}{t^\lambda + at^{\lambda-1} + bt^{\lambda-2} + \cdots}$$

for $t = m$, $m + 1$, $m + 2$,... . Gauss proved that "the sum of the series $M + M' + M'' + M'''+$ and so forth, continued to infinity, will certainly be finite" provided $A - a < -1$. In the opposite case, which Gauss divided into subcases, the series diverges (in the modern sense of the term). Thus we have here, completely proved, the *Gaussian convergence criterion for series*, to which, however, Gauss himself attached so little value that he did not even include it in the German abstract of the paper, which was written in Latin. The abstract mentions only (and only in passing) that it is proved with geometric rigor that when $x = 1$ the series converges to a finite sum only when $\gamma - \alpha - \beta > 0$.

Relying on the results of his study of the coefficients of the hypergeometric series as functions of the parameters α, β, and γ, Gauss established the formula

$$F(\alpha, \beta, \gamma, 1) = \frac{\Pi(\gamma - 1)\Pi(\gamma - \alpha - \beta - 1)}{\Pi(\gamma - \alpha - 1)\Pi(\gamma - \beta - 1)}$$

for $\gamma - \alpha - \beta > 0$, where $\Pi(z)$ is precisely the gamma function of Euler $\Gamma(z+1)$. We remark that in deriving this relation Gauss made repeated use of termwise passage

---

29) Riemann, B. *Werke*, S. 99–115.

30) Weierstrass, K. *Mathematische Werke*. Bd. 1, 1894, S. 153–221.

31) Gauss, C. F. *Werke*. Bd. 3.

to a limit in a power series at an endpoint of the interval of convergence. Gauss seems to have believed that such a passage to the limit required no justification. Actually the first justification was given by Abel in 1826 (Abel's second theorem, see above).

Gauss defined the function $\Pi(z)(= \Gamma(z+1))$ as the limit of the product

$$\Pi(z) = \lim_{k \to \infty} \Pi(k, z) = \lim_{k \to \infty} \frac{1 \cdot 2 \cdots \cdot k}{(z+1)(z+2)\cdots(z+k)} k^z,$$

Euler had defined $z!$ in the form of just such a limit in 1729, in a letter to Goldbach. But neither then nor later had he proved that the product converges. Gauss only sketched the proof of convergence, writing the difference $\ln \Pi(h+n, z) - \ln \Pi(h, z)$ (h and n are natural numbers) in the form

$$\sum_{k=2}^{\infty} \frac{z}{k}[1 + (-1)^k z^{k-1}]\left[\frac{1}{(h+1)^k} + \frac{1}{(h+2)^k} + \cdots + \frac{1}{(h+n)^k}\right]$$

(we are using modern notation here). He asserted that it is easy to prove that the increment "always remains finite as n increases without bound,"[32] i.e., that it tends to a finite limit. Considering that Gauss had just carried out a termwise passage to the limit in a power series (under the hypothesis that the series converged) as a thing taken for granted, there can be no doubt that here also he had in mind only a proof that the series obtained by replacing the finite sums in brackets by the corresponding infinite series

$$\sum_{n=1}^{\infty} \frac{1}{(h+n)^k}$$

converges.

Gauss also wrote the function $\Pi(z)$ as the infinite product

$$\Pi(z) = \frac{1}{z+1} \frac{2^{z+1}}{1^z \cdot (2+z)} \frac{3^{z+1}}{2^z(3+z)} \cdots;$$

the analogous notation for $z!$ was given in the letter of Euler mentioned above and in his *Institutiones calculi differentialis*:

$$m! = \frac{1 \cdot 2^m}{1+m} \frac{2^{1-m} \cdot 3^m}{2+m} \frac{3^{1-m} \cdot 4^m}{3+m} \cdots .$$

The remaining part of Gauss' paper is devoted to the study of the function $\Pi(z)$ and its applications to the computation of integrals. In particular Gauss introduced and studied the function

$$\Psi(z) = \frac{d}{dz} \ln \Pi(z).$$

---

32) *Ibid.*

In relation to $\Pi(z)$ Gauss wrote in his abstract:

> This function, which is of the highest importance for all of analysis, is essentially none other than the inexpressible Eulerian function $\Pi(z) = 1 \cdot 2 \cdot 3 \cdots z$; however this method of generating or defining it, in the opinion of the author, is entirely inadmissible, since it has a clear meaning only for positive integers z. The method of proof chosen by the author is applicable under general hypotheses and has the same clear meaning for imaginary values of z as for real values... .[33]

This passage, like the other place where Gauss exhibits an expression for the integral $\int_0^1 x^{\lambda-1}(1-x^\mu)^\nu \, dx$ (the Eulerian integral of first kind) in terms of the gamma function as a new result, is evidence that at the time Gauss was insufficiently acquainted with the published results of Euler or at least that he incorrectly judged Euler's role in the foundation and development of the elements of the theory of the gamma function.

The paper under consideration had a continuation: "Determinatio seriei nostrae per aequationem differentialem secundi ordinis," published posthumously in 1866.[34]

In this paper the hypergeometric differential equation was obtained immediately as a special case of the linear relations of the second section of the fundamental paper: if $F(\alpha, \beta, \gamma, x) = P$, then

$$\frac{dP}{dx} = \frac{\alpha\beta}{\gamma} F(\alpha+1, \beta+1, \gamma+1, x)$$

and

$$\frac{d^2P}{dx^2} = \frac{\alpha\beta(\alpha+1)(\beta+1)}{\gamma(\gamma+1)} F(\alpha+2, \beta+2, \gamma+2, x).$$

Gauss studied the transformations of the independent variable x and the function P that map the hypergeometric equation into itself with different values of α, β and γ. Thus he introduced the transformation $y = 1-x$, which made it possible to find a second integral of the same equation, $F(\alpha, \beta, \alpha+\beta+1-\gamma, 1-x)$, independent of $F(\alpha, \beta, \gamma, x)$, so that the general integral could be written as

$$M F(\alpha, \beta, \gamma, x) + N F(\alpha, \beta, \alpha+\beta+1-\gamma, 1-x),$$

where M and N were arbitrary constants. The transformation $P = (1-x)^\mu P'$ exhibited earlier by Euler gives for $\mu = \gamma - \alpha - \beta$

$$F(\gamma - \alpha, \gamma - \beta, \gamma, x) = (1-x)^{\alpha+\beta-\gamma} F(\alpha, \beta, \gamma, x).$$

---

33) *Ibid.*
34) *Ibid.*, S. 207–229.

The analogous transformation $P = x^\mu P'$ leads to the relation

$$F(\alpha, \beta, \alpha + \beta + 1 - \gamma, 1 - x) = \frac{\Pi(\alpha + \beta - \gamma)\Pi(-\gamma)}{\Pi(\alpha - \gamma)\Pi(\beta - \gamma)} F(\alpha, \beta, \gamma, x) +$$
$$+ \frac{\Pi(\alpha + \beta - \gamma)\Pi(\gamma - 2)}{\Pi(\alpha - 1)\Pi(\beta - 1)} x^{1-\gamma}(1-x)^{\gamma - \alpha - \beta} F(1 - \alpha, 1 - \beta, 2 - \gamma, x)$$

when $\mu = 1 - \gamma$. This last relation makes it possible to reduce the computation of the sum of the hypergeometric series in the interval $(0.5, 1)$, in which the convergence is slow, to the computation in the interval $(0, 0.5)$, where the convergence is more rapid. But the formula just obtained is not suitable in the case when $\alpha + \beta - \gamma$ is an integer, since it then contains the meaningless hypergeometric series with the third parameter equal to a nonpositive integer and infinite values of Π. Gauss got around this difficulty by using a device of d'Alembert known in the theory of linear differential equations. He applied the result just obtained to compute $F(1/2, 1/2, 1, 1 - x)$, in which we recognize $1/M(1, \sqrt{x})$. The result is

$$F\left(\frac{1}{2}, \frac{1}{2}, 1, 1 - x\right) = -\frac{1}{\pi}\left[\ln\left(\frac{1}{16}x\right) F\left(\frac{1}{2}, \frac{1}{2}, 1, x\right) + \frac{1}{2}x + \frac{21}{64}x^2 + \frac{185}{768}x^3 + \cdots\right],$$

a relation whose content Gauss had obtained in terms of the arithmetic-geometric mean as early as 1799, when he had not yet mastered the theory of hypergeometric functions. It is remarkable that Gauss explained one of the relations derived in this paper, which seems to be contradictory, using the multivaluedness of the hypergeometric function, here understood not as the sum of the hypergeometric series (where the definition was restricted by the condition $|x| < 1$ for convergence of the series) but as the solution of the hypergeometric equation studied for general values of x. In doing this Gauss referred explicitly to a variable x assuming both real and imaginary values traversing a circuit about the points 0 and 1. This explanation is in agreement with the information on the nature of multivaluedness of functions of a complex variable that Gauss had communicated to Bessel in his letter of 19 December 1811 (see above).

The actual study of the hypergeometric functions $F(\alpha, \beta, \gamma, x)$ as functions of a complex variable x, which could have been carried out on the basis of Euler's integral representation

$$F(\alpha, \beta, \gamma, x) = B^{-1}(\beta, \gamma - \beta) \int_0^1 z^{\beta - 1}(1 - z)^{\gamma - \beta - 1}(1 - xz)^{-\alpha}\, dz$$

(nowhere used explicitly by Gauss) would have required, however a significantly more profound and detailed development of the foundations of the theory of functions of a complex variable than Gauss was had at his disposal. It was for that reason that he did not continue the systematic study of the theory of hypergeometric functions, and even the second part of the work just discussed was not completed. Only after the work of Kummer, Weierstrass, Riemann, Fuchs, Schwarz, and Klein did the last-mentioned author succeed, in his posthumously published *Vorlesungen über die hypergeometrische Funktion* (Berlin 1933), in carrying out the synthesis of the analytic and geometric parts of the theory and actually realizing Gauss' plan of creating a comprehensive work embracing all the results and methods of Gauss relating to transcendental function.

However the development of the theory of hypergeometric functions did not end there. Hypergeometric functions were generalized in the work of P. Appel to the case of several complex variables and in this form found and continue to find numerous applications in the problems of mathematical physics.

The First Approach to Modular Functions

As early as 1794, when Gauss was only 17 years old, he studied power series of the form

$$P(x) = 1 + 2x + 2x^4 + 2x^9 + \cdots,$$
$$Q(x) = 1 - 2x + 2x^4 - 2x^9 + \cdots,$$
$$R(x) = 2x^{1/4} + 2x^{9/4} + 2x^{25/4} + \cdots \quad (|x| < 1),$$

in connection with his study of the arithmetic-geometric mean. These are the so-called zeroth theta functions, formally obtained from the trigonometric expansions of the Jacobi theta functions Θ_1, Θ, and H_1 by setting the argument equal to zero and denoting $q = e^{-\pi k'/k}$ by x. Similar series had been encountered before Gauss in the work of Jakob and Daniel Bernoulli and Euler. Exactly what Gauss knew about them in 1794 cannot be established at present. We shall use his (considerably later) exposition, written in 1818 at the earliest, which seems to have been intended for publication. This is his "Hundert Theoreme über die neuen Transcendenten."[35] Here Gauss established the following identities:

$$P^2(x) + Q^2(x) = 2P^2(x^2), \quad Q^2(x^2) = P(x)Q(x), \tag{7}$$
$$P(x^2)R(x^2) = \frac{1}{2}R^2(x), \quad P^4(x) - Q^4(x) = R^4(x). \tag{8}$$

It follows from formulas (7) that $P^2(x^2)$ is the arithmetic mean between $P^2(x)$ and $Q^2(x)$, while $Q^2(x^2)$ is the geometric mean between the same functions. For that reason if we set $P^2(x) = m$, $Q^2(x) = n$, $m^{(k)} = (m^{(k-1)} + n^{(k-1)})/2$, $n^{(k)} = \sqrt{m^{(k-1)}n^{(k-1)}}$, $m^{(0)} = m$, $n^{(0)} = n$, the application of the arithmetic-geometric mean algorithm yields

$$m' = (m+n)/2 = P^2(x^2), \quad n' = \sqrt{mn} = Q^2(x^2), \ldots,$$
$$m^{(k)} = P^2(x^{2^k}), \quad n^{(k)} = Q^2(x^{2^k}), \ldots,$$

so that

$$M(m,n) = \lim_{k \to \infty} P^2(x^{2^k}) = \lim_{k \to \infty} Q^2(x^{2^k}) = 1 \quad (|x| < 1).$$

---

35) Gauss, C. F. *Werke*. Bd. 3, S. 461–469.

If now m and n are any positive numbers, $m > n$, and $M(m, n) = \mu$, it is natural to pose the question of finding x such that $m = \mu P^2(x)$, $n = \mu Q^2(x)$. Then

$$Q^2(x)/P^2(x) = n/m \tag{9}$$

and the question reduces to solving this last equation for x.

To explain why the function $Q^2(x)/P^2(x)$ came to be known subsequently as the *modular function* we turn to the terminology and notation of Jacobi. Let z and w be related by

$$z = \int_0^w \frac{dt}{\sqrt{(1-t^2)(1-k^2t^2)}}$$

(the elliptic integral of first kind). The parameter k (not equal to 0 or 1) is called the *modulus* of the integral and $k' = \sqrt{1-k^2}$ the *complementary modulus*. The inverse of the integral — the Gaussian "most general lemniscatic sine" — is in this case Jacobi's elliptic sine: $w = \operatorname{sn}(z; k)$. Its fundamental periods can be taken as $4K$ and $2iK'$, where K and K' are the complete elliptic integrals with moduli k and k' respectively:

$$K = \int_0^1 \frac{dt}{\sqrt{(1-t^2)(1-k^2t^2)}}, \quad K' = \int_0^1 \frac{dt}{\sqrt{(1-t^2)(1-k'^2t^2)}}.$$

Here the values of the roots under the integral sign are chosen so that the condition $\operatorname{Re}(K'/K) > 0$ (if $k^2 < 1$, it obviously suffices to take positive values of the roots in both integrals). From the relations between the arithmetic-geometric means and the elliptic integrals of first kind established around 1800, it follows that the roots x of Eq. (9) have the form $x = e^{-\tau}$, where

$$\tau = K'/K \quad \text{and} \quad k^2 = (m^2 - n^2)/m^2, \quad \text{i.e., } k' = n/m.$$

Thus the function $[Q(x)/P(x)]^2$, regarded as a function of τ, is analytic in the half-plane $\operatorname{Re}\tau > 0$; it expresses the value of the (complementary) modulus of the Jacobi function $\operatorname{sn}(z; k)$ and can be called the *modular* function.

The problem of studying the set of solutions of Eq. (9) for an arbitrarily given n/m ($= k'$) is consequently the same as the problem of studying the distribution of values of one of the simplest modular functions. And Gauss actually did study such a problem. Thus, in fragments dating to 1827 he stated the following theorem:

If the imaginary parts of t and $1/t$ lie between $-i$ and $+i$, then the real part of $Q^2(t)/P^2(t)$ is positive.[36]

Here $Q(t)$ and $P(t)$ are the values of the previous functions $Q(x)$ and $P(x)$ when $x = e^{-\pi t}$, so that these functions are considered in the half-plane $\operatorname{Re} t > 0$.

---

36) *Ibid.*, S. 477–478.

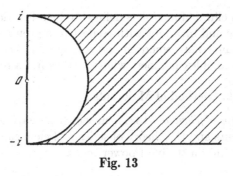

Fig. 13

Gauss' statement was accompanied by a drawing showing the region of the complex plane occupied by the values of t satisfying the conditions of the theorem (Fig. 13).

The idea that there exist many, in fact infinitely many, solutions of Eq. (9) corresponding to a given value of n/m had occurred to Gauss as early as 1800. Judging from his later notes (1825), it was a question of taking one of the two values of the square root in the computation of the geometric means in the arithmetic-geometric mean algorithm. Accordingly we find him making the following assertion:

> To solve the equation $Q(t)/P(t) = A$, set $A^2 = n.m$ and find the arithmetic-geometric mean between m and n; let it be μ. Then find the arithmetic-geometric mean between m and $\sqrt{m^2 - n^2}$ or, what is the same, between $m + n$ and $m - n$; let it be λ. Then you will have $t = \mu/\lambda$. In this way only one value of t is obtained; all the others are contained in the formula $\frac{\alpha t - 2\beta i}{\delta - 2\gamma t i}$, where α, β, γ, and δ denote all the integers satisfying the equation $\alpha\delta - 4\beta\gamma = 1$.[37]

This result, as Fricke remarked,[38] is not best-possible: the regularity noted by Gauss actually applies to the roots of the more general equation $Q^4(t)/P^4(t) = A^4 = n^2/m^2$. However what is important for our purposes is that Gauss, always relying on his arithmetic-geometric mean algorithm, gradually arrived at the discovery of the basic property of the modular function he had introduced: *invariance with respect to a certain special group of fractional-linear automorphisms of its domain of definition.*

The subsequent development of mathematics led to similar results both in relation to the theory of the modular function and in relation to the theory of elliptic functions. In the definition of both classes the historical sources are completely eradicated, and only the most general characteristic properties remain. To be specific, a modular function $\varphi(t)$ came to be defined as an analytic function in a half-plane (it is most convenient to talk of the upper half-plane $\operatorname{Im} t > 0$), which is invariant with respect to a certain subgroup of the group of linear automorphisms: $t' = (\alpha t + \beta)(\gamma t + \delta)^{-1}$, where α, β, γ, and δ are integers satisfying the condition

---

37) *Ibid.*, Bd. 8, S. 101.
38) *Ibid.*, S. 105.

$\alpha\beta - \gamma\delta = 1$. An elliptic function is defined to be a function $f(z)$ that is meromorphic in the plane and invariant with respect to a group of transformations of the form $z' = z + m\omega_1 + n\omega_2$, where m and n range over the integers, and ω_1 and ω_2 are *periods*, two complex numbers whose ratio is not a real number.

Power Series. The Method of Majorants

Cauchy's *Analyse algébrique* (1821), in which he gave a thorough exposition of the theory of limits and the theory of series and elementary functions based on it, contains in particular a study of power series

$$a_0 + a_1 x + a_2 x^2 + \cdots + a_n x^n + \cdots ,$$

in which x assumes both real and complex values. Cauchy established that such a series will converge or diverge according as the absolute value of x is less than or greater than

$$\rho = \frac{1}{\varlimsup_{n\to\infty} \sqrt[n]{a_n}}.$$

Of course Cauchy did not have the definition of the upper limit or the corresponding notation. But, as has already been stated, he speaks directly of the largest limit of the number $\sqrt[n]{a_n}$. Denoting this quantity by A, Cauchy introduced this number into the statement of the theorem.[39]

This theorem (the Cauchy-Hadamard Theorem — restated and proved by Hadamard in his 1892 doctoral dissertation), from which it follows in particular that for real x the series convefges inside the interval $(-\rho, \rho)$ and diverges outside this interval, gives complete information about the domain of convergence of a power series.

The problem of expanding functions in power series was placed in the correct light for the first time in the history of analysis in Lectures 36–38 of *Résumé des leçons données à l'École Polytechnique sur le calculu infinitésimal* (1823, T. 1). The value Cauchy attached to this problem is shown by an excerpt from his preface to this book, which he later reproduced word for word in the preface to his *Leçons sur le calcul différentiel* (1829). It will be quoted at the corresponding section of Book 3 of the present series.

In the study of power series the baton passed from Cauchy to Abel. The first issue of Crelle's *Journal* in 1826 contained his paper "Untersuchung über die Reihe $1 + \frac{m}{1}x + \frac{m(m-1)}{1\cdot2}x^2 + \frac{m(m-1)(m-2)}{1\cdot2\cdot3}x^3 +$ usw. Abel's problem was "to find the sum of the series

$$1 + \frac{m}{1}x + \frac{m(m-1)}{1\cdot2}x^2 + \frac{m(m-1)(m-2)}{1\cdot2\cdot3}x^3 + \text{etc.}$$

for all real or complex values of x for which the series converges." In doing this Abel assumed that m is an arbitrary, generally complex number. He gave an exhaustive

---

39) Cauchy, A.-L. *Œuvres complètes*. Sér. 2, T. 3, pp. 239–240.

solution of this problem. As preliminary theorems he established some important propositions that are of value not only for the general theory of series, but also in particular for the subsequent development of the elements of analytic function theory.

Theorem IV of this classic paper says:

If the series $f(\alpha) = v_0 + v_1\alpha + v_2\alpha^2 + \cdots + v_m\alpha^m + \cdots$ converges for a certain value of α equal to δ, then it also converges for any smaller value of α, and for steadily decreasing values of β the function $f(\alpha - \beta)$ approaches arbitrarily closely to the limit $f(\alpha)$ provided α is less than or equal to δ.

This theorem breaks up into two propositions joined by the conjunction "and." The first is usually called Abel's first theorem; the second, which asserts that the sum of a convergent power series is a (left-) continuous function of its argument, is called Abel's second theorem. We have seen above that Gauss relied on this proposition as something self-obvious.

In 1831 Cauchy obtained the theorem that a function of a complex variable can be expanded in a power series. This theorem says: *A function can be expanded in a convergent series of increasing powers of x according to Maclaurin's formula if the absolute value of the real or complex variable x remains less than any value at which the function ceases to be finite or continuous. Let X be this last value or a smaller value and $\bar{x}$ a complex expression whose absolute value is X. The absolute values of the general term and the remainder of the Maclaurin series will be less than the absolute values of the general term and remainder of the geometric progression having the sum*[40]

$$\frac{X}{X - \bar{x}} \Lambda f(\bar{x}).$$

Here $\Lambda f(\bar{x})$ denotes the "limit of the absolute value of $f(x)$" for the fixed absolute value $|\bar{x}| = X$ (or, as we now say, the upper bound of $|f(\bar{x})|$ at the points of the circle $|\bar{x}| = X$). This theorem established not only the simplest criterion for a function to be expandable in a power series, i.e., analyticity of the function, but also gives a majorant of the expansion, in the form of a geometric progression, which is extraordinarily important for the applications of power series. It is crucial to emphasize that these majorants and the corresponding estimate of the remainder of a power series provide the investigator with much more than the mere fact of uniform convergence of the power series in each disk concentric with the circle of convergence and having radius less than the radius of convergence. In other words, with his theorem Cauchy essentially removed the question of uniform convergence for power series,[41] proposing a simple quantitative characterization of it.

---

40) Cauchy, A.-L. *Exercices*, Paris 1841, T. 2, pp. 54–55. This theorem and some related results were published as lithographs in 1832, as will be discussed later.

41) The introduction of the concept of uniform convergence will be discussed in Book 3.

To prove his theorem Cauchy, using the fundamental relation (5), represented $f(x)$ as an integral

$$f(x) = \frac{1}{2\pi} \int_{-\pi}^{\pi} \frac{\bar{x} f(\bar{x})}{\bar{x} - x} \, dp \quad (\bar{x} = X e^{p\sqrt{-1}}),$$

which later came to be called the Cauchy integral.

Having done this, he had only to expand the fraction $\dfrac{\bar{x}}{\bar{x} - x} = \dfrac{1}{1 - \frac{x}{\bar{x}}}$ in a geometric series (using the fact that $|x/\bar{x}| < 1$) and integrate termwise after substituting that series under the integral sign. As a result he arrived at the required expansion with the majorant appearing in the form of a geometric progression as a consequence of the method by which the expansion was obtained.

To apply this theorem correctly one must keep in mind that Cauchy was convinced of the existence and continuity of the derivative of the function $f(z)$, although he did not state this at first in the hypothesis of the theorem. The reasons for this omission have already been examined in connection with the statement of the hypotheses of the Cauchy integral theorem in its original form. Here we must emphasize that the hypotheses of continuity and differentiability must hold for the function $f(x)$ regarded as a function of a complex variable. Cauchy explained all this by examples. The functions $\cos x$, $\sin x$, e^x, e^{x^2}, $\cos(1 - x^2), \ldots$ are continuous (and differentiable) for any complex number x. Therefore they can be expanded in a series of powers of x that converges everywhere. The functions $(1 + x)^{1/2}$, $\frac{1}{1-x}$, $\frac{x}{1+\sqrt{1-x^2}}$, $\ln(1 + x)$, and $\arctan x$ cease to be continuous in Cauchy's sense of the term at some x of absolute value 1. For us it suffices to remark that the derivative of the first function becomes infinite for $x = -1$ (Cauchy also considered the function $(1 + x)^{1/2}$ itself discontinuous at this point since no single-valued branch of it can be exhibited in a neighborhood of the point $x = -1$). The second function itself becomes infinite at $x = 1$. For the third function the derivative, which is $\frac{1}{(1+\sqrt{1-x^2})\sqrt{1-x^2}}$, becomes infinite at $x = \pm 1$ (again Cauchy considers the function itself discontinuous at these points). The function $\ln(1+x)$ becomes discontinuous for $x = -1$; and, finally, the derivative of $\arctan x$, which is $1/(1 + x^2)$, becomes infinite for $x = \pm i$. (Cauchy, limiting himself to the consideration of the function itself and not its derivative, expressed $\arctan(\pm X\sqrt{-1})$ in terms of logarithms:

$$\arctan(\pm X\sqrt{-1}) = \frac{\ln(1 \mp X) - \ln(1 \pm X)}{2\sqrt{-1}},$$

from which he concluded that $\arctan x$ has a discontinuity at $X = 1$, i.e., at $x = \pm\sqrt{-1}$.) It then follows from Cauchy's theorem that the power series for all these functions will converge only for $|x| < 1$. Finally, the functions $e^{1/x}$ are discontinuous at the origin, and therefore these functions cannot be represented as a convergent series of nonnegative powers of x.

These results were obtained by Cauchy in the years 1831–1832; they were first communicated to the Turin Academy at its session of 11 October 1831, and then published as lithographs in Turin in 1832. Cauchy informed his readers of all this in his *Exercices* (T. 1, 1840; T. 2, 1841), where he reproduced these results together with proofs in the articles "Note sur l'intégration des équations différentielles des mouvementes planétaires" (*Exercices*, 1840, T. 1, pp. 27–32), "Résumé d'un mémoire sur la mécanique céleste et sur un nouveau calcul appelé calcul des limites" (*Exercices*, 1841, T. 2, pp. 41–49), and "Formules pour le développement de fonctions en séries" (*Ibid.*, pp. 50–96). This last article was merely the lithographed Turin memoir of 1832, or rather an excerpt from it. However, this time in his statements he introduced the additional condition of continuity not only for the function itself but also for its derivative.

Cauchy had introduced this correction after a dispute with Liouville (to whom he had communicated his result in a letter of 27 January 1837, published in the *Comptes Rendus*) and Sturm. It was in the first clause of the statement given above that Cauchy replaced the words "the function ceases to be finite or continuous" with "the function $f(x)$ and its derivative $f'(x)$ cease to be finite and continuous."[42] Pointing out that in all the examples the function and its derivative become infinite or discontinuous for the same values of x, Cauchy added, "If one could be sure that such is always the case, one could eliminate the mention of the derivative in the theorem just stated. However, since one cannot be sufficiently confident on this account, it will be more rigorous to state the theorem in the terms we have used above."[43] Later on, in 1851, Cauchy again removed the requirement that the derivative be continuous, regarding it as unnecessary.

The absence of clarity and distinctness in Cauchy's statements and proofs attracted the attention of Chebyshev. In a short paper entitled "Note sur la convergence de la série de Taylor," (*J. für Math.*, **28** (1844), S. 279–283) he noted in passing that, "Cauchy assumes that the value of a definite integral can be expanded in a convergent series if the derivative can be expanded in a convergent series between the limits of integration; this is true only in special cases."[44] Thus Chebyshev here points out the inadmissibility of integrating series without some special proof of the validity of such an operation. We add that in the case of power series the legality of termwise integration follows immediately from two general results of Cauchy himself: his general criterion for the existence of a limit (the Cauchy convergence criterion) and the geometric progression majorant he had established for a power series.

In the paper "Résumé d'un mémoire sur la mécanique céleste..." quoted above and in several other papers Cauchy defines his "calcul des limites" (known in English as the method of majorants) to be not only the condition for convergence of power series for explicit and implicit functions of one and several variables (the most important of them being provided by the theorem given above), but also the method of estimating the remainder. In essence this "calculus of limits" reduces

---

42) Cauchy, A.-L. *Exercices*, 1840, T. 1, p. 29.

43) *Ibid.*, p. 32.

44) Chebyshev, P. L. *Complete Works* [in Russian]. Moscow/Leningrad 1947, T. 2, p. 13.

to the rules for finding an upper bound on the absolute value of a function of a complex variable at points of an arbitrary circle in the complex plane. As shown above, the limit of the absolute value of the function $f(x)$, where $\bar{x} = Xe^{p\sqrt{-1}}$, was defined by Cauchy as

$$\Lambda f(\bar{x}) = \max_{|\bar{x}|=X} |f(\bar{x})|.$$

Here are several of the simplest formulas of the "calculus of limits" (a is a positive number):

$$\Lambda(a + \bar{x}) = \Lambda(a - \bar{x}) = a + X, \quad \Lambda(a\bar{x}) = aX, \quad \Lambda(a/\bar{x}) = a/X,$$
$$\Lambda e^{\bar{x}} = \Lambda e^{-\bar{x}} = \Lambda e^{\bar{x}\sqrt{-1}} = \Lambda e^{-\bar{x}\sqrt{-1}} = e^X,$$
$$\Lambda \sin \bar{x} = \frac{1}{2}|e^X - e^{-X}|, \quad \Lambda \cos \bar{x} = \frac{1}{2}(e^X + e^{-X}),$$

Elementary as they are from our point of view, the introduction of relations of this type into the practice of analysis was a necessary condition for mathematicians to gain further mastery of the behavior of analytic functions in the complex plane.

We have already noted that for Cauchy the method of majorants was not an end in itself. He needed it in order to make use of the fundamental regularities contained in his theorem on the expansion of a function in a power series: every power series is majorized by some geometric progression for which the estimate of the remainder reduces to estimating the "limit" of the function. Cauchy applied the method of majorants to prove the convergence of power series obtained formally by the method of undetermined coefficients for expanding implicit functions (in particular algebraic functions) and solutions of differential equations, both in one and several variables. The development of this method later led S. V. Kovalevskaya to a general existence theorem for solutions of a system of partial differential equations with analytic data.

In the papers devoted to the method of majorants and its application to the solution of functional equations, Cauchy also laid the foundations of the theory of analytic functions of several variables. He showed that the coefficients of the power series that represent them, and also the remainders of these series, can be expressed and estimated by formulas analogous to the case of one variable. Long before Weierstrass he obtained the essential part of the fundamental theorem of the theory of analytic functions of several variables, the so-called preparation theorem (see Cauchy's paper cited above, "Formules pour le développement de fonctions en séries," *Exercices*, 1841, T. 2, pp. 41ff.).

Weierstrass wrote his first papers on the theory of functions in the early 1840's. He did not publish them at the time, and they first came to light in 1894 (though he included his results in his lectures on analytic function theory delivered at the University of Berlin starting in the late 1850's). His generalization of Cauchy's theorem on the expansion of a function in a power series to the case of a function of a complex variable that is continuous and differentiable in an annulus dates from 1841 (Weierstrass, not knowing of Cauchy's work, derived the theorem anew). In this case the result is a series of integer powers (including negative powers). these results are contained in his paper "Darstellung einer analytischen Funktion einer komplexen Veränderlichen deren absoluter Betrag zwischen zwei gegebenen Grenzen liegt,"[45] However, the first to publish a result of this type (independently of Weierstrass, of course) was the French military engineer and learned mathematician Pierre Alphonse Laurent (1813–1854), a graduate of the École Polytechnique. He was the author of a number of papers, mainly in mathematical physics, but the only paper that earned him a place in the history of mathematics was entitled "Extension du théorème de A. Cauchy, relatif à la convergence du développement d'une fonction suivant les puissances ascendantes de la variable," (*Comptes Rendus*, **17** (1843)). This theorem and a doubly infinite series of the form $\sum\limits_{-\infty}^{\infty} c_n (z - a)^n$ are now called respectively Laurent's theorem and a Laurent series.

In an 1842 paper, also not published at the time, entitled "Definition analytischer Funktionen einer Veränderlichen vermittelst algebraischer Differentialgleichungen"[46] Weierstrass expounded the idea of analytic continuation of power series, which later played a vital role in his construction of analytic function theory. However, independently of Weierstrass, the same idea occurred to V. Puiseux, who published a method of analytic continuation by means of power series (and even more general series) in 1850.

Of the papers on power series dating to the 1840's we note Cauchy's theorem that an entire function that is bounded in the entire plane is identically constant (*Comptes Rendus*, **19** (1844)).[47] However, this simple but very important theorem was only a generalization of a proposition stated by J. Liouville in the same year of 1844: *an entire doubly-periodic function is a constant.* E. Neuenschwander[48] notes that Cauchy gave the more general theorem, nowadays known as Liouville's theorem, at the session of the Paris Academy following the session at which Liouville stated his proposition. The latter made extensive use of his theorem in lectures on the theory of elliptic functions (1847, see below). The name of Liouville was attached to the general Cauchy theorem rather than the special case by the inattentive K. Borchardt, who attended both sessions.

Elliptic Functions in the Work of Abel

We have already mentioned that Gauss, having become acquainted with Abel's first paper on elliptic functions (1827, see above), found that Abel's results coincided with his own, explaining the coincidence by the fact that Abel, not knowing

45) Weierstrass, K. *Werke*. Bd. 1, 1894, S. 52–66.

46) *Ibid.*, S. 75–84.

47) Cauchy, A.-L. *Œuvres complètes*. Sér. 1, T. 8, pp. 378–383.

48) Neuenschwander, E. "The Casorati-Weierstrass theorem" (Studies in Complex Function Theory. I), *Hist. Math.*, **5** (1978), pp. 139–166.

of Gauss' work, nevertheless chose the same route. However in this paper and in
later papers, some published posthumously, Abel managed to develop the foundations of the theory much more completely and systematically than Gauss had
done. Moreover Abel has the honor of the first publication of a theory in which
elliptic functions are defined as the functions inverse to elliptic integrals, while his
predecessors and contemporaries (except for Gauss and, as we shall relate below,
Jacobi) had not resorted to this inversion.

The significance of the step taken by Abel can be judged by imagining that
mathematicians had contented themselves, up to a certain point in time, with
the study of the integrals $u = \int_0^x \dfrac{dx}{\sqrt{1-x^2}}$ and $u = \int_0^y \dfrac{dy}{1+y^2}$, and only at
that point had it occurred to one of them to base the study on the inverses of
these integrals: $x = \sin u$, $y = \tan u$. This gives an approximate description of
of the relation between the treatise of Legendre on elliptic integrals (*Traité des
fonctions élliptiques et des intégrales eulériennes, avec de tables pour en faciliter
le calcul numérique*, Paris 1825–1826, T. 1–2) and Abel's papers "Recherches sur
les fonctions élliptiques," (*J. für Math.*, **2** (1827); **3** (1828)) and "Précis d'une
théorie des fonctions élliptiques," (*J. für Math.*, **4** (1829)).

Abel began with the elliptic integral of first kind in the form

$$\alpha = \int_0^x \frac{dx}{\sqrt{(1-c^2x^2)(1+e^2x^2)}}$$

and, besides the function $x = \varphi(\alpha)$ inverse to this integral (which differs only
slightly from Gauss' most general sine), he introduced the two additional functions
$f(\alpha) = \sqrt{1 - c^2\varphi^2(\alpha)}$ and $F(\alpha) = \sqrt{1 + e^2\varphi^2(\alpha)}$, which are also elliptic functions.
He wrote out the corresponding addition theorems, citing general results of Euler.
Thus the values of $\varphi(\alpha + \beta)$, $f(\alpha + \beta)$, and $F(\alpha + \beta)$ can be expressed rationally
in terms of the values of these same functions of α and β. After defining the
function φ for purely imaginary values of the argument ($\varphi(\beta i)$) by inverting the
corresponding integral with x replaced by xi, and consequently defining $f(\beta i)$ and
$F(\beta i)$, Abel used the addition theorem to extend the definition of all three elliptic
functions to the case of an arbitrary complex argument $\alpha + \beta i$. When this is done,
their double periodicity reveals itself immediately. (The fundamental periods can
be taken to be

$$2\omega = 4 \int_0^{1/c} \frac{dx}{\sqrt{(1-c^2x^2)(1+e^2x^2)}}$$

and

$$2\omega i = \int_0^{1/c} \frac{dx}{\sqrt{(1+c^2x^2)(1-e^2x^2)}},$$

and all the zeros and poles can be found, etc.)

The addition theorem also immediately implies formulas for multiplication
of the argument, expressing $\varphi(n, \alpha)$, $f(n, \alpha)$, and $F(n, \alpha)$ rationally in terms of

$x = \varphi(\alpha)$, $y = f(\alpha)$, and $z = F(\alpha)$. In particular, Abel showed that for odd n

$$\varphi(n\alpha) = x\frac{P_n(x)}{Q_n(x)},$$

where $P_n(x)$ and $Q_n(x)$ are even polynomials of degree $n^2 - 1$, and for even n

$$\varphi(n\alpha) = xyz\frac{P_n(x)}{Q_n(x)} = x\sqrt{1 - c^2x^2}\sqrt{1 + e^2x^2}\frac{P_n(x)}{Q_n(x)},$$

where $P_n(x)$ and $Q_n(x)$ are also even polynomials, $P_n(x)$ has degree $n^2 - 4$, and Q_n has degree n^2. Hence for even n

$$\varphi^2(n\alpha) = x^2(1 - c^2x^2)(1 + e^2x^2)\frac{P_n^2(x)}{Q_n^2(x)}.$$

For comparison with the trigonometric functions $x = \sin\alpha$ and $y = \cos\alpha$ we derive from De Moivre's formula for odd n

$$\sin n\alpha = xP_n(x) = x\left[\frac{n}{1}(1 - x^2)^{\frac{n-1}{2}} - \frac{n(n-1)(n-2)}{3!}(1 - x^2)^{\frac{n-3}{2}}x^2 + \cdots\right],$$

and for even n

$$\sin n\alpha = xyP_n(x) = xy\left[\frac{n}{1}(1 - x^2)^{\frac{n-2}{2}} - \frac{n(n-1)(n-2)}{3!}(1 - x^2)^{\frac{n-4}{2}}x^2 + \cdots\right].$$

Here $P_n(x)$ is an even polynomial of degree $n - 1$ when n is odd and of degree $n - 2$ when n is even.

It follows from these formulas that in the problem of dividing the argument by a positive integer n, i.e., finding the function of α given the function of $n\alpha$, in the case of elliptic functions it is necessary to solve an equation of degree n^2, and if n is even, an equation of degree $2n^2$. These equations are

$$xP_n(x) - \varphi(n\alpha)Q_n(x) = 0,$$

and

$$x^2(1 - c^2x^2)(1 + e^2x^2)P_n^2(x) - \varphi^2(n\alpha)Q_n^2(x) = 0,$$

respectively. In the case of trigonometric functions one needs to solve only an equation of degree n, or $2n$ if n is even. This fact had been noted by Legendre as early as 1793, but Abel was the first to give an explanation of it. The explanation reduces to the fact that the elliptic functions are doubly periodic and the trigonometric functions singly periodic. Suppose, for example, n is an odd number. The relation $\varphi(n\alpha) = \varphi(\alpha_0)$ implies that $n\alpha$ differs from α_0 by a period, and consequently

$$x = \varphi(\alpha) = \varphi\left(\frac{\alpha_0 + 2k\omega + 2l\bar{\omega}i}{n}\right),$$

where k and l range over the values $0, 1, \ldots, n - 1$ independently of each other. Thus n^2 values of x are obtained. In the case of the sine the relation $\sin n\alpha = \sin \alpha_0$ implies that $x = \sin \alpha = \sin \frac{\alpha_0 + 2\pi k}{n}$, where $k = 0, 1, \ldots, n - 1$, i.e., only n values are obtained.

Abel remarked that the division equations (with $\varphi(n\alpha) = 0$) satisfied a criterion he had discovered for solvability of an equation in radicals, and deduced an expression for $\varphi(\alpha)$ in terms of $\varphi(n\alpha)$, in a rather complicated way at first, to be sure. However, using a note of Jacobi ("Addition au mémoire de M. Abel sur les fonctions élliptiques," 1828), which appeared in the same volume of Crelle's *Journal* as the second part of the "Recherches," Abel reached this goal by a very simple and elegant route in his paper "Théorèmes sur les fonctions élliptiques," (*J. für Math.*, 4 (1829)).

Finally, in the first part of his "Recherches," by representing $\varphi(\alpha)$ in terms of $\varphi(\alpha/(2n - 1))$ (and similarly for the other functions) using the multiplication formula without any further justification, Abel passed to the limit as $n \to \infty$. The idea of this derivation is completely analogous to the derivation used by Euler in his *Introductio* to expand the sine and cosine in power series. To be specific one replaces α by $\alpha/(2n + 1)$ in the formula given above for $\sin(2n + 1)\alpha$ and notes that as $n \to \infty$

$$x = \sin \frac{\alpha}{2n + 1} \sim \frac{\alpha}{2n + 1}, \quad 1 - x^2 \sim 1.$$

Then a termwise passage to the limit in each term of the right-hand side of the formula yields

$$\sin \alpha = \alpha - \frac{\alpha^3}{3!} + \cdots ,$$

i.e., we obtain formally the power series expansion of $\sin \alpha$. As a result of Abel's computations representations for the functions $\varphi(\alpha)$, $f(\alpha)$, and $F(\alpha)$ as ratios of certain entire functions and the infinite product expansions of these functions were obtained.

Almost simultaneously with the publication of the first part of the "Recherches" Jacobi's note "Extraits de deux lettres à Shoumacher" appeared in in Schumacher's journal *Astronomische Nachrichten* (No. 123, 1827). In this note Jacobi communicated without proof (which he was not in possession of at the time he submitted the note) an essentially new result on the transformation of elliptic integrals. There are grounds for supposing that Jacobi was assisted in finding the proof by becoming acquainted with Abel's paper, specifically with the idea of inverting the integral. However that may be, Abel was unpleasantly surprised at the end of March 1828 when he read in the same *Astr. Nachr.* the proofs of the results announced by Jacobi ("Demonstratio theorematis ad theoriam functionam ellipticarum spectantis," *Astronomische Nachrichten*, No. 127, 1827). A rivalry began between these two gifted mathematicians which was unfortunately not fated to last long: only one year of life remained to Abel. Abel hurriedly wrote up his theory of transformations of elliptic functions, which included Jacobi's results as a special case, and published them in Schumacher's journal ("Solution d'un problème général concernant la transformation des fonctions élliptiques," *Astr. Nachr.*, **6**

(1828); **7** (1829)). "My dispatch of Jacobi has been printed," he wrote to his friend Holmboe on 29 July 1828.[49] To Jacobi's credit it can be said that he considered this "dispatch" to be one of the most beautiful masterpieces of analysis.

The second, shorter part of Abel's "Recherches" was devoted mainly to two questions: the division of the lemniscate using compass and straightedge, and the transformation of elliptic functions. The Italian mathematician Fagnano (1682–1766) had found explicit algebraic formulas for the coordinates of the points of division of the lemniscate of Bernoulli into 2, 3, and 5 equal parts. Abel, who had read Gauss *Disquisitiones arithmeticae*, could not help being struck by Gauss' words on the principles he had developed in the theory of division of a circle into n equal parts: "They can be applied not only to the circular functions, but also to many other transcendental functions, for example, those that depend on the integral $\int \frac{dx}{\sqrt{1-x^4}} \ldots$."[50] This integral represented a special case of an elliptic integral well-known to Abel, in terms of which the arc length of the lemniscate could be expressed. Now Abel had established as early as 1826 that for the lemniscate the cases when the division into n equal parts can be carried out using a compass and straightedge are the same as for the circle (i.e., under the condition that $n = 2^m p_1 p_2 \cdots p_\mu$, where $m \geq 0$, and p_j ($j = 1, \ldots, \mu$) are distinct primes of the form $2^{2^{q_j}} + 1$); he therefore had grounds for believing that Gauss had also arrived at the same result (letter from Abel to Crelle, 4 December 1826).[51] For Abel this result signified only an application of his general theory of division of the argument of an elliptic function along with the criteria he had developed for solvability of an algebraic equation by radicals (square roots in the present case) in a particular problem. "I have at one stroke removed the mystery that shrouded Gauss' theory of division of the circle," he wrote to Holmboe in December of that same year of 1826. "I see clearly as day how he arrived at this result. What I said about the lemniscate is one of the results derived from my research on the theory of equations,"[52] These words, though spoken in reference to a special situation, have a much more general meaning: Abel's research on the theory of transcendental functions is intimately interwoven with his algebraic investigations.

The problem of transformation in the theory of elliptic functions, in the study of which Abel and Jacobi clashed, had occupied mathematicians since the second half of the eighteenth century (Euler, Landen, Lagrange, and later Legendre and Gauss). In terms of integrals it is a question of which changes of variable transform one elliptic integral into another (cf. Gauss' transformation). The significance of the work of Abel and Jacobi on this problem is that they stated the problem in its most general form and thereby obtained the most general results, which included as special cases all those known before. Here, for example, is how the problem

---

49) Abel, N. H. *Mémorial publié à l'occasion du centenaire de sa naissance*, Oslo, 1902. *Correspondance d'Abel...*, p. 68.

50) Gauss, C. F. *Werke*. Bd. 1, S. 412–413.

51) See the edition of Abel's correspondence cited above in Abel, N. H. *Mémorial*, pp. 52–53.

52) *Ibid.*, p. 54.

is stated in the paper of Abel mentioned above, "Solution d'un problème général concernant la transformation des fonctions élliptiques":

Find all cases in which the differential equation

$$\frac{dy}{\sqrt{(1 - c_1^2 y^2)(1 - e_1^2 y^2)}} = a \frac{dx}{\sqrt{(1 - c^2 x^2)(1 + e^2 x^2)}}$$

can be satisfied by substituting an algebraic function of x for y, rational or irrational.

Here Abel was making a slight change in the form of the canonical polynomial of degree four under the radical. It is not difficult to see that the problem just posed contains as a special case the problem of multiplication (or division) of the argument: it suffices to set $c_1 = c$, $e_1 = e$, and a equal to an integer.

The generality of this formulation of the problem is remarkable. From the outset Abel sought y in the class of algebraic functions. He proved that this problem can be reduced to the case when y is a rational function and gave a complete solution of it. It was this generality, combined with the elegance of his method, that elicited Jacobi's admiration. In developing his theory in the second part of the "Recherches" Abel gave in particular the following theorem, which gives the first case of so-called complex multiplication in the theory of elliptic functions: *If the equation*

$$\frac{dy}{\sqrt{(1 - y^2)(1 - c^2 y^2)}} = a \frac{dx}{\sqrt{(1 - x^2)(1 - c^2 x^2)}}$$

admits an algebraic integral and a is a complex number, then a necessarily has the form $a = \pm i \sqrt{n}$, where m and n are rational numbers and the parameter c^2 must satisfy a certain additional condition.

C. G. J. Jacobi. *Fundamenta nova functionum ellipticarum*

We have given brief biographical information on Jacobi above (see Book 1, p. 69). We reproduce that information here in somewhat expanded form.

Carl Gustave Jakob Jacobi (1804–1851) was born in Potsdam to the family of a banker. He entered the University of Berlin at an early age (he was not yet 16). He was not particularly diligent in attending mathematical lectures, preferring to study independently. Jacobi made a profound study of the works of Euler, in which he retained an unvarying interest throughout his life. In addition he found the time and the ability to take an active part in the seminar on classical philology conducted by the renowned philologist August Böckh (1785–1867). Jacobi graduated in 1825, having defended a dissertation on the expansion of rational functions in partial fractions. In 1827 he became professor extraordinarius at Königsberg University, and in 1829 he was made ordinarius, a position he retained until 1843. In 1829 he published his *Fundamenta nova functionum ellipticarum*, written in Latin. Because of this work he shares with Abel the glory of having founded the theory of elliptic functions. Of particular value is the theory of the theta functions named

C. G. J. Jacobi

after him, which he developed in this work and in some later ones, along with the representation of the three elliptic functions sn u, cn u, and dn u as quotients of theta functions. Jacobi originated the idea of inverting a system of Abelian integrals (1832, 1834), which led to the introduction of theta functions of several variables into mathematics and gave rise to the later theory of Abelian functions (see below).

In his fundamental lectures on mechanics Jacobi developed methods of integrating first-order partial differential equations and the equations of the calculus of variations (his "last-multiplier" method and the Hamilton-Jacobi theory). A course that he gave at Königsberg in 1842–1843, which contains the main results in this area, was recorded at the time by Borchardt, but not published until Clebsch undertook to do so in 1866, under the title *Vorlesungen über Dynamik* (Leipzig 1866). In this book a number of problems were studied and solved using the general methods developed by Jacobi: the two-body problem in space, the problem of attraction to two fixed centers, the determination of the geodesics on the surface of a triaxial ellipsoid, and others. His brief paper of 1837 entitled "Zur Theorie der Variations-Rechnung und der Differentialgleichungen" (*J. für Math.*, **17**, (1837)) is devoted entirely to the calculus of variations. In studying the so-called simplest problem of calculus of variations, he derived a supplement to the classical Lagrange-Euler equation — the Jacobi auxiliary differential equation —

and used it to establish a new necessary condition for an extremal that minimizes an integral (the Jacobi condition).

In studying the equilibrium shapes of a rotating fluid, Jacobi showed that under certain conditions a triaxial ellipsoid (Jacobi ellipsoid) can be such a shape (the case of ellipsoids of revolution had been known to Maclaurin). In algebra he is the author of a development of the theory of determinants entitled "De formatione et proprietatibus determinantum" (*J. für Math.*, **22** (1840)), whose significance is surpassed only by the fundamental work of Cauchy (1815), and also a number of results in the theory of quadratic forms.

Jacobi introduced into analysis the functional determinant now known as the Jacobian in his honor and showed its significance in his paper "De determinantibus functionalibus," (*J. für Math.*, **22** (1841)). An important class of orthogonal polynomials, which generalize the Legendre polynomials, bears his name.

On the advice of his physician Jacobi spent six months of 1843 in Italy. Upon his return he settled in Berlin as a professor of the University of Berlin and a member of the Prussian Academy of Sciences. As an honorary member of the Petersburg Academy of Sciences (after 1839) Jacobi participated actively in the (abortive) plan of the Academy to publish the complete works of Euler. This is attested by his correspondence with P. N. Fuss (1798–1859), the permanent secretary of the Academy, by his personal meeting with Fuss during a trip abroad by the latter, and by some letters to his brother, the prominent physicist and electrical engineer Moritz Hermann (Boris Semyonovich) Jacobi, who was elected to membership in the Petersburg Academy of Sciences in 1837 and settled permanently in the Russian capital.

To characterize Jacobi one should also note his radical political views. In March of 1848 he became a member of the opposition Constitutional Convention and agreed to be a candidate for the Prussian National Assembly. In a speech before the Convention he stated, among other things, that he was not frightened by the prospect of a republic. This proclamation frightened the leaders of the Bund, however, and his nomination as a candidate for deputy was dropped. After the triumph of reaction, Jacobi was reminded of his behavior in the days of the revolution. His tenure was taken away and an attempt (unsuccessful) was made to remove him from Berlin back to Königsberg. Jacobi died of smallpox in 1851.

As an outstanding creative personality Jacobi exerted great influence on his contemporaries during his lifetime. In Germany his followers included Richelot, Kirchhoff, Hesse, and Clebsch, in France Hermite and Liouville, in England Cayley.

In this section we shall give only a brief summary of the content of the basic paper *Fundamenta nova...* (1829) mentioned above.

Jacobi chose as his starting point the elliptic integral of first kind, just as Abel and Gauss had done, only in the Legendre normal form, writing it as

$$u = \int_0^\varphi \frac{d\varphi}{\sqrt{1 - k^2 \sin^2 \varphi}},$$

where k is a parameter $(0 < k < 1)$ called the *modulus*. Jacobi called the variable φ the *amplitude*, and set $\varphi = \operatorname{am} u$. Since the substitution $x = \sin \varphi$ yields

$$u = \int_0^x \frac{dx}{\sqrt{(1-x^2)(1-k^2x^2)}},$$

for x as a function of u — the inverse of the elliptic integral of first kind — we obtain $x = \sin \operatorname{am} u$ (read "sine-amplitude of u"). Along with this function he defined two more functions

$$\cos \operatorname{am} u = \sqrt{1 - \sin^2 \operatorname{am} u}, \quad \Delta \operatorname{am} u = \sqrt{1 - k^2 \sin^2 \operatorname{am} u}.$$

In 1838 C. Gudermann (who will be discussed below) gave the simpler notation generally used nowadays for these three functions — the Jacobi elliptic functions — $\operatorname{sn} u$, $\operatorname{cn} u$, and $\operatorname{dn} u$ in his paper "Theorie der Modular-Functionen und der Modular-Integrale," (*J. für Math.*, **18** (1838)). The addition theorem, the double periodicity, the zeros, and the poles were all established following the same route taken by Abel and earlier by Gauss. (For $\operatorname{sn} u$ the fundamental periods are

$$4K = 4 \int_0^1 \frac{d\varphi}{\sqrt{1 - k^2 \sin^2 \varphi}} \quad \text{and} \quad 2iK' = 2i \int_0^1 \frac{d\varphi}{\sqrt{1 - k'^2 \sin^2 \varphi}},$$

where $k' = \sqrt{1 - k^2}$ is the complementary modulus; they are also periods, but not fundamental periods, for $\operatorname{cn} u$ and $\operatorname{dn} u$.) Jacobi went on to study the problem of transformation of elliptic functions, multiplication and division of arguments, the relations between elliptic integrals of second and third kinds, and so forth. He used the theta functions as auxiliary functions. The Jacobi elliptic functions can be expressed in terms of the fundamental theta functions (the theta function $\Theta(u)$ and the eta function $H(u)$) by the formulas

$$\operatorname{sn} u = \frac{1}{\sqrt{k}} \frac{H(u)}{\Theta(u)}, \quad \operatorname{cn} u = \sqrt{\frac{k'}{k}} \frac{H(u+K)}{\Theta(u)}, \quad \operatorname{dn} u = \sqrt{k'} \frac{\Theta(u+K)}{\Theta(u)}.$$

The functions $\Theta(u)$ and $H(u)$ can be represented by the following everywhere-convergent trigonometric series:

$$\Theta(u) = 1 - 2q \cos 2v + 2q^4 \cos 4v - 2q^9 \cos 6v + \cdots,$$
$$H(u) = 2q^{1/4} \sin v - 2q^{9/4} \sin 3v + 2q^{25/4} \sin 5v - \cdots,$$

where $v = \pi u / 2K$ and $q = e^{-\pi K'/K}$. Both functions are entire and periodic, and the fundamental period is $2K$ for $\Theta(u)$ and $4K$ for $H(u)$. Moreover, when the imaginary number $2iK'$ is added to u, both functions acquire the same exponential multiplier $q^{-1}e^{-i\pi K'u/K}$. This is the reason that the quotients

$$H(u)/\Theta(u), \quad H(u+K)/\Theta(u), \quad \Theta(u+K)/\Theta(u)$$

turn out to be doubly periodic functions. The valuable function-theoretic properties of the theta functions combine with their convenience for computation in applications, for example in mechanics. The reason is that when u and k are real and k satisfies the condition $0 < k < 1$ (so that K and K' are also positive real numbers), the series given above converge very rapidly, due to the factors of the form $q^{n^2/4}$ (or $q^{(2n+1)^2/4}$), where $q = e^{-\pi K'/K} < 1$.

The Jacobi Theta Functions

In his lectures on the theory of elliptic functions given at the University of Königsberg in 1835–1836 and published after his death by his student K. Borchardt (1817–1880)[53] the basis of the development consists of four theta functions, whose definition and properties are given independently of elliptic integrals. The starting point is a series of the form $\sum\limits_{n=-\infty}^{\infty} e^{an^2+bn+c}$, where b and c are certain linear functions of x and a is a complex constant whose real part is negative. This last condition guarantees, in modern terminology, the absolute and uniform convergence of the series in any disk of the finite plane. Consequently the sum of the series is an entire function.

For convenience we rewrite the series in the form proposed somewhat later by Hermite:

$$\Theta_{\mu,\nu}(x) = \sum_{n=-\infty}^{\infty} \exp\left\{\pi i\left[n\nu + x(2n+\mu) + \tau\frac{(2n+\mu)^2}{4}\right]\right\}.$$

Since the coefficient of n^2 in the exponent here is $\pi i\tau$, it follows that the imaginary part of τ must be assumed positive in order to assure convergence. We shall give the parameters μ and ν only the two values 0 and 1. In accordance with the various combinations of these values we obtain Jacobi's four theta functions. We shall write them out in the notation of J. Tannery and J. Molk, which differs only slightly from the notation of Jacobi himself (here q denotes $e^{\pi i\tau}$, $|q| < 1$):

$$\vartheta_1(x) = -i\Theta_{1,1}(x) = -i\sum_{-\infty}^{+\infty}(-1)^n q^{\frac{(2n+1)^2}{4}} e^{(2n+1)\pi ix}$$

$$\sum_0^{\infty}(-1)^n 2q^{\frac{(2n+1)^2}{2}} \sin(2n+1)\pi x;$$

$$\vartheta_2(x) = \Theta_{1,0}(x) = \sum_{-\infty}^{+\infty} q^{\frac{(2n+1)^2}{4}} e^{(2n+1)\pi ix}$$

$$= \sum_0^{\infty} 2q^{\frac{(2n+1)^2}{4}} \cos(2n+1)\pi x;$$

---

53) Jacobi, C. G. J. *Gesammelte Werke*. Berlin 1881, Bd. 1, S. 479.

$$\vartheta_3(x) = \Theta_{0,0}(x) = \sum_{-\infty}^{+\infty} q^{n^2} e^{2n\pi i x} = 1 + \sum_{1}^{\infty} 2q^{n^2} \cos 2n\pi x;$$

$$\vartheta_4(x) = \Theta_{0,1}(x) = \sum_{-\infty}^{+\infty} (-1)^n q^{n^2} e^{2n\pi i x} = 1 + \sum_{1}^{\infty} (-1)^n 2a^{n^2} \cos 2n\pi x.$$

These functions can be expressed in terms of one another by replacing x by $x+1/2$, $x + i\tau/2$, and $x + 1/2 + i\tau/2$:

$$\vartheta_2(x) = \vartheta_1\left(x + \frac{1}{2}\right), \quad \vartheta_1(x) = -\vartheta_2\left(x + \frac{1}{2}\right), \quad \vartheta_3(x) = \vartheta_4\left(x + \frac{1}{2}\right),$$

$$\vartheta_4(x) = \vartheta_3\left(x + \frac{1}{2}\right), \quad \vartheta_3(x) = q^{\frac{1}{4}} e^{-\pi x i} \vartheta_2\left(x + \frac{1}{2} i\tau\right),$$

$$\vartheta_4(x) = -i q^{\frac{1}{4}} e^{\pi x i} \vartheta_2\left(x + \frac{1}{2} + \frac{1}{2} i\tau\right),$$

etc. (these functions originally appeared in the *Fundamenta nova...* under the guise of the Θ and H functions).

We remark that the functions $\vartheta_2(x)$ and $\vartheta_3(x)$ obtained from $\Theta_{\mu,0}(x)$ with $\mu = 1$ and $\mu = 0$ respectively can be represented as

$$\Theta_{\mu,0}(x) = e^{-\frac{\pi i}{\tau} x^2} \sum_{n=-\infty}^{\infty} \exp\left\{\frac{\pi i}{\tau}\left[\tau\left(\frac{2n+\mu}{2}\right) + x\right]^2\right\} \quad (\mu = 1, 0).$$

Therefore for any quadruple of complex numbers w, x, y, z we find by multiplying the series that

$$\vartheta_2(w)\vartheta_2(x)\vartheta_2(y)\vartheta_2(z) + \vartheta_3(w)\vartheta_3(x)\vartheta_3(y)\vartheta_3(z) =$$

$$= \exp\left[-\frac{\pi i}{\tau}(w^2 + x^2 + y^2 + z^2)\right] \sum_{(m,n,p,q)}^{+\infty} \exp\left\{\frac{\pi i}{\tau}\left[\left(\frac{\tau m}{2} + w\right)^2 + \right.\right.$$

$$\left.\left. + \left(\frac{\tau n}{2} + x\right)^2 + \left(\frac{\tau p}{2} + y\right)^2 + \left(\frac{\tau q}{2} + z\right)^2\right]\right\},$$

where the indices m, n, p, and q range over all quadruples of integers of the same parity. Setting

$$w' = \frac{1}{2}(w + x + y + z), \quad x' = \frac{1}{2}(w + x - y - z),$$

$$y' = \frac{1}{2}(w - x + y - z), \quad z' = \frac{1}{2}(w - x - y + z)$$

(when this is done, the identity

$$w'^2 + x'^2 + y'^2 + z'^2 = w^2 + x^2 + y^2 + z^2$$

holds), and carrying out a similar transformation of the indices of summation m, n, p, and q, Jacobi arrived at the fundamental relation

$$\vartheta_2(w)\vartheta_2(x)\vartheta_2(y)\vartheta_2(z) + \vartheta_3(w)\vartheta_3(x)\vartheta_3(y)\vartheta_3(z) =$$
$$= \vartheta_2(w')\vartheta_2(x')\vartheta_2(y')\vartheta_2(z') + \vartheta_3(w')\vartheta_3(x')\vartheta_3(y')\vartheta_3(z'),$$

from which he deduced dozens of other identities by pure algebra. Among these other identities are the algebraic relations among theta functions of the same argument, such as the following:

$$\vartheta_3^2(0)\vartheta_3^2(x) = \vartheta_4^2(0)\vartheta_4^2(x) + \vartheta_2^2(0)\vartheta_2^2(x), \tag{10}$$
$$\vartheta_2^2(0)\vartheta_4^2(x) = \vartheta_3^2(0)\vartheta_1^2(x) + \vartheta_4^2(0)\vartheta_2^2(x), \tag{11}$$
$$\vartheta_3^2(0)\vartheta_4^2(x) = \vartheta_2^2(0)\vartheta_1^2(x) + \vartheta_4^2(0)\vartheta_3^2(x), \tag{12}$$

and algebraic addition theorems for the following quotients of theta functions:

$$\vartheta_1(x)/\vartheta_4(x), \quad \vartheta_2(x)/\vartheta_4(x), \quad \vartheta_3(x)/\vartheta_4(x).$$

Finally (by passing to the limit in the identities that express the addition theorem), he obtained formulas for the derivatives of these last functions in the form

$$
\begin{aligned}
\frac{d}{dx}\frac{\vartheta_1(x)}{\vartheta_4(x)} &= \frac{\vartheta_4(0)}{\vartheta_2(0)}\frac{\vartheta_1'(0)}{\vartheta_3(0)}\frac{\vartheta_2(x)}{\vartheta_4(x)}\frac{\vartheta_3(x)}{\vartheta_4(x)}, \\
\frac{d}{dx}\frac{\vartheta_2(x)}{\vartheta_4(x)} &= -\frac{\vartheta_3(0)}{\vartheta_4(0)}\frac{\vartheta_1'(0)}{\vartheta_2(0)}\frac{\vartheta_1(x)}{\vartheta_4(x)}\frac{\vartheta_3(x)}{\vartheta_4(x)}, \\
\frac{d}{dx}\frac{\vartheta_3(x)}{\vartheta_4(x)} &= -\frac{\vartheta_2(0)}{\vartheta_4(0)}\frac{\vartheta_1'(0)}{\vartheta_3(0)}\frac{\vartheta_1(x)}{\vartheta_4(x)}\frac{\vartheta_2(x)}{\vartheta_4(x)},
\end{aligned}
\tag{13}
$$

It follows from this, by virtue of relations (10)–(12), that

$$\vartheta_1(x)/\vartheta_4(x), \quad \vartheta_2(x)/\vartheta_4(x), \quad \text{and} \quad \vartheta_3(x)/\vartheta_4(x)$$

satisfy simple first-order algebraic differential equations.

Jacobi needed these relations to show that the elliptic functions he had introduced earlier starting from the elliptic integral $\displaystyle\int_0^\varphi \frac{d\varphi}{\sqrt{1 - k^2 \sin^2\varphi}}$ could easily be expressed using his theta functions. First of all, setting $x = 0$ in (10), he verified that the complementary moduli k and $k' = \sqrt{1 - k^2}$ could be defined by the equalities

$$\sqrt{k} = \vartheta_2(0)/\vartheta_3(0) \quad \text{and} \quad \sqrt{k'} = \vartheta_4(0)/\vartheta_3(0). \tag{14}$$

Relation (11) then enabled him to introduce the auxiliary variable φ defined by the formulas

$$\frac{1}{\sqrt{k}}\frac{\vartheta_1(x)}{\vartheta_4(x)} = \sin\varphi, \quad \frac{\sqrt{k'}}{\sqrt{k}}\frac{\vartheta_2(x)}{\vartheta_4(x)} = \cos\varphi; \tag{15}$$

it then followed from (12) that

$$\sqrt{k'}\frac{\vartheta_3(x)}{\vartheta_4(x)} = \sqrt{1 - k^2 \sin^2 \varphi} = \Delta(\varphi). \qquad (16)$$

Finally, formulas (13)–(16) enabled him to conclude that

$$\frac{\vartheta_3(0)\vartheta_1'(0)}{\vartheta_4(0)\vartheta_2(0)} x = \int_0^\varphi \frac{d\varphi}{\sqrt{1 - k^2 \sin^2 \varphi}}. \qquad (17)$$

Consequently

$$\varphi = \mathrm{am}\left[\frac{\vartheta_3(0)}{\vartheta_4(0)}\frac{\vartheta_1'(0)}{\vartheta_2(0)} x\right] \qquad (18)$$

and the functions (15) and (16) are none other than the Jacobi elliptic functions sn, cn, and dn.

It now remained only to express the coefficient of x in formula (18) (i.e., $\frac{\vartheta_2(0)}{\vartheta_4(0)}\frac{\vartheta_1'(0)}{\vartheta_2(0)}$)) in terms of quantities that have a direct relation to the elliptic integral (17). To do this Jacobi first of all derived the relation

$$\vartheta_1'(0) = \pi\vartheta_2(0)\vartheta_3(0)\vartheta_4(0),$$

which is valid for $q = e^{\pi i \tau}$. It follows from this that

$$\frac{\vartheta_3(0)}{\vartheta_4(0)}\frac{\vartheta_1'(0)}{\vartheta_2(0)} = \pi\vartheta_3^2(0).$$

Jacobi carried out the subsequent computations assuming that q is a positive real number less than 1 (i.e., τ is a pure imaginary number). The theta functions then assume real values when x is real. Using this fact, he found that

$$\pi\vartheta_3^2(0) = 2\int_0^{\pi/2} \frac{d\varphi}{\sqrt{1 - k^2 \sin^2 \varphi}} = 2K.$$

The expressions for his elliptic functions in terms of the theta functions then assume their final form

$$\mathrm{sn}\, x = \frac{1}{\sqrt{k}}\frac{\vartheta_1(x/2K)}{\vartheta_4(x/2K)}, \mathrm{cn}\, x = \frac{\sqrt{k'}}{\sqrt{k}}\frac{\vartheta_2(x/2K)}{\vartheta_4(x/2K)}, \mathrm{dn}\, x = \sqrt{k'}\frac{\vartheta_3(x/2K)}{\vartheta_4(x/2K)}.$$

Jacobi also studied the inverse problem, obviously not considering it to have been solved, or even stated, in his *Fundamenta Nova...*: *starting from the elliptic functions* sn x, cn x, *and* dn x, *construct the theta-functions corresponding to them*. In doing this he restricted himself to the case when the modulus k is a real number satisfying the condition $0 < k < 1$. For him this problem reduced to determining the number q. Remarking that K'/K, regarded as a function of q, has the same property as $\ln q$, namely the property of acquiring a multiplier n when q is replaced by q^n, Jacobi concluded that $(K/K')\ln q$ is independent of q and equal to $-\pi$. Therefore

$$q = e^{-\pi K'/K},$$

which enabled him to construct the corresponding theta-functions from the formulas given above.

Jacobi's new approach to the theory of elliptic functions just discussed is quite remarkable from the historical point of view. Indeed, a decade had not yet passed since the idea of inverting the elliptic integrals had been announced, and already in constructing the foundations of the theory of elliptic functions mathematicians were trying to get rid of the conditions of the particular problem that had led them to these functions and to introduce a corresponding class based on more general considerations. The elements of the construction were taken to be single-valued entire functions defined by corresponding everywhere-convergent series. This was an important step not only toward the construction of the theory of elliptic functions soon proposed by Liouville and later by Weierstrass, but, what is much more significant, toward the later theory of Abelian functions. It should be emphasized, however, that neither in this paper nor in his later papers did Jacobi use the general methods of investigation of analytic functions that Cauchy had developed by this time (the Cauchy integral theorem, the residue theorem, and its corollaries).

Elliptic Functions in the Work of Eisenstein and Liouville.

The First Textbooks

Jacobi's *Fundamenta nova functionum ellipticarum* greatly promoted the expansion of this area of mathematical analysis. The clarity and the systematic nature of his exposition, his happy choice of functions connected with the mathematical tradition (the starting point being the elliptic integral of first kind in the Legendre form), his notation, which brought elliptic functions into a close relationship with trigonometric functions, and the expression of them in terms of trigonometric series — all this brought about an increasing popularity of the Jacobi functions among mathematicians and especially those who specialized in mechanics, who applied them to solve various problems of mechanics. In his *Dictionary of Pure and Applied Mathematics* (St. Petersburg 1839; unfortunately the volume A–D was omitted) V. Ya. Bunyakovskiĭ had no article on "Elliptic functions." That article was supposed to appear in the second volume. But the article "Amplitude (of an elliptic function). Amplitude, longitude, capacity of an elliptic function" leaves no doubt with respect to the interpretation of the question noted by Bunyakovskiĭ. He writes: "Mathematicians often use the following notation: representing the elliptic function $\int_0^\varphi \dfrac{d\varphi}{\sqrt{1 - k^2 \sin^2 \varphi}}$ by u, they write $\varphi = \operatorname{am} u$, $\sin \varphi = \sin \operatorname{am} u$,

$\varphi' = \text{co am}\, u$, $\sin \varphi' = \sin \text{co am}\, u$, etc., See Elliptiques (Fonctions)."[54] As can be seen, this was the notation of Jacobi ($\text{co am}\, u = \text{am}\,(K - u)$ is the co-amplitude.)

Even the first course of elliptic functions in Russian *Foundations of the Theory of Elliptic Functions* (St. Petersburg 1850) was basically an exposition of the work of Jacobi. This book came from the pen of I. I. Somov (1815–1876), a professor at St. Petersburg University, a member of the Academy of Sciences, and the author of many original works in the theory of elliptic functions and their applications in mechanics. In their review of this book Academicians Ostrogradskiǐ and Bunyakovskiǐ noted that it was the "first complete, systematic work in Russian on one of the most remarkable and difficult areas of integral calculus," but they also pointed out certain deficiencies of the work: the absence of "ultra-elliptic functions" (apparently ultra-elliptic integrals) and the proofs of a number of theorems.

In the French *Dictionnaire des sciences mathématiques pure et appliquées* of A. S. de Montferrier, which was published about the same time (1838–1840) as Bunyakovskiǐ's *Dictionary*, we find the article "Transcendantes elliptiques," only in the supplementary (third) volume. Here this term means elliptic integrals, but nevertheless along with Legendre and his treatise Jacobi and his *Fundamenta nova* are also mentioned; Abel's name is absent.

However by this time (the late 1830's) and especially during the 1840's, new elements sprang up in the theory of elliptic functions, destined as they developed to force Jacobi's functions into the background.

Around 1836 Jacobi's *Fundamenta nova* came into the hands of the young Weierstrass, who was teaching himself mathematics. The book turned out to be too difficult for him. After all, Jacobi assumed that his readers had a nodding acquaintance with Legendre's treatise, which was as yet unfamiliar to Weierstrass. Having heard that a professor at the Münster Theological/Philosophical Academy, C. Gudermann (1798–1852), was announcing a course of "modular functions" (Gudermann's name for elliptic functions), Weierstrass immediately headed there to become the only member of the audience for these lectures. In 1840 he presented as an examination paper for the teacher's license a composition on the theme he had chosen for himself: "Über die Entwicklung der Modular-Funktionen,"[55] which earned him a high mark from Gudermann, although the author had severely criticized Gudermann himself. This work already, as Felix Klein noted, contained the germ of many of Weierstrass' later investigations and represented an important step forward in the theory of elliptic functions. To be specific, using a remark of Abel, Weierstrass obtained a representation for the Jacobi functions as ratios of entire functions whose power series expansions have coefficients that are polynomials in k^2 (where k is the modulus of the elliptic integral). He later denoted these functions by Al_1, Al_2, Al_3, and Al in Abel's honor. However, starting from his

54) Bunyakovskiǐ, V. Ya. *Dictionary of Pure and Applied Mathematics* [in Russian]. St. Petersburg 1839.

55) Weierstrass, K. *Mathematische Werke*. Bd. 1, S. 1–49.

Berlin lectures (see below) the Weierstrass sigma-function, which differs from Al by an exponential factor, took the place of these functions.

After the mid-1840's in the work of the young Berlin Privatdocent Eisenstein (1823–1852) there appeared series (one may say in finished form) that later entered the Weierstrass theory of elliptic functions as its analytic machinery. These papers were essentially combined in a collection entitled *Mathematische Abhandlungen besonders aus dem Gebiete der höheren Arithmetik und der Elliptischen Funktionen von Dr. G. Eisenstein* (Berlin 1847) published with a foreword by Gauss. Among these papers the long paper "Genaue Untersuchung der unendlichen Doppelprodukte, aus welchen die elliptische Funktionen als Quotienten zusammengesetzt sind, und der mit ihnen zusammenhängenden Doppelreihen," (*J. für Math.*, **35** (1847)), which first appeared in Crelle's *Journal*, is of particular significance. Eisenstein remarked that if ω_1 and ω_2 are complex numbers whose ratio is not real (periods of an elliptic function), then the value of the double product $u \prod \left(1 - \dfrac{u}{m_1\omega_1 + m_2\omega_2} \right)$, where m_1 and m_2 range over all pairs of integers that are not simultaneously zero, depends on the order in which the factors are arranged in the product, i.e., the product does not converge absolutely. We recall that in the case when $\omega_2 = i\omega_1$, corresponding to the lemniscatic sine, this product had already occurred in Gauss' manuscripts (see above). In studying this product Eisenstein arrived at the series by means of which Weierstrass later introduced his functions $\wp(z)$ and $\wp'(z)$ and their invariants g_2 and g_3, and derived the basic relations between them. In comparison with the Weierstrass theory Eisenstein was lacking the factors of the form

$$\exp\left[\frac{u}{m_1\omega_1 + m_2\omega_2} + \frac{u^2}{2(m_1\omega_1 + m_2\omega_2)^2} \right]$$

in the products, whose presence guarantees absolute convergence (and uniform convergence in any finite disk) and turns them into sigma-functions.

It is interesting to note that very recently André Weil has returned to Eisenstein's approach to the theory of elliptic functions, which does not rely on any preliminary construction of the general elements of the theory of analytic functions, and has used it both to produce an exposition of the theory of elliptic functions which is quite complete and at the same time does not require very much space and (invoking the Kronecker series, which are closely related to those of Eisenstein) for applications to some very delicate questions of modern number theory. Here we have an example showing the usefulness of rereading old, seemingly outdated papers.[56]

But Weierstrass' first paper and the paper of Eisenstein just mentioned served at the time merely as a sort of proclamation of the future basic machinery of the theory of elliptic functions; they did not alter the essence of the theory being developed.

---

56) Weil, A. *Elliptic Functions According to Eisenstein and Kronecker*, Berlin 1976.

G. Eisenstein

A fundamentally new step was taken by J. Liouville in his lectures on the theory of doubly periodic functions, which were given in 1847 before an audience containing Borchardt and F. Joachimsthal (1818–1861). Although Borchardt did not publish his notes until 1880 ("Leçons sur les fonctions doublement périodiques (faites en 1847)," *J. für Math.*, **88** (1880)), the Liouville treatment of the subject did not escape the notice of contemporaries. A detailed description of the lectures was published in the *Comptes Rendus*, **32** (1851), p. 450). Moreover Ch. Briot and J. Bouquet wrote their book *Théorie des fonctions doublement périodiques et, en particulier, des fonctions elliptiques*, Paris 1859), which we shall discuss later, under the influence of these lectures. The very title of Liouville's lectures characterizes the structure of the course: the object of study is the entire class of doubly periodic single-valued functions of a complex variable that are continuous and differentiable everywhere except at poles, i.e., meromorphic functions.

Remarking that any period of a single-valued nonconstant doubly periodic function can be written as $m\omega+m'\omega'$, where m and m' are integers and ω and ω' are fundamental periods whose ratio is a complex number (the corresponding general theorem was first established by Jacobi in 1834), Liouville covered the plane by a grid of parallelograms with sides ω and ω', and reduced the general problem to the study of functions in any one parallelogram, for example, the parallelogram with vertices 0, ω, $\omega+\omega'$, ω'. In doing this he did not use Cauchy's results on functions of a complex variable, at least not explicitly. Here, for example, is how Liouville's theorem that an entire doubly periodic function is a constant was established. Expand $f(z)$ in a Fourier series, regarding it as a function with period ω. We obtain

$$f(z) = \sum_{-\infty}^{\infty} A_n e^{\frac{2\pi i n}{\omega} z}.$$

If ω' is also a period of this function such that ω'/ω is a complex number, then

$$f(z + \omega') = \sum_{-\infty}^{\infty} A_n q^n e^{\frac{2\pi i n}{\omega} z} = f(z),$$

where $q = e^{2\pi i \omega'/\omega}$, and $|q| \neq 1$ by virtue of the assumption on ω and ω'. From the uniqueness of the expansion, it follows that $A_n = A_n q^n$, so that $A_n = 0$ when $n \neq 0$, i.e., $f(z) = A_0$.

Liouville proved that a doubly periodic function (not identically constant) must have at least two simple poles $h + h'$ and $h - h'$ (or one pole of order two) in a period parallelogram, and he constructed the simplest function with such poles as the sum of a series

$$f(z) = \sum_{-\infty}^{\infty} \left[\cos \frac{2\pi(z + n\omega' - h)}{\omega} - \cos \frac{2\pi h'}{\omega} \right]^{-1}.$$

What is important, however, is that he obtained theorems establishing simple and elegant properties of the entire class of doubly periodic meromorphic functions (all having the same periods). It was functions of this class that came to be called elliptic functions (the inverses of elliptic integrals are only a subclass of them), and the corresponding general theorems came to be called Liouville's theorems. However these theorems can be proved (and in some cases stated) most simply using the theory of residues. Hermite laid the foundation for the application of the theory of residues to the properties of elliptic functions when he proved in an 1848 paper[57] that the sum of the residues of an elliptic function with respect to the poles inside a period parallelogram is zero. Hence in particular it follows, as Liouville had asserted, that an elliptic function cannot have only one simple pole in a period parallelogram.

We shall now give a brief list of certain other theorems of Liouville on elliptic functions, using the concept of the order of a function, introduced by Weierstrass.

Following Weierstrass, we call the number n of poles of an elliptic function in a period parallelogram (counted according to multiplicity) the *order* of the function. The result of Liouville given above says that $n \geq 2$. Cauchy's residue theorem, applied to $\frac{f'(z)}{f(z)-A}$ and $z\frac{f'(z)}{f(z)-A}$ respectively (where A is any complex number), makes it possible to assert that the general number of roots of the equation $f(z) -$

---

57) Hermite, C. Œuvres. Paris 1905, T. 1, pp. 75–83.

$A = 0$ in a period parallelogram is equal to n independently of A, and that the sum of the roots of this equation can differ from the sum of the poles only by the value of some period. We note further that each elliptic function can be expressed as a rational function of any function of second order and its derivative (having the same poles), and that only functions of second order are inverses of elliptic integrals. Nevertheless, as already noted, the name elliptic functions later came to be attached to the whole class of doubly periodic meromorphic functions.

In 1856 there appeared the paper "Étude des fonctions d'une variable imaginaire," (*J. Éc. Polyt.*, **21** (1856)) written by Briot and Bouquet, students of Cauchy and Liouville. This paper contained in simple and clear form a systematic exposition of the fundamental results obtained mainly by Cauchy in analytic function theory. This small memoir (consisting of 48 pages) can be regarded in terms of its content and manner of exposition as the first textbook in the theory of analytic functions.

The French mathematicians Briot and Bouquet did much to propagate the ideas of analytic function theory in the third quarter of the nineteenth century. Charles Auguste Albert Briot (1817–1882), a graduate of the École Normale Supérieure, taught mathematics in Orleans and Lyons, and (after 1855) at the École Normale Supérieure. In his later years he was professor of mathematical physics at the Sorbonne. Jean-Claude Bouquet (1819–1885), who studied at the École Polytechnique and the École Normale Supérieure in Paris from 1839 to 1841, taught mathematics in Marseilles and Lyons, and (after 1852) in the *Lycées* of Paris. After 1873 he was professor of mathematics at the Sorbonne and professor of mechanics and astronomy at the École Normale Supérieure. In 1875 he was elected a member of the Paris Academy.

After the paper just mentioned Briot and Bouquet co-authored the book *Théorie des fonctions elliptiques* mentioned above, which in its second edition (Paris 1879) grew into a fundamental work. The only book that Briot wrote by himself, *Théorie des fonctions Abéliennes* (Paris 1879), won the Poncelet Prize.

Weierstrass' first lectures on the analytic function theory, "Allgemeine Theoreme über die Darstellung der analytischen Funktionen durch convergente Reihe" also appeared at this time. Starting in 1861 he gave a course entitled "Allgemeine Theorie der analytischen Funktionen." For Weierstrass the point of departure was provided by power series expansions of functions. But where his predecessors had usually regarded each function as the sum of only one series, Weierstrass considered along with that series all of its analytic continuations, defining the function by means of the totality of all possible organically interconnected series.

If to all this we adjoin the famous dissertation of Riemann, "Grundlagen für eine allgemeine Theorie einer veränderlichen komplexen Grösse" (1851), which will be discussed below, we can assert that during the 1850's three basic concepts of the general theory of analytic functions of a complex variable, which were to determine the future development of the theory for a century, in a certain sense acquired a definitive incarnation: the concepts of Cauchy, Weierstrass, and Riemann. The basis of the first was complex integration, which revealed that the integral was independent of the path of integration; the basis of the second was the representability of a function by power series interconnected via the process of analytic continuation; the basis of the third was the concept of an analytic function of a complex variable as the solution of a certain partial differential equation (in complex form): $i\dfrac{\partial f}{\partial x} = \dfrac{\partial f}{\partial y}$.

From the point of view of the historian of mathematics it is instructive to observe that this entire treasure trove of general ideas was valued by contemporaries not for its own sake, but most of all for its usefulness in the study of special classes of analytic functions, which had been the focus of all the interest of analysts of the time, from the greatest to the least, since the beginning of the second quarter of the nineteenth century: elliptic functions (and the theta and modular functions connected with them) and Abelian functions. However, in the last quarter of the nineteenth century these functions faded into the background, yielding its place to automorphic functions.

It is characteristic that in one of the early attempts to survey the history of mathematics in the nineteenth century[58] the "theory of functions of double or multiple periodicity" appears as a separate rubric, while the general theory of functions is not accorded a separate section. A convincing example of the longevity of an old tradition is provided by the course in analytic and elliptic functions given by A. Hurwitz (1859–1919), a student of Weierstrass and Klein and professor at the University of Königsberg and later in the Technische Hochschule in Zürich (*Vorlesungen über allgemeine Funktionentheorie und elliptische Funktionen*, Berlin 1922). This book, which has become a classic in our era, consists of two parts: "General theory of functions of a complex variable," and "Elliptic functions," the second part being only slightly shorter than the first.

One of the external manifestations of the situation just described was that in the 1850's and long afterwards (say in courses given in Russian universities in the early twentieth century) the elements of the general theory of analytic functions were regarded mainly as an introduction to the theory of elliptic or Abelian functions. As an example we may cite the book *Théorie des fonctions doublement périodiques, et, en particulier, des fonctions elliptiques* of Briot and Bouquet (1st ed., 1859). Their paper "Étude des fonctions d'une variable imaginaire" was wholly incorporated into this work as the introductory part. This book was enormously successful, as shown by the mere fact that a German edition of it was published in 1862. By the time of the second, expanded French edition of 1875 the general theory occupied about 200 pages. But the greater part of this work (consisting altogether of 700 quarto pages) is devoted to elliptic functions, and the work itself is entitled, *Théorie des fonctions elliptiques*. Another example is provided by a book written entirely in the spirit of Riemann's ideas. This is the book *Vorlesungen über Riemann's Theorie der Abel'schen Integrale* (Leipzig 1865) of K. Neumann (1832–1925), who occupied professorships successively at the Universities of Halle, Basel, Tübingen, and Leipzig. This, by the way, is the first textbook in which a sphere

---

58) Ball, W. W. R. *Histoire des mathématiques*. Éd. française, avec des additions de R. de Montessus, Paris 1907, T. 2.

is used to give a geometric representation of complex numbers (the Riemann or Neumann sphere). Among the early textbooks of a more elementary character we mention also the book of J. Durège (1821–1893), a professor at the University of Prague. He published his *Theorie der elliptischen Funktionen* (Zürich 1861), which gained wide acceptance (the fifth edition was issued in 1908). However, in his exposition he did not rely on the theory of analytic functions of a complex variable. Three years later Durège issued a textbook devoted entirely to the general theory of analytic functions. This book, *Elemente der Theorie der Funktionen einer complexen veränderlichen Grösse. Mit besonderer Berücksichtung der Schöpfungen Riemanns bearbeitet*, (Leipzig 1864), also went through a number of editions.

One of the earliest monographs on the theory of analytic functions of a complex variable was also devoted entirely to the Riemann construction of the theory. That was the doctoral dissertation of Professor M. E. Vashchenko-Zakharchenko (1825–1912) of the University of Kiev, and it was entitled *The Riemann Theory of Functions of a Complex Variable* (Kiev 1861). Finally, among the early original textbooks one must mention the broadly conceived work of the Italian mathematician F. Casorati (1835–1890), a professor at the University of Pavia: *Teorica delle funzioni di variabili complesse* (Pavia 1868). Casorati was strongly influenced by Riemann. He failed to publish the second volume of his work only because he had not been able to prove rigorously the Dirichlet principle, which was intended as the foundation of the planned exposition.[59] However Casorati was not one-sided, and attempted to combine the concepts of Riemann and Cauchy in his textbook, which he prefaced with an extensive historical essay (143 pages!) on the development of the theory from Gauss to Riemann.

Abelian Integrals. Abel's Theorem

"Abel," wrote Hermite, "left mathematicians enough work for 150 years."[60] In saying this he had in mind not only the theory of elliptic functions, which had completely engrossed him personally, but also the famous theorem of Abel on integrals of algebraic functions. Abel had discovered this theorem as early as 1825 and devoted his long paper "Sur une propriété générale d'une classe étendue de fonctions transcendantes"[61] presented to the Paris Academy in 1826. Through the fault of Cauchy, who lost the manuscript sent to him for review, it was not published until 12 years after Abel's death, in 1841 (*Mém. sav. étrang.*, **7** (1841)). However, for the special case of hyperelliptic integrals Abel did manage to expound and publish his discovery at the end of 1828 in the paper "Remarques sur quelques propriétés générales d'une certaine sorte de fonctions transcendantes" (*J. für Math.*, **3** (1828)), and this special case was enough to cause the bedaz-

59) Bottazzini, U. "Riemann's Einfluss auf E. Betti und F. Casorati," *Arch. Hist. Exact Sci.*, **18** (1977).

60) Quoted in: *Histoire de la science (Encyclopédie de la Pléiade)*. Paris 1957, p. 630.

61) Abel, N. H. *Œuvres complètes*. Christiania 1881, T. 1, pp. 145–211.

zled Legendre to call this theorem Abel's "monumentum ære perennius," quoting Horace.[62]

In accordance with a suggestion of Jacobi integrals of the form $\int R(x,y)\,dx$, in which R is a rational function and $y = y(x)$ an algebraic function defined by an equation $f(x,y) = 0$ (f being an irreducible polynomial in x and y), are called *Abelian integrals*. In the particular case when $y^2 = P(x)$, where $P(x)$ is a polynomial of degree n having no multiple roots, the integral is called *elliptic* if $n = 3$ or $n = 4$ and *ultra-elliptic* or *hyperelliptic* if $n \geq 5$.

For the following discussion we shall have need of certain details about these integrals: We shall discuss them in a geometric language whose elements were revealed in the letter of Gauss to Bessel quoted above. However, neither Abel nor Jacobi used this language, taking for granted a great deal that we now state explicitly. To determine the value of the Abelian integral $\int R(x,y)\,dx$ it is not sufficient merely to indicate the limits of integration x_0 and x. It is also necessary to fix the path of integration l, i.e., the curve joining x_0 to x and, in addition, to distinguish a single-valued branch of the algebraic function $y = y(x)$ (which in general is multi-valued) on l. We shall assume that all the roots of the equation $f(x_0, y) = 0$ are distinct (there are only finitely many points x_0 at which the equation $f(x_0, y) = 0$ has multiple roots; they are called the *critical points* of the corresponding algebraic function). Having chosen one of the roots y_0, we define a single-valued branch of the function $y(x)$ in a neighborhood of x_0 by the condition $y(x_0) = y_0$. It can then be continued along l by maintaining its continuity. If we assume further that l does not pass through any critical points of $y(x)$ or through points at which $R[x, y(x)]$ becomes infinite, the integral $\int R(x,y)\,dx$ will be uniquely determined. Moreover, in some neighborhood of the curve l — a simply connected domain containing no critical points — its value will be independent of the path of integration. Here it is a single-valued and continuous (even analytic) function of x, a finite point of the path l. But if there are critical points (or poles of the function $R[x, y(x)]$) between two paths l and l_1 having the same initial and terminal points, the values of the integral along l and l_1 will in general be different.

The set of all distinct values of the Abelian integral between two points x_0 and x can be expressed using integer linear combinations of the periods of this integral. The simplest example of a period caused by a pole of the integrand, is the polar period $2\pi i$ that occurs in the case $\displaystyle\int_1^x \frac{dx}{x} = \ln x$. Both Gauss (1811) and Poisson (1820) had known this. It is obtained by integrating the function $1/x$ in the complex plane over a circle of arbitrarily small radius with center at the pole $z = 0$ of this function. Periods of a different type, the so-called *cyclic periods*, are observed, for example for the integral $\displaystyle\int_0^x \frac{dx}{\sqrt{1 - x^2}}$. Here the difference of two integrals over paths l and l_1 joining the points 0 and z (Fig. 14) equals the integral over a closed contour enclosing the two branch points -1 and $+1$ of the function $1/\sqrt{1 - z^2}$. Without changing the value of the latter integral this curve can be

---

62) Quoted in: *Histoire de la science*, p. 631.

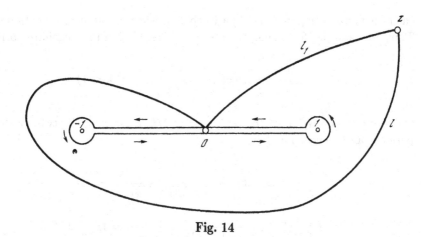

Fig. 14

replaced by two circles with centers at these points and the interval of the real axis joining them, traversed twice. The value of the integral — the period under a positive traversal of this contour — is 2π.

Let us consider in particular an elliptic or hyperelliptic integral of first kind (the general name for Abelian integrals having a finite value in a neighborhood of any point x, finite or infinite). Its general form is: $\displaystyle\int \frac{Q(x)\,dx}{\sqrt{P(x)}}$, where $P(x)$ and $Q(x)$ are polynomials, the degree of $Q(x)$ being less than $p = [(n-1)/2]$ (n is the degree of the polynomial $P(x)$; under this condition the integral converges at the point $x = \infty$). Depending on the configuration of the paths l and l_1, one of the following two formulas must hold:

$$\int_{l_1} \frac{Q(x)\,dx}{\sqrt{P(x)}} = \int_l \frac{Q(x)\,dx}{\sqrt{P(x)}} + 2m_1\omega_1 + \cdots + 2m_{2p}\omega_{2p}$$

or

$$\int_{l_1} \frac{Q(x)\,dx}{\sqrt{P(x)}} = 2\omega_0 - \int_l \frac{Q(x)\,dx}{\sqrt{P(x)}} + 2m_1\omega_1 + \cdots + 2m_{2p}\omega_{2p},$$

where $m_1, \ldots, m_{2p}$ are integers and $2\omega_0,\ 2\omega_1, \ldots,\ 2\omega_{2p}$ are complex numbers depending only on the integrand. The $2p$ numbers $2\omega_0,\ 2\omega_1, \ldots,\ 2\omega_{2p}$, any integer multiples of which may occur in the right-hand sides of the last formulas, are called *periods* of the integral $\displaystyle\int \frac{Q(x)\,dx}{\sqrt{P(x)}}$. If $\alpha_1, \ldots, \alpha_n$ are the roots of the equation $P(x) = 0$, one can, for example, set

$$2\omega_j = 2\int_{\alpha_j}^{\alpha_n} \frac{Q(x)\,dx}{\sqrt{P(x)}} \quad (j = 1, 2, \ldots, 2p),$$

each time integrating over fixed paths that satisfy the conditions mentioned above. Thus, for example, for an elliptic integral of first kind in the Legendre normal form

$$\int_0^x \frac{dx}{\sqrt{(1-x^2)(x-k^2x^2)}} \quad (0 < k < 1),$$

where $p = 1$ and $\alpha_1 = -1$, $\alpha_2 = 1$, $\alpha_3 = -1/k$, $\alpha_4 = 1/k$, the two corresponding periods can be chosen as follows:

$$2\omega_1 = 2\int_{-1}^1 \frac{dx}{\sqrt{(1-x^2)(1-k^2x^2)}} = 4K,$$

$$2\omega_2 = 2\int_{-1}^{1/k} \frac{dx}{\sqrt{(1-x^2)(1-k^2x^2)}} = 4K + 2iK',$$

where

$$. K = \int_0^1 \frac{dx}{\sqrt{(1-x^2)(1-k^2x^2)}} \quad \text{and} \quad K' = \int_0^1 \frac{dx}{\sqrt{(1-x^2)(1-k^2x^2)}}$$

$(k'^2 = 1 - k^2)$. If $n = 5$ or $n = 6$, then $p = [(n-1)/2] = 2$, and we obtain the four periods of a hyperelliptic integral. These periods were discovered by Jacobi following a very peculiar route, in any case having no connection with the complex plane, in 1832 (see below). Abel himself knew as early as 1824 that the transition from elliptic integrals to the more general Abelian integrals leads to an increase in the number of periods. However there is no indication that he had any more precise concept of this regularity, which was discovered in its general form by Riemann.

Naturally interest in the periods was connected with the problem of inverting the integrals: the periods of the integrals must be periods (in the usual sense of the term) for the inverse function. That is why the presence of more than two independent periods for an Abelian integral implied that the inverse function must have more than two (independent) periods. We shall return to this question below; just now we state Abel's theorem. In doing so, for the sake of brevity, we use a geometric language (curves, points of intersection) that Abel did not employ.

The sum of the values of any Abelian integral from a general initial point (x_0, y_0) to the N points of intersection of the curve $f(x, y) = 0$ with a variable algebraic curve $F(x, y) = 0$ (whose equation contains parameters) is a rational function of the coefficients of the polynomial $F(x, y)$ added to a finite number of logarithms of analogous rational functions, multiplied by constants.

The proof of this general theorem proposed by Abel himself is quite simple. In essence it follows from two elementary facts: 1) any symmetric rational function of the roots of an algebraic equation is a rational function of its coefficients; and 2) the integral of a rational function is the sum of a rational function and a finite number of logarithms of rational functions. This circumstance caused Picard to say[63] that perhaps nowhere else in the history of science has a proposition of such importance been obtained by such simple means.

In analyzing the consequences of his general theorem Abel came to the conclusion that for each algebraic curve $f(x,y) = 0$ there must exist a smallest positive integer p such that for the sum of any number q $(q > p)$ of Abelian integrals

$$S = \sum_{j=1}^{q} \int_{(x_0,y_0)}^{(x_j,y_j)} R(x,y)\,dx,$$

$$(f(x_0,y_0) = f(x_j,y_j) = 0, \quad j = 1,2,\ldots,q)$$

the following relation holds:

$$S = \sum_{k=1}^{p} \int_{(x_0,y_0)}^{(X_k,Y_k)} R(x,y)\,dx + V_0 + A_1 \ln V_1 + \cdots + A_m \ln V_m.$$

Here X_k $(k = 1,\ldots,p)$ are algebraic functions of $x_1,\ldots,x_q$ and moreover are such that each rational symmetric function of $X_1,\ldots,X_p$ can be expressed rationally in terms of $x_1,\ldots,x_q$, and V_0, $V_1,\ldots,$ V_m are rational symmetric functions of $x_1,\ldots,x_q$. Finally $A_1,\ldots,A_m$ are constants (the addition theorem for Abelian integrals). In the case when $\int R(x,y)\,dx$ is an integral of first kind one can assert that all the numbers $A_1,\ldots,A_m$ are zero and V_0 is a constant.

The integer characteristic p for any algebraic curve $f(x,y) = 0$, which occurs here for the first time in the history of mathematics, was later studied by Riemann, and in his paper "Theorie der Abelschen Funktionen" (1857, see below) he denoted it by the letter p. Clebsch laid the foundations for a systematic use of it in the theory of algebraic curves in a cycle of papers including "Über die Anwendungen der Abel'schen Funktionen in der Geometrie" (*J. für Math.*, **63** (1864)), "Über diejenigen ebenen Curven, deren Coordinaten rationale Funktionen eines Parameters sind," (*J. für Math.*, **64** (1865)) and others. He called the number p the *genus* of the curve. In the case $y^2 = (x - \alpha_1) \cdots (x - \alpha_n)$, $(\alpha_1,\ldots,\alpha_n$ being distinct numbers and $n \geq 3$), the genus is $p = [(n-1)/2]$. In particular, for elliptic integrals, where $n = 3$ or $n = 4$, the genus is 1; if these are also integrals of first kind, the theorem just quoted makes it possible to assert that the sum of any number q of such integrals is, up to an additive constant, equal to just one integral whose upper limit X is an algebraic function of the upper limits of the terms $x_1, x_2,\ldots,x_q$ (more precisely — a rational function of $x_j, y_j, j = 1,2,\ldots,q$). This special case had been known to Euler.

---

63) Picard, É. *Traité d'analyse.* Paris 1893, T. 2. We have used the third edition (1925), where these words are given on p. 464.

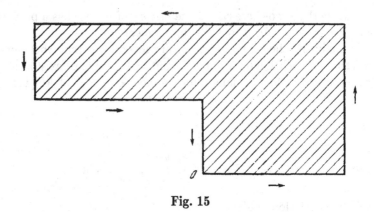

Fig. 15

Quadruply Periodic Functions

As noted in the preceding section, the attempt to invert Abelian integrals, in particular, hyperelliptic integrals, necessarily leads to the consideration of functions having more than two independent periods. But what kind of functions can these be, and what is their nature? The attention of mathematicians had been attracted to these questions by a short paper of Jacobi "De functionibus duarum variabilium quadrupliciter periodicis, quibus theoria transcendentium Abelianarum innititur" (*J. für Math.*, **13** (1835)).

Noting that the periods of functions studied in mathematics can be expressed either as integer multiples of one period or as integral linear combinations of two periods whose ratio is not real (the case of elliptic functions), Jacobi asked: Can a function (of one variable, of course) have three periods not reducible to two periods? By a thorough examination of a purely arithmetic nature he arrived at a negative answer (Jacobi's theorem). More precisely, he proved that the assumption that such a function exists leads to the conclusion that it must have periods of arbitrarily small (non-zero) absolute value, which he pronounced absurd without further explanation.

On this basis he proclaimed as absurd any attempt to invert an integral $u = \int_0^x \dfrac{(\alpha + \beta x)\, dx}{\sqrt{X}}$, where X is a polynomial of degree 5 with distinct real roots. Indeed he showed that the function $x = \lambda(u)$ would have to have four distinct periods (he had noted this fact as early as 1832). The last assertion is correct, but is completely explained by the fact that the function $x = \lambda(u)$ is an infinitely many-valued analytic function for which equal values (belonging, however, to different branches of the function) are assumed in an arbitrarily small neighborhood of any point $u = u_0$. All this could have become intuitively clear through the remark that the integral $u = \int_0^x \dfrac{(\alpha + \beta x)\, dx}{\sqrt{X}}$ gives a one-to-one mapping of the upper half-plane Im $z > 0$ onto a hexagon with right angles. Each side represents one of the half-periods of the integral, which, as can be seen from Fig. 15, can be reduced to four. If the hexagon is reflected through each of its six sides and the reflection process is repeated ad infinitum for all of the new hexagons obtained, the z-plane is covered by an infinite set of layers of a hexagonal parquet. In each tile a single-valued branch of $z = z(w)$ is defined, mapping the hexagon onto the upper or lower half of the w-plane Here in each neighborhood of any point $w = w_0$ there is an infinite number of points whose images are represented by the same number z_0, but which must be located on different tiles of the multi-layered parquet. Of course all these considerations were alien to Jacobi, and he insisted on the "absurdity" of a function such as $z = z(w)$.

Hermite later attempted to buttress Jacobi's position, explaining that the latter had implicitly assumed single-valued (and of course also continuous and nonconstant) functions in his theorem on periods. However, such an explanation, which was shared by H. Weber,[64] the editor of the German translation of Jacobi's paper (written originally in Latin), does not seem to us to be sufficiently supported by the facts.

In fact, as becomes clear from the ensuing text of this paper, Jacobi does not exclude a priori multi-valued functions such as $\arcsin x$ or $\ln x$ or an elliptic integral of first kind, which for a given value of the upper limit x (the Jacobi sine of the amplitude) assumes an infinite set of values, differing by the values of the periods. In all these examples, the different values of the multi-valued function at a given point form, in modern language, a discrete set: none of them is a limit of the others. It was this condition that Jacobi seems to have considered necessary for analytic functions (for him, the functions studied in analysis). He regarded as beyond mathematical analysis and its possibilities those multi-valued functions that, in his words, "lead to such strong multiplicity of values that... among these values there are always some that differ from any given real or imaginary value by less than any given small quantity."[65] But, as explained above, the functions inverse to hyperelliptic integrals, and even the hyperelliptic integrals themselves, must have precisely such a "strong multiplicity of values."

Only by following the routes later marked out by Riemann and Weierstrass, were mathematicians able to verify that the "strong multi-valuedness" that disturbed Jacobi is not at all inconsistent with analyticity in the precise sense of the concept. Essentially "Jacobi's paradox," as one might call it, was fully explained by Riemann in 1857. In fact in his paper "Theorie der Abel'schen Funktionen" (*J. für Math.*, **54** (1857)) he showed in particular that an arbitrary Abelian integral corresponding to an algebraic curve of any genus p, maps the corresponding Riemann surface T onto some Riemann surface S in such a way that the function inverse to the integral is $2p$-periodic on S.[66] As a result all doubts concerning the case of an hyperelliptic integral also vanished.

---

64) Jacobi, C. G. J. *Über die vierfach periodischen Functionen zweier Variablen*, Herausgegeben von H. Weber. Leipzig 1895, S. 37 (*Ostwald's Klassiker der Exakten Wissenschaften*, No. 64).

65) *Ibid.*, S. 28.

66) We shall study this paper of Riemann's below.

Nevertheless Riemann's contemporaries had difficulty making sense of his results, comprehension of which requires mastery of the idea of a Riemann surface. This is attested, for example, by the fact that even Riemann's student F. Prym (1841–1915), a professor at the Technische Hochschule in Zürich and at the University of Würzburg, asserted in 1864 in relation to the hyperelliptic integral studied by Jacobi, that in general the integral could not be regarded as a function of its upper limit, nor could the upper limit be regarded as a function of the integral. F. Casorati, a professor at the University of Pavia, adhered to a different view, defending his views in a number of papers (1863–1866). But he too required 20 years to acquire a definitive picture of the mapping of the corresponding Riemann surfaces.[67] For the time being it is necessary to note that the desire to escape from this imaginary cul-de-sac turned out to be quite fruitful for the subsequent development of the subject. It forced Jacobi to give up his first study of individually considered Abelian integrals and his attempts to invert them in analogy with the inversion of elliptic integrals and led him to the discovery of Abelian functions, which it would probably be more accurate to call Jacobian functions.

Jacobi's ideas had been first expounded three years before the paper we are studying in his "Considerationes generales de transcendentibus Abelianis," (*J. für Math.*, **9** (1832)). This idea, using the example of the hyperelliptic integral of first kind $\int_a^x \dfrac{(\alpha + \beta x)\, dx}{\sqrt{X}}$, where X is a polynomial of degree 5 or 6 (here the genus of the curve $y^2 = X$ is 2), amounts to considering the sum of two values of the integral with independent upper limits x and y instead of one such integral considered individually:

$$u = \int_a^x \frac{(\alpha + \beta x)\, dx}{\sqrt{X}} + \int_b^y \frac{(\alpha + \beta x)\, dx}{\sqrt{X}}.$$

We further set

$$u' = \int_a^x \frac{(\alpha' + \beta' x)\, dx}{\sqrt{X}} + \int_b^y \frac{(\alpha' + \beta' x)\, dx}{\sqrt{X}}$$

(the integrals $\int_a^x \dfrac{(\alpha + \beta x)\, dx}{\sqrt{X}}$ and $\int_a^x \dfrac{(\alpha' + \beta' x)\, dx}{\sqrt{X}}$ are assumed to be linearly independent, for which it suffices that $\alpha\beta' - \alpha'\beta \neq 0$). The problem proposed by Jacobi is to regard x and y as functions of the two independent variables u and u':

$$x = \lambda(u, u') \ \text{ and } \ y = \lambda'(u, u').$$

It is remarkable that the functions so defined (Abelian functions, in terminology suggested by Jacobi himself) are no longer infinitely many-valued: they are only

---

67) One can read about all of the interesting details of the discussion of the "Jacobi paradox," whose basis, in our view, is more a psychological problem than a mathematical one, in the paper of E. Neuenschwander "Der Nachlass von Casorati (1835–1890) in Pavia" (*Arch. Hist. Exact Sci.*, **19** (1978), No. 1.

two-valued, which, however, Jacobi himself did not prove. That he believed in their two-valuedness, or more precisely in the single-valuedness of the symmetric functions $x + y = \lambda + \lambda'$ and $xy = \lambda\lambda'$, is shown by his proposal to study x and y as the roots of a quadratic equation of the form $Ux^2 - U'x + U'' = 0$, where U, U', and U'' are functions of the two variables u and u'. The single-valuedness of the latter was established in later investigations.

As his "fundamental theorem" in the paper "De functionum duarum variabilium quadrupliciter periodicis..." discussed above Jacobi proved that $\lambda(u, u')$ and $\lambda'(u, u')$ are quadruply periodic functions. He introduced them for the particular case

$$X = x(1 - x)(1 - \varkappa^2 x)(1 - \lambda^2 x)(1 - \mu^2 x).$$

As fundamental periods here one can take the numbers

$$\omega_0 = 2i \int_{-\infty}^{0} \frac{\alpha + \beta x}{\sqrt{X}}\, dx, \quad \omega_1 = 2 \int_0^1 \frac{\alpha + \beta x}{\sqrt{X}}\, dx,$$

$$\omega_0 = 2i \int_{1/\lambda^2}^{1/\mu^2} \frac{\alpha + \beta x}{\sqrt{X}}\, dx, \quad \omega_1 = 2 \int_{1/\mu^2}^{\infty} \frac{\alpha + \beta x}{\sqrt{X}}\, dx,$$

for the variable u and an analogous quadruple ω_0', ω_1', ω_2', ω_3' computed using the analogous formulas with α and β replaced by α' and β' for the variable u'. Jacobi demonstrated that the two systems ω_0, ω_1, ω_2, ω_3, ω_0', ω_1', ω_2', ω_3' are periods for λ (and λ') in the following sense:

$$\lambda(u + m\omega_0 + m'\omega_1 + m''\omega_2 + m'''\omega_3, u' + m\omega_0' + m'\omega_1' + m''\omega_2' + m'''\omega_3') =$$
$$= \lambda(u, u')$$

and similarly for λ'.

It is essential that the integral linear combinations of the systems of periods ω_j and ω_j' ($j = 0, 1, 2, 3$) be taken for u and u' *with the same respective coefficients* (ω_j and ω_j' are *joint* systems of periods).

Jacobi demonstrated that for the functions λ and λ' the presence of four periods does not contradict the nature of the concept of a function, for here there are two independent variables u and u', not just one. To be specific, when $\omega = m\omega_0 + m'\omega_1 + m''\omega_2 + m'''\omega_3$ assumes values that are arbitrarily small (in absolute value), the quantity $\omega' = m\omega_0' + m'\omega_1' + m''\omega_2' + m'''\omega_3''$ with the same coefficients assumes values that are arbitrarily large (in absolute value). In modern language, one can say that the distance between the points (u, u') and $(u + \omega, u' + \omega')$ tends to infinity.

Jacobi's research received an important continuation in the work of Adolph Göpel (1812–1847), who settled for a modest post on the staff of the Royal Berlin Library. His main work, "Theoriae transcendentium Abelianarum primi ordinis adumbratio levis," (*J. für Math.*, **35** (1847)) appeared posthumously with a brief afterword by Jacobi and Crelle. In this paper Göpel carried out for Abelian functions work analogous to that which Jacobi himself had carried out for elliptic functions in the years 1835–1836 in a course at the University of Königsberg. Briefly stated, Göpel constructed 16 single-valued functions P, P', P'', P'''', Q, Q', Q'', Q''',..., R,..., S'''' of the two variables u and u' analogous to the theta functions in their structure (sums of series of exponential functions whose exponents are polynomials of degree 1 in u and u' and degree 2 with respect to the integer indices of summation) and in their properties. It follows from these properties that the ratios of these functions taken in pairs are single-valued quadruply periodic functions of u and u'. Göpel went on to establish the algebraic relations among these functions and derive differential equations that they satisfy. In transforming the latter, he showed finally which formulas must be used to compute the coefficients of Jacobi's quadratic equation $Ux^2 - U'x + U'' = 0$. They were given by Göpel as ratios of the entire functions of two variables u and u' that he studied, which are generalizations of the Jacobi theta functions, i.e., they turn out to be meromorphic quadruply periodic functions of u and u'. Nowadays it is precisely meromorphic functions of p complex variables $u_1,...,u_p$ having $2p$ independent systems of periods that are called Abelian functions. In the particular case when $p = 1$, the Abelian functions coincide with the elliptic functions. When $p \geq 1$ they contain as a proper subclass the functions by means of which the suitably stated Jacobi inversion problem is solved, as, for example, in the paper of Göpel (special Abelian functions). However, in its full extension the concept of Abelian integrals goes beyond this problem. Abelian functions have widespread applications in the problems of the algebraic geometry of surfaces.

Having a high opinion of the paper of Göpel, whom he did not know personally, Jacobi nevertheless stated that his former student Georg Rosenhain (1816–1887) had arrived at analogous results in working on the problem in more detail. (After 1857 Rosenhain was a professor at the University of Königsberg.) Rosenhain's long paper, which emphasized its dependence on Jacobi in various places (in particular by beginning with the problem of inverting elliptic integrals of third kind, in the formulation proposed by Jacobi for hyperelliptic integrals), was presented for a prize competition announced by the Paris Academy, and in 1846 it was awarded the Grand Prix. Not until 1851, however, did the Academy publish Rosenhain's paper "Sur les fonctions de deux variables et à quatre périodes qui sont les inverses des intégrales ultraelliptiques de la première classe" (*Mém. sav. étrang.*, **11** (1851)).

Thus in the work of Göpel and Rosenhain, carried out under the influence of Jacobi, classes of single-valued quadruply periodic functions of the variables u and u' were made the object of study and represented as ratios of functions of the same variables analogous to the Jacobi theta functions. The inversion problem for hyperelliptic integrals of first kind (the case when a polynomial of degree 5 or 6 stands under the square root) in the form stated by Jacobi was solved by their use. These results were a prefiguration of the general solution of the problem of inverting Abelian integrals obtained later in the work of Riemann and Weierstrass. We remark that by 1847 Weierstrass had surpassed Göpel and Rosenhain. Having chosen a different route, which may perhaps be compared with Gauss' first approach to the theory of elliptic functions (which Weierstrass himself did not know), he solved the inversion problem in the Jacobi formulation for hyperelliptic integrals of any degree. However, he did not begin to publish a complete exposition of these results until 1856. Weierstrass' method will be discussed below.

Results of the Development of Analytic Function Theory over the First Half of the Nineteenth Century

The famous dissertation of Riemann "Grundlagen einer allgemeine Theorie der Funkionen einer komplexen Veränderlichen" (1851) divides the nineteenth century into equal halves. To be specific, by that time internal contradictions had accumulated in the historical process of laying the foundations of analytic function theory, and new ideas were urgently needed to resolve them.

We have seen that as early as 1811 Gauss was in possession of the concept of a complex variable in the form of a point moving on a plane and was able to explain the multi-valuedness of elementary functions ($\ln z$) by the presence of different paths of integration joining two points of the complex plane. This concept was not stated fully and explicitly either in the published work of Gauss or in the work of any of his contemporaries. Cauchy came closer to it than anyone else; but until the second half of the 1840's even he, as noted above, was still looking for an interpretation of a complex number that he could accept. In doing this he was striving to get away from the purely hypothetical understanding of a complex number that he had announced in 1821, as a symbol having no meaning by itself. Only after considerable vacillation did he settle on a geometric interpretation. We recall that such an interpretation had been supported by the authority of Gauss a decade and a half earlier ("Theoria residuorum biquadraticorum," 1832) and that the term "complex numbers" appeared for the first time in this paper of Gauss.

Cauchy's own papers and those of his younger contemporaries contained a collection of powerful methods for investigating various problems of analysis. We have already mentioned the most important of these methods: the Cauchy integral theorem, the Cauchy integral, the power series expansion, the estimates of the coefficients of the series in terms of the maximum modulus on a circle, the concept of residue and the residue theorem. Using the method of majorants, Cauchy showed the value of these methods for establishing the convergence of the power series expansions of implicit functions or functions defined by differential equations. The properties just enumerated were theoretically completely sufficient, for example, for the systematic development of the foundations of the theory of entire and meromorphic functions. But this area remained "unknown territory" on the map of the theory of functions until the work of Weierstrass and Mittag-Leffler. Only one portion of it was conquered through the labor of Liouville — the theory of elliptic functions as the class of meromorphic functions having double periodicity. Of course, such an advance of a partial theory ahead of the general theory occurred because elliptic functions continued to occupy one of the central places in the field of interest of mathematicians of the time. On the other hand, this center had already shifted toward Abelian (in particular, hyperelliptic) integrals and the problem of inverting them. But Jacobi had imposed a sort of veto on the direct inversion of hyperelliptic integrals by proclaiming the absurdity or nonexistence of functions of one variable having more than two independent periods.

We have already said that the point here was not that he tacitly required such functions to be single-valued, as Hermite and Weber later explained it, but that for such an inversion the very concept of a function ceased to have any meaning or sense in Jacobi's eyes, as a consequence of the "excessively strong multivaluedness." In short, using modern language, Jacobi did not see the possibility of decomposing functions of this type into single-valued branches.

To be sure, at the same time Jacobi had made a brilliant discovery in showing that the analogy with the inversion of elliptic integrals could be resurrected by inverting a suitable number of sums of several (linearly independent) hyperelliptic integrals of first kind rather than just one individual integral. But when this is done, functions of several complex variables, rather than one, necessarily arise; and knowledge of them was still quite fragmentary. Moreover to carry out Jacobi's idea even in the simplest cases of integrals of first kind depending on the square root of a polynomial of degree 5 or 6 involved serious difficulties. To begin with these integrals caused a significant complication of the computations in comparison with the problem of inverting elliptic integrals.

To gain an appreciation for these complications it suffices merely to turn the pages of the papers of Göpel and Rosenhain, which are solidly filled with cumbersome formulas. But the matter did not end there. There were also deep-rooted difficulties in the reluctance of the authors of these papers, beginning with Jacobi, to rely on the representation of a complex variable and integration over paths in the complex plane suggested by Cauchy's work. Figuratively speaking, one can say that these authors were afraid to leave the real line, although they could not get by without imaginary numbers. Thus, for example, in the classical work of both Jacobi and Rosenhain on quadruply periodic functions all the zeros of the polynomial X under the square root sign are real numbers, and the variable x ranges only over the intervals into which these zeros divide the real axis. In this situation complications arise due to the double-valuedness of $\sqrt{X}$.

It is amazing that Jacobi nevertheless managed to find the periods of the integral. He did this by using an auxiliary variable φ in terms of which both x and

$$u = \int_0^x \frac{\alpha + \beta x}{\sqrt{X}}\, dx$$ can be expressed (here X is a polynomial of degree 5). Let a and b $(a < b)$ be two adjacent zeros of the polynomial X. Jacobi made the substitution $x = a + (b - a)\sin^2 \varphi$, so that x increases from a to b as φ increases from 0 to $\pi/2$. When this is done, the integrand assumes the form $F(\varphi)$, where $F(\varphi)$ is an even function of period π (and moreover, as we now say, it is analytic on the real axis). Therefore its Fourier series expansion $F(\varphi) = \gamma_0 + \gamma_2 \cos 2\varphi + \gamma_4 \cos 4\varphi + \cdots$ converges (uniformly) to the function on any interval. Rewriting u in the form

$$u = \int_0^a \frac{\alpha + \beta x}{\sqrt{X}}\, dx + \int_a^x \frac{\alpha + \beta x}{\sqrt{X}}\, dx = C + \int_0^\varphi F(\varphi)\, d\varphi$$

and integrating the expansion of $F(\varphi)$ termwise, we find

$$u = C + \gamma_0 \varphi + (\gamma_2/2)\sin 2\varphi + \cdots .$$

Hence, in particular, when $\varphi = \pi/2$ $(x = b)$, we obtain

$$2 \int_a^b \frac{\alpha + \beta x}{\sqrt{X}} \, dx = \gamma_0 \pi.$$

But it follows from the expansion just found for u that when φ changes by π the value of the integral changes by $\gamma_0 \pi$, while $x = a + (b-a) \sin^2 \varphi$ remains unchanged. It follows from this that $\gamma_0 \pi = 2 \int_0^b \frac{\alpha + \beta x}{\sqrt{X}} \, dx$ is a *period of the function* $x = x(u)$ or, what is the same, a *period of the integral*

$$\int_0^x \frac{\alpha + \beta x}{\sqrt{X}} \, dx.$$

Since the zeros of the polynomial X divide the real axis into six intervals (for the infinite extreme intervals the substitution for x described above must be suitably modified), Jacobi thereby obtained six periods, three real numbers and three purely imaginary. But he reduced them to only four periods, remarking that their sum is

$$\int_{-\infty}^{+\infty} \frac{\alpha + \beta x}{\sqrt{X}} \, dx = 0,$$

from which it followed that the sums of the real and purely imaginary periods must be separately equal to zero.

For a mathematician equipped with Cauchy's theorem (which had been published in 1825) the vanishing of this last integral follows from Cauchy's theorem applied to a single-valued branch of the function $(\alpha + \beta x)/\sqrt{X}$ in the upper half-plane; here, of course, a passage to the limit is assumed under which the branch points of $\sqrt{X}$ are bypassed over semicircles in this domain. But such was not Jacobi's reasoning. He assumed that the vanishing of the integral simply followed from the fact that when x is replaced by $1/x$ the two limits of the integral coincide, becoming zero! This same argument, without any changes, was given by Rosenhain. Moreover Rosenhain was writing 12 years after Jacobi in a paper submitted to a prize competition of the Paris Academy, of which Cauchy was a member! This gives an indication of the lack of appreciation of Cauchy's ideas on the part of Jacobi and mathematicians directly influenced by him.

In this connection the opinion on the validity of mathematical proof expressed in a letter of 21 December 1846 from Jacobi to Humboldt, cited in Book 1, is curious: "When Gauss says that he has proved a thing, it seems very probable to me; when Cauchy says that, it is a fifty-fifty bet; but when Dirichlet says it, you can count on it."[68]

As for Cauchy, we have seen previously that one of his theorems raised doubts in Abel, who gave a counterexample. Chebyshev had doubts on another score and

---

68) Biermann, K.-R. *Die Mathematik und ihre Dozenten an der Berliner Universität 1810–1820.* Berlin 1973, S. 31.

asked for proof of the validity of termwise integration of a series. At the beginning of the preface of the second edition of their *Traité des fonctions elliptiques* (1875, p. 1) Briot and Bouquet expressed regret that in some works devoted to the theory of functions of a complex variable (which they called an imaginary variable) "on ne rend pas à Cauchy la justice qui lui est due."

Looking ahead, we note that Weierstrass, in the courses that he gave at the University of Berlin for several decades, hardly ever cited Cauchy. And when the Paris Academy presented him with the first volume of the recently-published collected works of Cauchy, Weierstrass wrote not a line in response.[69]

How can one explain this neglect of the ideas and particular results of Cauchy on the part of people who would seem to have been better situated than anyone else to appreciate their exceptional value for science? Let us leave aside considerations of sympathy and antipathy, friendship and enmity, in which we have no basis for judgment. There is nevertheless no doubt that this situation was to some extent explained by the character and direction of the research of Cauchy himself. Being unusually productive, he filled the pages of the *Comptes Rendus* with hundreds of his notes, not always substantive, sometimes repeating what he had said earlier, occasionally disputing it. Just recall his vacillation on the question whether one needed to assume the existence of a continuous derivative in the hypothesis of his theorem on the power series expansion of a function! It was as if there were not enough scholarly journals at the time and he was creating his own, in which he was the only author.

However, the main reason for the absence of serious interest in Cauchy's ideas on the part of people who were eagerly developing the new problematics generated in the investigations of Abel and Jacobi (elliptic functions, Abelian integrals, and the problem of inverting them), was, in our opinion, that Cauchy himself, despite the great variety of his scholarly interests, was indifferent to these themes. They were alien to his interests, as shown by his cavalier attitude toward Abel's paper, which contained the famous theorem of Abel in full generality. This paper was published by the Paris Academy in 1841 under the influence of an official request by Norway. Cauchy himself, as a researcher did very little for the direct application of his ideas to multi-valued functions, except for the elementary ones — the logarithm and the inverse trigonometric functions, with respect to which he was not sparing of explanations.

However, it is precisely in the example of the elementary functions that one can demonstrate the difficulties presented to Cauchy (and of course also to his contemporaries) by the idea of multi-valuedness.

In his "Mémoire sur les fonctions de variables imaginaires," published in the 36th issue of his *Exercices d'analyse* (1846), Cauchy defined what we would now call the principal branches of the multi-valued functions of a complex variable $x^{1/n}$, x^q, and $\ln x$, subjecting the argument p of the complex number x to the condition $-\pi < p \leq \pi$. He noted that this condition could be otherwise, for

---

69) Dugac, P. "Eléments d'analyse de Karl Weierstrass," *Arch. Hist. Exact Sci.*, **10** (1973), p. 61.

example $\varphi - \pi < p \le \varphi + \pi$. However, if we identify p with the polar angle, for example, and measure it from 0 to 2π $(0 \le p < 2\pi)$ we encounter the inconvenience that these functions will have a discontinuity at every point of the positive real axis. It is important to emphasize here that this paper seems to be the first place where cuts in the plane were used (along the real axis) that happened to be lines of discontinuity. In Riemann's investigations such lines were accorded primary attention. In doing this Cauchy indicated that the position of these lines in the plane depends on the conventions adopted (i.e., on the way in which a single-valued branch of a multi-valued function was singled out). We can discern in all this a sort of preparation of construction materials for the future Riemann surfaces.

Modest as the scientific contribution in this paper seems to us, one cannot avoid the evaluation due to Cauchy himself: "In my *Analyse algébrique*," he wrote, "in this paper (p. 287), and in my earlier work, when using this notation [i.e., ln x, x^q], I confined myself to the case where the real part of the variable x was positive." Thus it was a question of passing from a single-valued branch in a half-plane to a branch in the cut plane, not a large step, but a necessary one on the route to the Riemann surface!

The last numbers of the fourth volume of the *Exercises d'analyse*, whose title page is dated 1847, were actually published after 1850. In this volume, in the article "Sur les fonctions continues de quantités algébriques ou géométriques" Cauchy continued to insist that a branch point (the term was introduced shortly afterwards by Riemann) is a point of discontinuity of the function (for example, $z = 0$ for $z^{1/p}$). We have already stated that he had grounds for such an interpretation. Indeed for functions of a complex variable the continuity requirement introduced by Cauchy played the role of necessary and sufficient condition for expandability of the function in a power series. It implicitly included single-valuedness, finiteness, and continuity not only of the function, but also of its first derivative, whose existence was assumed. Therefore in his terminology a point of discontinuity (literally a particular value of the variable in a neighborhood of which the function ceases to be continuous or at which it becomes discontinuous) means in essence the same thing as a singularity in the work of Puiseux (1850). In this interpretation Cauchy had mastered the notion of a singular point of an analytic function of a complex variable as early as 1832, in his lithographed Turin memoir.

Let us return to the paper of Cauchy mentioned above. In this paper he singled out the functions in a given domain (region) that arrive at the same value independently of way in which "the point labeled z runs here and there provided it does not leave the domain S and avoids the singular points corresponding to infinite values of the functions,"[70] calling such functions *monodromic*, from the Greek words μόνος (alone) and δρόμος (race). As one can see, Cauchy used the term *point singulier* here only in relation to a pole. He used the term *le chemin* several pages later to denote a continuous curve described in the plane by the point labeled z, and in a footnote mentioned "the remarkable paper (of Puiseux) entitled "Recherches sur les fonctions algébriques (1850)" in which, he claimed,

---

70) Cauchy, A.-L. *Exercices*, Paris 1847, T. 4, p. 325.

the term *path* was used in this sense for the first time.[71] In the next article of his *Exercices* "Sur les différentielles de quantités algébriques ou géométriques, et sur les dérivées de fonctions de ces quantités," which cannot be assigned a date earlier than 1850, as follows from the reference to Puiseux's memoir just cited, the class of *monogenic* functions appears, defined by the condition of having a monodromic derivative. An example of a monogenic but not monodromic function given by Cauchy is $\ln z$.

We see that Cauchy moved rather cautiously away from the requirement of single-valuedness, introducing after the single-valued functions only those multi-valued functions whose derivatives were single-valued. Algebraically irrational functions and their integrals remained outside of this extended class (of monogenic functions), and these were precisely the functions of the class on whose study the efforts of the mathematicians of the time were concentrated!

We remark that the term *monodromy* has kept its original meaning to the present time, in the monodromy theorem (which asserts that a function obtained by analytic continuation over all possible paths inside a simply connected region is single-valued). But the class of monogenic functions, as understood by Cauchy, turned out to be unviable. The term *monogeneity* is now used to mean simply differentiability of a function with respect to a complex variable (at points of a domain or at points of some nowhere dense sets satisfying certain conditions of a metric character; in the latter case one speaks of *functions monogenic in the sense of Borel*).

Let us return to algebraic functions. Without making them the subject of a special study Cauchy nevertheless created all the necessary means used by Puiseux to construct the theory of them. These means were, first of all, complex integration, which led to the power series expansion of a function via the Cauchy integral, and the (more special) Cauchy implicit function theorem. We have already mentioned this theorem above as an early form of the so-called Weierstrass preparation theorem. Translated into modern language, Cauchy's theorem would sound as follows:

If $f(x, y)$ is an analytic function of two variables in a neighborhood of the point (a, b), b is an n-fold root of the equation $f(a, y) = 0$, and $F(y)$ is a function analytic in a neighborhood of the point $y = b$, then there exists a neighborhood of the point $x = a$ in which

a) *the equation $f(x, y) = 0$ has exactly n roots $y_1(x)$, ..., $y_n(x)$ close to b for each $x \neq a$ (this expresses the continuity of the roots);*

b) *$F(y_1) + \cdots + F(y_n)$ is a single-valued analytic function of x.*

In particular when $n = 1$, the function $y = y(x)$ happens to be analytic and single-valued in a neighborhood of $x = a$.

This last theorem can be applied in the case when $f(x, y)$ is a polynomial in x and y, which for definiteness we shall assume irreducible, i.e., not decomposable into a product of polynomials of lower degree. If the degree of $f(x, y)$ with respect

---

71) *Ibid.*, p. 328.

to y is m, the equation $f(x,y) = 0$ defines an *algebraic function* $y = y(x)$ that is in general m-valued. Suppose that for a given $x = a$ the coefficient of y^m (which is a polynomial in x) is non-zero and that the equation $f(a, y) = 0$ has no multiple roots. Then by the preceding theorem we obtain for y in a neighborhood of $x = a$ exactly m distinct series of nonnegative integer powers of $x - a$. This conclusion does not hold for those points $x = a$ of the plane where either the coefficient of y^m vanishes or the equation $f(a, y) = 0$ has multiple roots, but there are only finitely many such points.

It turns out that the same implicit function theorem (more precisely the assertion that the roots are continuous) based on complex integration made it possible to establish that in a neighborhood of an exceptional (singular) point $x = a$ the function admits an expansion into a generalized power series in which the exponents of the powers of $x - a$ can be fractions (all having the same denominator), positive or negative.

We repeat that Cauchy himself did not draw these conclusions from his theory. He did not single out the class of algebraic functions as an object of study in analysis. The credit for this formulation of the question and the first results relevant to it belongs to the French mathematician and astronomer V. Puiseux, who considered it necessary, however, to emphasize in his investigations that he was relying on the work of Cauchy and developing his ideas.

V. Puiseux. Algebraic Functions

Victor Alexandre Puiseux (1820–1883) studied at the École Normale, where he attended the lectures of Charles Sturm (1803–1855). He then taught at Reims and later at Besançon, where he wrote his most important paper "Recherches sur les fonctions algébriques" (*J. math. pures et appl.*, **15** (1850)), which was soon followed by "Nouvelles recherches sur les fonctions algébriques," (*J. math. pures et appl.*, **16** (1851)). In this paper the concept of an algebraic function is formulated precisely, and its behavior in a neighborhood of singular points is studied in detail, so that the group-theoretic character of the corresponding regularities became clear. The process of analytic continuation by means of power series was studied using the example of an algebraic function, and the significance of the topological (more precisely, homotopic) properties of closed curves in a multiply connected domain for the theory of functions was established. A general definition of the periods of Abelian integrals was given, and their properties were studied in a number of important special cases (in particular, for hyperelliptic integrals). This paper and others related to it immediately gained wide fame for Puiseux. The rest of his career was spent in Paris: in the École Normale, at the Sorbonne, at the Bureau des Longitudes, and at the Observatoire, where he studied mainly astronomical questions (observations of the transit of Venus, celestial mechanics). He was elected a member of the Academy of Sciences in 1871.

The fundamental importance of Puiseux's main paper, mentioned above, lay in the fact that it was the first time in the history of mathematics that an extensive class of multi-valued analytic functions was subjected to study, functions for whose

V. Puiseux

analytic representation the mathematician had only local methods at his disposal: power series expansions (ordinary or generalized) that converged only in some neighborhood of the point in question.

Since the classical machinery of analysis (the derivative and integral) had been constructed for single-valued functions, Puiseux was forced to begin by determining conditions under which he was able to reduce the multi-valued function being studied to single-valued functions. But the local character of the analytic representation of the function required that he establish connections and ways of passing from one of them to another, i.e., that he clarify the continuation process, which in the final analysis was analytic continuation.

Puiseux studied an irreducible algebraic equation of the form

$$f(u, z) = Au^m + Bu^{m-1} + \cdots + K = 0,$$

where A, B, ..., K are polynomials in z (we are using his notation). Let b be one of the roots of the equation $f(b, c) = 0$ Based on the property of continuity of the roots of the equation $f(u, z) = 0$ established by Cauchy, Puiseux asserted that at a point $z = c'$ near c there will be a root b' of the equation $f(u, z) = 0$ near b. This situation gives rise to a problem: determine the value of u at some point $z = k$ if, in moving from $z = c$ to $z = k$ along a curve CMK joining them, we maintain the condition that the values of the functions at sufficiently near points of the curve must be arbitrarily close to one another. In other words, here at the outset it was a question of *continuing an algebraic function* along a given curve CMK as a *single-valued continuous function*, i.e., continuous extension.

In order for such a continuation to be possible the path must avoid those points at which u may become infinite, i.e., the poles (Puiseux does not use this term). But such points correspond to roots of the equation $A = 0$. Suppose, for example, that at some point $A = 0$, but $B \neq 0$; let us set $u = v/A$. Then for v we have the equation

$$v^m + Bv^{m-1} + \cdots + KA^{m-1} = 0.$$

When $A = 0$, this equation has the root $v = -B$, corresponding to the value $u = \infty$. Moreover, as follows from the same implicit function theorem of Cauchy, one must avoid the points $z = c$ at which the equation $f(u, c) = 0$ has multiple roots (these points satisfy the two equations $f(u, z) = 0$ and $\partial f / \partial u = 0$, or the single algebraic equation obtained from them by eliminating u). For points c' arbitrarily close to c there exist at least two distinct values $u = b'_1$ and $u = b'_2$ arbitrarily close to b and consequently the condition for uniqueness of the choice of values of the function near b is violated.

In this way Puiseux distinguished as singular points of an algebraic function (not using the term "singular") the finite set of points of the plane at which either u becomes infinite or two or more values of u become equal. The first points later came to be known as poles, the second as critical points.[72] In the historical literature it is emphasized that when Cauchy spoke of points of discontinuity, he meant only poles; critical points escaped his notice (Puiseux himself started this version of events in a remark on p. 375 of his first paper). But we have already seen that Cauchy's interpretation of the concept of a point of discontinuity also encompassed branch points (for example, $z = 0$ for the case $u = \sqrt{x}$).

In accordance with what has been said above Puiseux went on to study the continuous extension of algebraic functions along paths not passing through any singular points. Let CMK be such a path and u_1 a function defined at points of the path by the initial condition $u = b_1$ at $z = c$. As a fundamental proposition of his theory he proved that the value of the function at the endpoint of the path

---

72) The classification of the singularities of an algebraic function as poles and critical points was carried out, for example, in the *Théorie des fonctions elliptiques* of Briot and Bouquet (1875). However, the term "pole," as pointed out by E. Neuenschwander ("The Casorati-Weierstrass theorem," *Hist. Math.*, **5** (1978), pp. 139–166) was first used in this sense by K. Neumann in his *Vorlesungen über Riemann's Theorie der Abelschen Integrale* (1865) in connection with the fact that the point at infinity was depicted as the pole of the sphere in this book. The term "pole" has become firmly entrenched in the literature. As for the term "critical point," it is frequently replaced by the more general term "branch point" introduced earlier by Riemann ("Theorie der Abel'schen Functionen," 1857). For the sake of precision, the adjective "algebraic" is added to distinguish this kind of branch point from a logarithmic or transcendental branch point.

$z = k$ does not change if "the curve CMK passes to an infinitely near curve $CM'K$." This implied the following theorem:

If the point z moves from C to K once over the curve CMK and another time over CNK, then the function u_1 that assumes the value v_1 at C arrives at the same value h_1, provided it is possible to deform the curve CMK so as to coincide with CNK without passing through [points] at which the function becomes infinite or equal to some other root of the equation $f(u, z) = 0$.[73]

In particular it follows that if $CLMC$ is a closed curve, generally intersecting itself and winding about C an arbitrary number of times, the function u_1 having the value b_1 at the initial point of the path will have the same value at the terminal point provided that curve can be shrunk to a point without passing through any singular points. Thus continuous extension led Puiseux to distinguish a *single-valued branch of an algebraic function* in any (simply connected) domain containing no singular points.

The integral of an algebraic function $\int u_1\, dz$, where the integration extended over the path CMK, now made sense. Essentially repeating Cauchy's reasoning, and using the differentiability of the function (the point is that the differentiability of the implicit function is a consequence of its continuity and the fact that $\partial f/\partial u \neq 0$), Puiseux first established that the integral was independent of the path. He then established an integral formula for u_1, and finally derived the expansion of the function in powers of $z - c$, pointing out that this expansion holds in a disk with center c and radius equal to the distance from c to the nearest singularity. In a footnote (p. 375) that we have already mentioned he explained that all this had already been done by Cauchy under more general hypotheses. But Cauchy had required only that the path not pass through any points at which the function becomes discontinuous, while in the present case, where algebraic functions were being studied, one could make his statement and proof more precise by speaking of poles and critical points instead of points of discontinuity.

Having obtained the power series, Puiseux emphasized that for points inside the disk mentioned above this series gives the same value for u_1 that could have been obtained by continuous extension. We would say that he had established here *the identity (under the hypotheses of his study) of continuous extension with analytic continuation.* He went on to use this fact, showing how to use power series to distinguish the value of u_1 at a point k outside the circle of convergence (p. 379). The process proposed by Puiseux for this purpose coincides exactly with the process of *analytic continuation along the curve CMK* nowadays discussed in textbooks of the theory of functions and connected with the name of Weierstrass. However, Puiseux did not restrict himself to power series as a means of analytic continuation. He replaced them by more general series

$$\sum_0^\infty c_k [\psi(z)]^k,$$

---

73) *J. math. pures et appl.*, **15** (1850), pp. 370–371.

where $\psi(z)$ is a rational function that vanishes at the point c. Let σ be a curve $|\psi(z)| = \lambda$ enclosing the point c, inside and on which ψ is one-to-one, and suppose all the singular points lie outside σ. Using an integral formula that is a natural generalization of Cauchy's integral formula (the latter corresponds to the special case $\psi(z) = z - c$), Puiseux derived an expansion of the required form, converging inside and on σ. He remarked that a function could also be continued along a curve by means of this kind of expansion. Thus, if one sets successively

$$\psi(z) = (z - c)(z - c'), \quad \psi_1(z) = (z - c')(z - c''), \ldots,$$

where the points c, c', c'',... are successive points on the curve CMK, we obtain a process of analytic continuation involving lemniscates with foci at the pairs of points (c, c'), (c', c''),... instead of disks.

This concludes the first part of the first paper, which can be said to be entirely devoted to the extension (originally continuous, later analytic) of a function.

The main problem of the second part is a detailed study of the analytic machinery, i.e., series representing an algebraic function in a neighborhood of a singular point. For simplicity it is first assumed that the coefficient A of u^m is independent of z, and hence cannot vanish. Then all the values of u at all the points of the plane are finite. Let $z = a$ be a critical point at which the equation $f(u, a) = 0$ has a p-fold root b. A circle $CLMC$ is described about a in such a way that except for the point a there are no critical points inside or on the circle, and p functions $u_1, \ldots, u_p$ are studied, satisfying the equation $f(u, z) = 0$ and having values at the points of a circle of sufficiently small radius that differ by an arbitrarily small amount from b. Setting $z = a + \alpha$, $u = b + \beta$, Puiseux studied the behavior of the infinitesimal quantities $\beta_1, \ldots, \beta_p$ as the point z traverses the circle $CLMC$ (in such a way that it returns to the same point C from which it started). Since the set of values u_j $(j = 1, \ldots, p)$, and consequently also β_j $(j = 1, \ldots, p)$ is recovered after a circuit (in a neighborhood of a there are p and only p roots of the equation near b), the only possible change is that some β_j maps to some β_{k_j}. Puiseux set himself the task of ascertaining the nature of this transition, i.e., determining the permutation of the indices:

$$\begin{pmatrix} 1 & 2 & \ldots & p \\ k_1 & k_2 & \ldots & k_p \end{pmatrix}.$$

The general result, amounting to the fact that this permutation is the product of a certain number of cycles, could have been foreseen by appealing to Cauchy's theorem on permutations, as Puiseux did at the end of the paper. But Puiseux was interested in the connection of the permutation with the nature of the analytic expression for u_j. For that reason he required a thorough and very minute investigation.

We shall give details here only in the simplest case, that in which $\partial f/\partial z \neq 0$ at the point $z = a$, $u = b$. Then the equation $f(b + \beta, a + \alpha) = 0$ assumes the form

$$A\beta^p + B\alpha + \sum C\beta^q \alpha^r = 0,$$

where A, B, $C,\ldots$, are complex numbers with $A \neq 0$, $B \neq 0$, and the nonnegative integer exponents q and r are such that $q > p$ if $r = 0$ and $r > 1$ if $q = 0$, so that $pr + q > p$ in all cases. If α and β are infinitesimals, then $A\beta^p + B\alpha \approx 0$, i.e., $\beta \approx h\sqrt[p]{\alpha}$ up to infinitesimals of order higher than $1/p$ (with respect to α). For a suitable enumeration of β_j it follows from this that β_j becomes β_{j+1} ($j < p$) and β_p becomes β_1 when z makes a circuit about the point z (i.e., when α makes a circuit about the origin) in the positive direction. Thus the required permutation is cyclic in this case:

$$\begin{pmatrix} 1 & 2 & \cdots & p \\ 2 & 3 & \cdots & 1 \end{pmatrix}.$$

In accordance with this Puiseux called the corresponding system u_1, u_2, ..., u_p circular.

To find a representation for each of its terms in a neighborhood of $z = 0$ Puiseux made the change of variables $\alpha = \alpha'^p$, $\beta = v\alpha'$ in the equation connecting β and α. The result is

$$Av^p\alpha'^p + B\alpha'^p + \sum Cv^q\alpha'^{pr+q} = 0$$

or after cancelling α'^p (we recall that $pr + q > p$ under these conditions)

$$Av^p + B + \sum Cv^q\alpha'^{r'},$$

where $r' > 0$. When $\alpha' = 0$, this algebraic equation has p distinct roots, representing p values of the root $\sqrt[p]{-B/A}$. Therefore by Cauchy's implicit function theorem there exist p distinct power series such that in a neighborhood of the point $\alpha' = 0$

$$v_j = \gamma_j + a_j\alpha' + b_j\alpha'^2 + \cdots$$

($j = 1, 2, \ldots, p$; γ_j is one of the p values of $\sqrt[p]{-B/A}$). Therefore for $u_j = b + \beta_j = b + v_j\alpha'$ we obtain

$$u_j = b + \gamma_j\alpha' + a_j\alpha'^2 + b_j\alpha'^3 + \cdots =$$
$$= b + \gamma_j(z-a)^{1/p} + a_j(z-a)^{2/p} + \cdots \quad (j = 1, \ldots, p).$$

Thus the character of the expansion of an algebraic function in a neighborhood of the critical point $z = a$ is established in the case when the system $u_1, \ldots, u_p$ is circular. We remark that the coefficients on the right-hand side are actually independent of the index j: the differences in values of the functions of the system are entirely determined by the differences in the values of $(z - a)^{1/p}$.

The case when $\partial f/\partial z = 0$ at the point $z = a$, $u = b$ is much more complicated and requires the study of various possibilities. The final conclusion was indicated above: the functions $u_1, \ldots, u_p$ are distributed in general among several circular systems some of which may contain only a single function. If the z functions

$u_1, \ldots, u_s$ form a circular system, then for any u_j $(1 \leq j \leq s)$ there is an expansion of the form

$$u_j(z) = b + A(z - a)^{r/s} + B(z - a)^{(r+1)/s} + \cdots$$

in a neighborhood of the point $z = a$. Puiseux proposed a simple geometric method (the idea of which goes back to *Newton's diagram*, see HM, T. 2, pp. 49–50), making it possible to determine all the circular systems corresponding to a given critical point $z = a$ (it is a matter of computing the corresponding s).

If the coefficient A of the highest power of u in the equation $f(u, z) = 0$ depends on z, so that there are poles among the singularities in the plane, the substitution $u = v/A$ reduces the study of the algebraic function u to the preceding case (the coefficient of v^m in the transformed equation is 1). Returning to u, we verify that in the expansion in a neighborhood of the singular point there may be negative as well as fractional powers of $z - a$. We remark that Puiseux omits entirely the expansion of the function in a neighborhood of the point $z = \infty$, which could easily have been obtained by the substitution $z = 1/\zeta$, applying the preceding theory at the point $\zeta = 0$.

If we trace the course of development of Puiseux's ideas using later terminology referring to the concept of a multi-sheeted Riemann surface extended over the complex z-plane, we can say that Puiseux had already explained in full the structure of the m-sheeted Riemann surface determined by an algebraic equation $f(u, z) = 0$. He had found the location of all its branch points and determined the nature of the transition from one sheet to another in a neighborhood of each of them. As it turned out, in such a neighborhood the sheets separate into cycles (some cycles may have only one sheet) so that for each of them a single circuit around the branch point is equivalent to passing from one sheet to the next sheet of the same cycle; only after a number of cycles equal to the number of sheets in the cycle does the point of the surface return to its original position. To be sure, the point at infinity was left out of this investigation; but every circuit around that point is equivalent to a circuit around the whole set of finite singularities, and the result can easily be deduced from knowledge of the structure of the surface in a neighborhood of the latter. For that reason the analysis carried out by Puiseux can be considered exhaustive. Indeed, he showed how one could trace the change in values of an algebraic function along any paths with a given initial point C and terminal point K by using the facts he had established, noting in the process that closed paths in the z-plane correspond in general to nonclosed curves in the plane of values of the function.

We have already seen that in the first part of this paper Puiseux had solved the problem of continuation of an algebraic function along any path CMK not passing through a branch point. It was now a matter of comparing the end results of continuation along different paths on a Riemann surface with a common initial point C and terminal points located on different sheets projecting to the same point of the z-plane without carrying out the continuation process, relying only on the information on the behavior of the function in a neighborhood of the singular points.

The elementary curves that entered the later theory of algebraic functions under the name of *loops* play a fundamental role. An *elementary contour* (Puiseux's term) corresponding to the initial point C and the singular point A consists of a path joining C to a point D in a neighborhood of A (usually a line segment) followed by a circle DNP with center at A and then by the same path DC, this time traced from D to C. Knowing how the values of the functions $u_1, u_2, \ldots, u_m$ determined by the initial values $b_1, b_2, \ldots, b_m$ at the point c permute among themselves in a neighborhood of the point A, one can determine how these values will permute at the point C when z describes the loop.

We denote such a loop by the symbol (A), and for all the singular points A, A', A'',... we construct a system of loops (A), (A'), (A''),..., any two of which have only the point C in common. In doing this we shall distinguish two orders of traversal — forwards and backwards — for each loop, by writing $(+A)$ for the loop traversed forwards and $(-A)$ for the same loop traversed backwards. We then claim that any closed contour passing through the point C, not passing through any of the points A, A', A'',..., and having a given direction of traversal can be deformed without passing through any of the singular points and without moving the point C in such a way as to reduce it to a sequence of loops, possibly traversed several times in either direction. This sequence is called the *characteristic* of the contour. Here is an example of the way in which Puiseux wrote the characteristic of a certain contour:

$$(-A')(+A)(+A')(-A'')(-A').$$

If a closed contour can be deformed to the point C without passing through any singular points, it is assigned characteristic (0). Under these conditions Puiseux claimed that two contours having the same characteristic could always be deformed into one another without passing through the singularities. Conversely, the characteristic does not change if a contour is deformed without passing through any singular points. In this way Puiseux established a purely topological (homotopic) invariant for closed curves in the plane.

Identity of the characteristics of two closed contours passing through the point C guarantees that the values of the algebraic functions reached at a point by describing the two contours starting from the same initial value will be the same. In other words, equality of the characteristics means that both paths lead to the same point of the Riemann surface over the point in the z-plane.

Now let the values of the functions $u_1, \ldots, u_m$ at a point C labeled c be $b_1, \ldots, b_m$, and let $h_1, \ldots, h_m$ be the values reached at the point K when z describes some path CMK. Then in order to determine which value each of the functions will assume when z reaches K over a different path CLK it suffices to determine the characteristic of the closed path composed of CMK and KLC (the path CLK tranversed from K to C). From this point of view the path CLK can be symbolically written as $(\Gamma) + CMK$, where (Γ) is the characteristic of the closed path mentioned above.

The concepts introduced in the first two parts of Puiseux's paper and the regularities established in them were directly applied in the third, concluding part of the paper to study the integral $\int_c^k u_1\, dz$ of an algebraic function. The process of analytic continuation described in the first part made it possible to express the value of the integral over a given curve CMK as the sum of a finite number of power series. In this way, in particular, the value of the integral over any loop and the value of each of the functions u_j defined by its own initial value $u_j = b_j$ at the point $z = c$ could be established. Puiseux called these integrals *elementary*. The elementary integral $\int u_j\, dz$ over the loop $(+A^{(k)})$ or $(-A^{(k)})$ was denoted by $A_{\pm j}^{(k)}$. Certain linear relations hold between elementary integrals. For example, if u_i becomes u_j after traversal of the loop $(A^{(k)})$, then $A_{-j}^{(k)} = -A_i^{(k)}$. Here is an important example of a different sort. Suppose there exists a closed contour Δ passing through the point C and enclosing all the points $A, A', A'', \ldots$, such that the function u_i returns to its original value at the point C after a traversal of Δ. If u_i becomes $u_{i'}$ after traversing $(A^{(i)})$, u'_i becomes $u_{i''}$ after traversing $(A^{(i')})$, etc., then

$$A_i + A'_{i'} + A''_{i''} + \cdots = 2\pi\lambda_i\sqrt{-1},$$

where λ_i is the residue of u_i with respect to the point at infinity.

The integral $\int u_1\, dz$ over any closed contour $CLMC$ can be expressed as an integer linear combination of elementary integrals. It follows from this that if

$$v_j = \int_c^k u_j\, dz \quad (j = 1, \ldots, m)$$

are the values of the integrals of u_j along the same path CMK, then the value of the integral $\int_c^k u_1\, dz$ along any other path joining C and K is an integer linear combination of the elementary integrals corresponding to the closed contour Γ composed of the paths CMK and KLC added to one of the numbers v_j. The index of the latter is the same as the index of the u_j to which u_1 passes after traversing the contour Γ.

Noting that not every integer linear combination of elementary integrals added to some v_j gives one of the values of $\int_c^k u_1\, dz$, Puiseux introduced the concept of a *period* of the integral $\int_c^k u_1\, dz$ (see p. 438 of the first paper). This was his name for a linear combination of elementary integrals that, first of all, is independent of the choice of the point $z = c$ and, second, has the property that any integer multiple of it added to one of the values of the integral $\int_c^k u_1\, dz$ again gives a value of the same integral. The most important problem that arises in connection with this concept is that of finding all the distinct periods, i.e., all linear independent integer combinations of periods. Puiseux solved this problem for a number of special cases, among them the case of hyperelliptic integrals. In the process some light was shed on the assertions made earlier by Jacobi in connection with his inversion problem for hyperelliptic integrals. We remark that Puiseux took Jacobi's point of view relative to the possibility of regarding z as a function of v in the equation

$$\int_c^z \frac{\alpha + \beta z}{\sqrt{(z-a)(z-a')(z-a'')(z-a''')(z-a^{iv})}}\, dz = v,$$

"since v may change by arbitrarily small amounts for a given value of z."[74]

In relation to the concept of a period itself Puiseux referred to Cauchy as his forerunner,[75] since Cauchy had made some remarks on the derivation of the periods of the integrals of a rational function in a note of 1846 entitled "Considérations nouvelles sur les intégrales définies qui s'étendent à tous les points d'une courbe fermée," (*Comptes Rendus*, **23** (1848), p. 698) and in an unpublished paper had explained the derivation of the periods of elliptic integrals by the same route (i.e., complex integration), but his interpretation of the question was not sufficiently complete and general.

Puiseux continued this research in the short paper of 1851 "Nouvelles recherches sur les fonctions algébriques" already mentioned. In this paper he established in particular that a single-valued algebraic function is necessarily rational and furthermore that in the case of an irreducible equation $f(u, z) = 0$ one can find paths starting from an arbitrary point c with any of the values u_j $(j = 1, \ldots, m)$ over which one arrives at the same point with value u_1. This completed the concept of the connectedness of all the branches of an algebraic function understood as the totality of the branches.

To summarize, one can say that in his work Puiseux gave a model for the study of multi-valued functions using the methods created by Cauchy. In doing so he discovered the importance of purely algebraic and topological methods for the further development of the theory. In particular, for progress in the study of the periods of Abelian integrals he lacked the concepts of order of connectivity and genus of a surface. Nevertheless he laid before his contemporaries the concept of an algebraic function in a form sufficiently complete and well articulated that interpreting it as a unified geometric image became both possible and necessary.

Bernhard Riemann

Despite his brief life — he died of tuberculosis before the age of forty — Riemann is beyond a doubt one of the most profound and penetrating mathematical geniuses known to the history of science. His works, essentially all dating to a single decade, the 1850's, encompass all the basic areas of mathematics, and they exerted and continue to exert a productive influence on its development.

Georg Friedrich Bernhard Riemann was born 17 September 1826 in the village of Breselenz (Lower Saxony) to the family of a Lutheran pastor, the second of six children. His brilliant mathemtical talent manifested itself while he was still in

---

74) *J. math. pures et appl.*, **15** (1850), p. 463.

75) *Ibid.*, p. 479.

the Gymnasium, where he went beyond the required course of study and took up the works of Euler and Legendre. Having entered the University of Göttingen in 1846, he first studied philology and theology at his father's insistence, but attended lectures on mathematics and physics at the same time. Thus in the summer he attended the lectures of Moritz Stern (1807–1894) on the numerical solution of equations and those of Karl Goldschmidt (1807–1851) on terrestrial magnetism, while in the winter semester of 1846–1847 he attended the lectures of Gauss on the method of least squares and those of Stern on definite integrals. His attraction to mathematics was so strong that his father was forced to yield, and Riemann devoted himself entirely to his favorite subject. But Gauss took very little interest in teaching at that period of his life, and by the spring of 1847 we find Riemann in Berlin, where Jacobi, Lejeune-Dirichlet, and Steiner were teaching. There he also attended the lecture of Eisenstein on the theory of elliptic functions. Riemann later spoke of his disagreements with Eisenstein over the concept of a function of a complex variable. Eisenstein defended the primacy of the formal computational machinery, while by that time Riemann was basing functions of a complex variable on partial differential equations (the d'Alembert-Euler equations).

In the spring of 1849 Riemann was again in Göttingen. Over the course of three semesters there he attended lectures on natural science and philosophy. The strongest influence on him was due to the lectures on experimental physics given by Wilhelm Weber (1804–1891), in whom Riemann found a true friend. A deep interest in physics and a striving to create a unified mathematical theory encompassing the entire complex of physical phenomena and closely connected with the properties of space are typical of Riemann. "My main work," he once wrote, "involves a new understanding of the known laws of nature, their expression in terms of other fundamental concepts intended to make it possible to use experimental data on the interaction between heat, light, electricity, and magnetism to obtain a connection between these phenomena."[76] In the fall of 1850 he participated in the seminar on mathematical physics conducted by Weber, the mathematician G. Ullrich (1798–1879), Stern, and J. B. Listing (1808–1882). As is well known, the last-mentioned author was the first to attempt an exposition of the prinicples of a nascent mathematical subject — topology. In terms of their actual contribution in mathematical research neither Listing nor Riemann was destined to become the father of topology. But it is very probable that Riemann's original ideas on the the unity of physical phenomena and the properties of space, which had not been studied before that time, could have found support in conversations with the participants in this seminar.

In November 1851 Riemann presented to the philosophical faculty of the University of Göttingen his famous doctoral dissertation "Grundlagen für eine allgemeine Theorie der Funktionen einer komplexen Veränderlichen," which marked the beginning of a new era in the development of the theory of analytic functions and included the basic ideas of the topology of surfaces. However his professional position was to remain modest for a long time. He worked as an assistant in Weber's

---

76) Riemann, B. *Werke*. Berlin 1990, p. 539.

B. Riemann

seminar; among his duties were to conduct exercises for the new students. The title of Privatdocent, which in no way altered his material situation, was not accorded to Riemann until 1854, when he presented, as required by the rules, his *Habilitationsschrift*, and gave his *Probevorlesung*, chosen by Gauss from three topics offered by Riemann. Riemann's *Habilitationsschrift*, entitled "Über die Darstellbarkeit einer Function durch eine trigonometrische Reihe," which was published posthumously in 1868, gave a powerful stimulus to the development of the theory of functions of a real variable; it has already been discussed. His *Probevorlesung*, entitled "Über die Hypothesen, welche der Geometrie zu Grunde liegen," which was published that same year and contained all the basic ideas of Riemannian geometry, was studied in detail in the first chapter of the present volume.

In the years 1855–1856 Riemann gave a course on the theory of Abelian functions, and a year later sent to the publisher a long paper that has become a classic, bearing the title "Theorie der Abel'schen Functionen" (*J. für Math.*, **54** (1857)). This paper contained a brilliant development of the ideas and methods of his 1851 dissertation in application to the theory of algebraic functions and their integrals and to the inversion problem posed by Jacobi and his students, which at the time was considered with good reason to be one of the most complicated problems of analysis. In the meantime he completed and in 1857 published his paper

"Beiträge zur Theorie der durch die Gauss'sche Reihe $F(\alpha, \beta, \gamma, x)$ darstellbaren Functionen" (*Abh. Ges. Wiss. Göttingen*, **7** (1857)), which served as the point of departure for the analytic theory of differential equations. Finally, to mention only Riemann's ground-breaking work, in 1859 he published the paper "Über die Anzahl der Primzahlen unter einer gegebenen Grösse," which was discussed in Book 1, pp. 189–192.

Only in November 1857 did this scientific genius receive the title of extraordinarius and, after the death of Dirichlet in 1859, the title of professor ordinarius. Among the courses he gave at the University of Göttingen, his lectures on the differential equations of mathematical physics were of particular significance. They were written up by his student Hattendorf and first published in 1869 (*Partielle Differentialgleichungen und deren Anwendung auf Physikalische Fragen. Vorlesungen von Bernhard Riemann. Für den Druck bearbeitet und herausgegeben von K. Hattendorf*, Braunschweig 1869). In 1862, shortly after his marriage, Riemann caught cold and fell ill. The exacerbation of the illness, which turned into tuberculosis, forced him to interrupt his intense work for a long time and take journeys to Italy for his health. But the disease progressed, and Riemann died 20 July 1866 in Selasca, near Lago Maggiore.

Riemann's Doctoral Dissertation. The Dirichlet Principle

In his papers Riemann is very sparing in his citations of the work of others. He, like other mathematicians of genius, seems to have preferred to think a question through by himself rather than searching on the shelves of libraries. An apparent exception is the introductory part of his *Habilitationsschrift*, devoted to the history of the representation of functions by trigonometric series. Here the literature of the question was presented thoroughly. But this exception may be explained by one of his letters to his father (autumn 1852), in which he reports a two-hour conversation with Dirichlet and rejoices that Dirichlet had supplied him with such a copious supply of the necessary notes that his work was significantly facilitated. Without those notes he would have had to undertake lengthy library searches.[77]

The absence of citations of a paper in Riemann's work does not mean that he was unfamiliar with that paper. He almost never cites Cauchy, but one can confidently assert that he knew Cauchy's papers on functions of a complex variable from the 1840's. Thus, for example, in the paper "Theorie der Abel'schen Functionen" (1857) mentioned above Riemann spoke of a function w of a complex variable for which it follows "from a well-known theorem, that w can be represented by a series of integer powers of $z - a$ in the form $\sum_{0}^{\infty} a_n(z - a)^n$, provided that in a neighborhood of the point a it always has one definite value that varies continuously with z, and that this representation is valid out to the distance from a or modulus of $z - a$ at which a discontinuity is encountered."[78] It is obvious that

---

77) Riemann, B. *Werke*. S. 578.

78) *Ibid.*, S. 120.

this is Cauchy's theorem even down to the terminology (including the term *modulus* that Cauchy had taken over from Argand) in a form in which Riemann could have read it either in the first volume of the *Exercices* (1840, pp. 29–31) or the second volume (1841, pp. 54–55). One may suppose that these books of Cauchy, who by this time was famous throughout Europe and a member of the Göttingen Scientific Society, were available in Göttingen and, of course, in Berlin, where Riemann lived in 1847–1849. The publisher of the *Exercices*, Bachelier issued them in installments on subscription not only in France but also abroad. In particular the subscriptions in Berlin were taken by the bookseller B. Behr. We remark that in the paper just quoted Riemann used along with the term single-valued (*einwertig*) also the term *monodromic* due to Cauchy. As we have mentioned, this last term occurs in No. 46 of the fourth volume of the *Exercices*, which was published after 1850 (but not later than 1853).

How much information on the work of Cauchy and Puiseux could Riemann have had at his disposal before he wrote his dissertation (basically from the autumn of 1850 to the autumn of 1851)? On this score one can only conjecture. We consider it possible that by that time Riemann was already familiar with three volumes of Cauchy's *Exercices*.[79]

Incidentally, is it possible that Cauchy's paper "Sur les fonctions d'une variable imaginaire" in No. 36 of the *Exercices* could have been, as early as 1847, one of the stimuli to the discussion between Riemann and Eisenstein on the fundamental principles of the use of complex quantities in the theory of functions, which Riemann later related to Dedekind? As reported by Dedekind, the conversation went as follows:

> Eisenstein maintained the position of formal computations, while he himself [i.e., Riemann] perceived the important thing in the definition of a function of a complex variable to be in partial differential equations.[80] The equations in question are, of course, the d'Alembert-Euler equations.

As for the "position of formal computations," it was represented not only in the papers of Eisenstein at that time on elliptic functions, for which the basis was infinite products and double series with complex terms, but in even more explicit and elementary form in the paper of Cauchy just mentioned. Here is how the latter began a paragraph on functions of complex variables:

---

79) We recall that the title pages of these volumes bear the dates 1840, 1841, and 1844, and only the fourth volume is dated 1847. However, as already mentioned, the complete publication of the *Exercices*, was done in separate *cahiers*, with consecutive numeration, 12 in each volume; the publication was delayed more and more as time passed. Thus in the last *cahier* of the fourth volume Cauchy referred to a paper of his from 1853. It follows that the subscribers to the fourth volume of the *Exercices* had to wait only 7 years for the last installment!

The last *cahier* (No. 36) of the third volume, which will be discussed below, was obviously published before 1847, when No. 37, which began the fourth volume, appeared.

80) Dedekind, R. "Bernhard Riemann's Lebenslauf," in: Riemann, B. *Werke.* S. 576.

Complex variables, just like real variables, can be subjected to various operations, whose results are functions of these variables. These functions are completely defined when the operations themselves are defined and when the meaning of the notation used in the computations is completely fixed.[81]

It is typical that in his dissertation Riemann seemingly continued his dialogue, no longer with Eisenstein, but with Cauchy (not naming either one, of course). Along with his new approach to functions of a complex variable, rigorously distinguishing the class of analytic functions (see below), he also kept constantly in mind the traditional approach, in which the functions are defined by operations on quantities (Grössenoperationen). In doing this he distinguished two interpretations of the latter: a narrower one and a broader one. In the first interpretation we are dealing with elementary operations: addition, subtraction, multiplication, division, differentiation, and integration. Using a finite number of these operations, one could define all the functions used in analysis up to that time, Riemann claimed. The second interpretation he connected only with addition, subtraction, multiplication, and division, but now allowed these operations to be performed not only a finite number of times, but also *an infinite number of times*, i.e., *he implicitly admitted passage to a limit*. In doing this Riemann in no way denied the naturalness or legitimacy of the approach to functions based on operations on quantities, and admitted the possibility of a proof that "what is set down here [i.e., in his dissertation] as the basis of the concept of a function of a complex variable coincides completely with the dependence expressed by means of operations on quantities"[82] [in the second interpretation]. This hypothesis testifies to the stability of Riemann's interest in the problem of the analytic representation of a function (probably from 1847 on). From here it was a direct route to his *Habilitationsschrift* "Über die Darstellbarkeit einer Function durch eine trigonometrische Reihe" (1854). But this belongs to the area of functions of a real variable. As for functions of a complex variable, Riemann's hypothesis that the two concepts of a function coincide turned out to be true under an additional restriction: not just any limiting passage can be considered, but only uniform limits in a connected two-dimensional domain. Weierstrass was the first to prove this in his paper "Zur Funktionenlehre" (*Berl. Monatsber.*, 1880). Weierstrass, however, considered it necessary to emphasize that he had refuted Riemann's conjecture, since the latter did not include explicitly the requirement of connectivity (or uniform convergence, see below).

We may believe that Riemann did not mention Cauchy either because he did not share his approach to functions of a complex variable, in which the basic principles were not distinguished with sufficient clarity, or because he considered this approach too narrow and unsuitable for treating the problems involved with the study of algebraic functions and their integrals that he was occupied with. However that may be, Riemann's position in turn freed Cauchy from the need to

---

81) Cauchy, A.-L. *Exercises*, 1844, T. 3, p. 366.

82) Riemann, B. *Werke*. S. 71.

cite his papers, especially in cases when it might seem to Cauchy that Riemann had invaded territory already occupied by him. This may explain why Cauchy used the conditions for existence of a derivative of a function Z of a complex variable z in approximately the same way as Riemann but without mentioning him, in his paper "Sur les différentielles des quantités algébriques ou géométriques et sur les dérivées de ces quantités," which appeared in the 47th *cahier* of the fourth volume of the *Exercices* (1852?, see above). Here once more an essential fact appears, which had remained in the shadows in Cauchy's previous work: the requirement of continuity alone for the functions of a complex variable being studied is not sufficient grounds for the regularities he had discovered earlier; it was necessary that the d'Alembert-Euler equations hold. One may believe that the paper of Cauchy just mentioned was, so to speak, induced by Riemann's dissertation. However, we have no factual confirmation of this hypothesis.

On the other hand, we also have no direct proof that Riemann knew the paper of Puiseux on algebraic functions while he was working on his dissertation. Nevertheless we consider it highly likely that Riemann had read the paper by that time. The paper was published in one of the most influential and widely distributed mathematical journals of the time, a journal necessary to anyone working in the area of analysis, namely Liouville's *Journal des mathématiques pures et appliquées*. This publication of the paper coincided with the first stage of Riemann's work on his dissertation, from October to December 1850. Riemann could not have overlooked Puiseux's paper, which contained the latest work in the theory of algebraic functions and their integrals, if only because by his own definition the area of research in which variables were assigned complex values reduced almost entirely to the cases in which one variable is an algebraic function of the other or the integral of an algebraic function (see below). We add, finally, that the very appearance of the concept of a multi-sheeted surface at the beginning of Riemann's paper (§ 5) with the definition of branch points of the surface and the description of the connections of the sheets among themselves, decomposing into disjoint cycles in a neighborhood of each such point, would appear completely unexpected and extremely artificial only for those who had not read Puiseux's paper. But the attentive reader of that paper would be able to see in that concept an excellent geometric commentary and at the same time what appears to be a profound summary of Puiseux's memoir.

As for the achievements of Abel and those of Jacobi and his students in analytic function theory and the theory of Abelian integrals, Riemann had undoubtedly mastered them while he was still planning his dissertation. Both these achievements and especially the questions in this area that remained unresolved probably convinced him that any significant progress would require the creation of a solid basis for the study of multi-valued functions.

The main significance of Riemann's dissertation, however, we attribute to § 20. This section follows the main part of the paper, which is devoted to the construction of Riemann surfaces and the proof that there exist functions on them that are subject to certain conditions typical of algebraic functions and their integrals and precedes the two concluding sections, in which the problem of conformal mapping of simply connected domains is stated and solved. At the beginning of this section Riemann mentions that the harmony and regularity hidden in the theory of regular connections between variables had up to that time manifested itself when complex values were assigned to these variables. To be sure, the cases when this happened still encompassed only a small portion of the dependences in which one of the variables is an algebraic function of the other or has a derivative that is an algebraic function. But each step that had been taken had not only given a simpler and more finished form to results obtained without complex quantities, but had also pointed the way to new discoveries, an example of which was, "the history of the research devoted to algebraic, circular or exponential, elliptic, and Abelian functions."[83]

What is new in this research? To this question Riemann gave the following answer:

> The methods of studying these functions that have existed up to now had as their basis the definition of a function by means of a formula that made it possible to compute its value for each given value of the argument; in our study we show that by the properties intrinsic to functions of a complex variable some of the data in the definition are consequences of the rest, and we establish how the number of data can be decreased and reduced to those that are strictly necessary.[84]

As an example of the application of the principles on which his paper was based, which made it possible to determine the behavior of a function independently of any representation of it by a formula, he gave the following complete characterization of the concept of an algebraic function: a function for which the domain of the variable z covers the entire infinite plane once or several times, having as its only discontinuities a finite number of points at which it becomes infinite to finite degree. Naturally Riemann assumed that at all the remaining points there exists a derivative with respect to the complex variable z. This characterization is indeed complete. The mathematician of the following decades could have remarked, however, that the condition of finite order for the poles allowable for an algebraic function was superfluous: it follows from the other assumptions. But for such a conclusion it would have been necessary to rely, for example, on the analysis of Weierstrass based in the final analysis on a "formula": the representation of a function in a neighborhood of an isolated singular point by the Laurent-Weierstrass series.

At the end of § 20 Riemann stated the problem of proving that two concepts of a function of a complex variable are the same: the new concept (his own), and the traditional concept, which was based on operations on quantities (the analytic representation). This, however, was discussed above.

The dissertation begins by establishing the difference between functions of a real and complex variable. Without subjecting the former to any restrictions other

---

83) *Ibid.*, S. 70.
84) *Ibid.*, S. 70.

than possibly continuity, Riemann proposed studying $w = u + iv$ as a function of $z = x + iy$ if the expression

$$\frac{du + dv\,i}{dx + dy\,i} = \left[\frac{1}{2}\left(\frac{\partial u}{\partial x} + \frac{\partial v}{\partial y}\right) + \frac{1}{2}\left(\frac{\partial v}{\partial x} - \frac{\partial u}{\partial y}\right)i\right] +$$

$$+ \left[\frac{1}{2}\left(\frac{\partial u}{\partial x} - \frac{\partial v}{\partial y}\right) + \frac{1}{2}\left(\frac{\partial v}{\partial x} + \frac{\partial u}{\partial y}\right)i\right]e^{-2\varphi i}.$$

(where we have set $dx + dy\,i = \varepsilon e^{\varphi i}$) is independent of φ. We remark that the expressions on the right-hand side enclosed in brackets are the so-called formal derivatives of the function $w(z)$, which were thus first introduced by Riemann in 1851. The condition just discussed has the following form:

$$\frac{\partial u}{\partial x} = \frac{\partial v}{\partial y}, \quad \frac{\partial v}{\partial x} = -\frac{\partial u}{\partial y}. \tag{19}$$

These are the famous partial differential equations customarily called the Cauchy-Riemann equations or (with equal justification) the d'Alembert-Euler equations (see HM, T. 3, pp. 365–367).

It was these equations that led to the definition laid down by Riemann as the basis of his paper: "A variable complex quantity w is a *function* of another variable complex quantity z if it varies in such a way that the value of the differential quotient dw/dz is independent of the values of the differential dz."[85] If we omit the now-familiar details, this is the definition of an analytic function of a complex variable (in a two-dimensional domain). In the abovementioned paper of Cauchy "Sur les différentielles de quantités algébriques ou géométriques et sur les dérivées de ces quantités," published in the 47th *cahier* of the *Exercices*, after Cauchy had time to become acquainted with Riemann's dissertation, we see the same equations (19), but in the complex form

$$D_y Z = i D_x Z,$$

where Z is a function of z and D_x and D_y denote the partial derivatives with respect to x and y. Cauchy's reasoning was similar to that of Riemann, although Cauchy did not write out the expressions for the formal derivatives.

Let us return to Riemann. Representing the complex variables z and w by points A and B respectively in the complex plane, he showed that the dependence of w on z could be represented as a mapping (eine Abbildung) of the A-plane into the B-plane under which a point maps to a point, a curve to a curve, and a region (ein zusammenhängendes Flächenstück) to a region. If w is a function of z in the sense defined above, this mapping transforms small pieces of the A-plane to pieces of the B-plane that are similar to them (i.e., it is conformal, though Riemann did not use that term). In this place he referred to Gauss, who had studied the

---

85) *Ibid.*, S. 37.

problem of conformal mapping of one surface onto another in 1822. We remind the reader that the problem of conformal mapping of regions of a sphere into a plane had been studied in general form by Euler in a paper published in 1778 (see HM, T. 3, p. 170).

In proposing that the concept of a mapping of one region onto another be associated with each function of a complex variable, Riemann was taking an essentially new step in the history of analytic function theory. It can be compared in significance with the introduction of the graphs of functions originally represented by formulas. In a graph the derivative is expressed as the slope of the tangent line, in conformal mapping it is represented by the complex coefficient of similarity, to which the mapping under consideration reduces on the infinitesimal level (the absolute value of the derivative defines the similarity ratio or scale of the mapping, the argument defines the angle through which the transformed figure is rotated).

As a consequence of Eqs. (19) Riemann wrote out the equations

$$\frac{\partial^2 u}{\partial x^2} + \frac{\partial^2 u}{\partial y^2} = 0, \quad \frac{\partial^2 v}{\partial x^2} + \frac{\partial^2 v}{\partial y^2} = 0,$$

i.e., in modern language he proved that the real and imaginary parts of an analytic function of a complex variable are harmonic functions. Again this explicit statement has first-rank importance: it opens the possibility of transferring into the theory of functions the methods applied in the problems of physics connected with equilibrium or steady-state motion, in which these functions naturally arose.

But before discussing these methods and the integral representations of functions on which they are based Riemann extended the concept of a domain by introducing the surfaces T later named after him — the Riemann surfaces which cover the entire plane or part of it many times (§ 5 of the dissertation).

There is not space here to reproduce the details of his exposition, which subsequently passed into textbooks and monographs on the theory of functions of a complex variable in one form or another. We shall say only that it follows from Riemann's proposed explanations that a neighborhood of a point of a Riemann surface has the structure of either a simple disk or an m-fold disk ($m \geq 2$). In the latter case a point of the surface T distinct from the center O of the disk and revolving around O in a definite direction while remaining on the surface must rotate through the angle $2m\pi$ before returning to its original location. More precisely, if a and z denote the complex numbers represented by the point O and the moving point respectively, then the mapping $\sqrt[m]{z - a} = t$ is a one-to-one mapping of this neighborhood onto a disk in the t-plane with center at the origin. Riemann called such a point O of the surface T a *winding point* (*Windungspunkt*). Later, in his "Theorie der Abel'schen Functionen," he called it a *branch point* of order $m - 1$. Different branch points may lie above a given point of the complex plane, each having its own order, and there may also be ordinary points of the surface T above the point (for which one may consider that $m = 1$, or that the order is 0). Riemann added, quite correctly, that giving the location and direction of traversal of the boundary of T with respect to which the interior and exterior of T are determined along with the locations and orders of the branch points does not always uniquely determine the surface itself.

Multi-sheeted surfaces make it possible to extend the original definition of a function to variables assuming a definite value for each point of such a surface, varying continuously as the position of the point varies. Here it is allowed that there exist isolated points or lines on the surface at which the function is either not defined or has a discontinuity.

What has just been said reveals clearly Riemann's plan, which led him to introduce into mathematics a new kind of surface composed of planes and parts of planes spread out over one another and joined together in a particular way. This plan was, so to speak, to endow each function of a complex variable in the sense in which he understood the term (i.e., analytic function), a function that is in general multi-valued (recall the algebraic functions and Abelian integrals) with its own domain, in which it would be a single-valued function of a point. Thanks to Riemann each analytic function was fated henceforth to wear its Riemann surface as a snail wears its shell.

We repeat that the Riemann surface probably did not produce the impression of anything invented and unnatural to mathematicians capable of grasping the structure of algebraic functions and their integrals on the level, for example, of Puiseux's paper. The reasoning of the latter, which had been entangled in a complicated web of intersecting paths encircling the critical points many times, became lucid, and the knots on the paths untangled themselves when the whole structure was transferred to a suitable Riemann surface.

We recall that if we are correct in our assumption that in the argument between Riemann and Eisenstein in 1847 both parties must have known the paper of Cauchy "Sur les fonctions d'une variable imaginaire," which had recently been published in the third volume of the *Exercices*, then one could have read there both the idea of isolating a single-valued branch of a multi-valued function, which can be done in many ways, and the fact that doing so leads to the appearance of lines of discontinuity on the plane. Then, in 1847, Riemann's thought could have been attracted to these elements of the definition of multi-valued functions, which had not yet been assimilated into mathematics.

However, what has just been said does not mean that all the researchers in this area easily and gladly left the plane and moved to the Riemann surface, which of course represented a higher degree of abstraction. Thus nearly a quarter-century later, in 1875, Briot and Bouquet, in publishing their revised *Théorie des fonctions elliptiques*, in which multi-valued functions were introduced in the early pages under the name of polytropic functions, wrote in the preface,

> In Cauchy's theory the path of a complex variable is represented by the motion of a point in a plane. To represent functions assuming many values for the same value of the variables, Riemann regarded the plane as formed of many sheets lying on it and joined by welds (les soudures) in such a way that the variable can pass from one sheet to another by crossing one of the connecting lines. The concept of surfaces with multiple sheets presents certain

difficulties; despite the beautiful results Riemann achieved using this method, it does not seem to us that it has any advantages for the object we have in view. Cauchy's idea gives a very good approach to multi-valued functions: it suffices to adjoin to a value of the variable the corresponding value of the function, and when the variable describes a closed curve and the value of the function changes, indicate this by changing the subscript.[86]

In this passage from true believers not only in the spirit but also the letter of Cauchy one senses critical notes, most of all the conviction that the methods proposed by Cauchy (and developed by Puiseux) are equivalent to Riemann's methods. Of course this is unfair to Riemann, but such was the judgment of his contemporaries!

The content of §6 of Riemann's dissertation is entirely topological. Here the concept of order of connectivity of a surface is introduced by making cuts in the surface, i.e., curves that join one boundary point of the surface to another without self-intersections. Each such cut should be thought of as having two edges. When this is done, if several cuts have already been made in a surface, then the points of these cuts are regarded as boundary points when a new cut is made. A surface is called *simply connected* if any cut separates it into two disjoint parts, i.e., makes it disconnected. A surface that is not simply connected is called *multiply connected*.

The number of cuts needed to turn a multiply connected surface into a simply connected surface, increased by 1, is called the *order of connectivity* of the surface. Since each cut increases the number of boundary curves of the surface by 1, the number of boundary curves can differ from the order of connectivity only by an even number.

Sections 7 and 8 are devoted to the proof of the Green-Gauss formula in the form

$$\iint \left(\frac{\partial X}{\partial x} + \frac{\partial Y}{\partial y}\right) dT = -\int (X \cos \xi + Y \sin \eta)\, ds,$$

where the double integral extends over the whole surface T, and the line integral over its boundary, ξ and η being the angles between the inward normal p and the coordinate axes. In §9 the consequences of this formula are explored under the assumption that $\partial X/\partial x + \partial Y/\partial y = 0$ everywhere on the surface T except at isolated curves or points at which the derivatives may have discontinuities.

The analysis contained in these sections really provided nothing new to mathematicians of the time, except that the order of connectivity of the surface T was taken into account. Because of this fact Riemann obtained the full picture of the behavior of a line integral on T that is independent of the path. In particular he noted that if T is made into a simply connected surface T^* by means of cuts, then the integral

$$\int_{O_0}^{O} \left(Y \frac{\partial x}{\partial s} - X \frac{\partial y}{\partial s}\right) ds,$$

86) Briot, C. and Bouquet, J. *Théorie des fonctions elliptiques.* Deuxième éd., Paris 1875, pp. 1–11.

where O_0 is fixed and O is a moving point of T^*, is a single-valued function of O (still under the condition that $\partial X/\partial x + \partial Y/\partial y = 0$). On each edge of the cut the integral maintains a constant value so that a crossing of a cut is accompanied by a jump in this constant value.

In this way a basis formally independent of Cauchy's theorem was gradually built up for complex integration over a surface. In §10 Riemann carried out a study of a harmonic function u on this basis under the assumption that the surface T "covers the plane simply" everywhere (i.e., is one-sheeted), that the points of discontinuity of u, $\partial u/\partial x$, and $\partial u/\partial y$ are isolated, and that in a neighborhood of each of them $\rho\partial u/\partial x$ and $\rho\partial u/\partial y$ are infinitesimal, where ρ is the distance from (x,y) to the point of discontinuity (singular point), and finally that the points of discontinuity for u are not removable (i.e., u cannot be made continuous at these points by redefining its value). Under these hypotheses, using Green's formula, which expresses the value of a function at a point in terms of the values of the function and its normal derivative on the boundary (Riemann derived this formula for the case of a plane), he proved that u and its partial derivatives of all orders are finite and continuous at all points of T. Thus we encounter here the earliest theorem on the a priori "elimination" of possible singularities of a (harmonic) function.

Section 11 contains corollaries of this Green's formula. In particular, it was established there that the values of u and $\partial u/\partial p$ on a certain arc of a curve (belonging to either the boundary or the interior of T) determine u completely on the entire surface T, and also that (a non-constant) u cannot have either a maximum or minimum at any point of T.

In §§12–15 the preceding results were applied to a function $w = u + iv$ of the variable $z = x + iy$ which (except perhaps for isolated lines and points) is continuous at each point O of the surface T and satisfies the equations $\partial u/\partial x = \partial v/\partial y$, $\partial u/\partial y = -\partial v/\partial x$. Moreover it was assumed that this function may have discontinuities that can be removed by redefining the function at a suitable point. Riemann called such a function simply a *function of $z = x + iy$.*

He began by stating the following theorem, which he deduced from the theorem of § 10 on harmonic functions.

If a function w of z has no discontinuities along lines and in addition for any point O' of the surface at which $z = z'$ the product $w(z - z')$ is infinitely small as the point O approaches infinitely near to the point O', then the function w is necessarily finite and continuous together with all its derivatives at all points of the surface.[87]

This time we have a quite general theorem on the "removal" of the singularities of an analytic function. Riemann's proof cannot be considered sufficient; however, the same can be said of the theorem of § 10 and a number of major points of this dissertation, as well as his other papers on analysis. In making this

---

87) Riemann, B. *Werke*, S. 55.

judgment, however, one must keep in mind that we are talking about the mid-nineteenth century, with the possibilities, means and requirements with respect to rigor and completeness of proof characteristic of that time. If we consider this theorem from the modern point of view, we must say that it was only in 1932 (the Lehman-Men'shov theorem) that a completely rigorous proof was achieved that a function $w(z)$ is analytic if it is continuous in a domain and except at a finite or countably infinite set of points there exist finite derivatives $\partial u/\partial x$, $\partial u/\partial y$, $\partial v/\partial x$ and $\partial v/\partial y$ satisfying the equations $\partial u/\partial x = \partial v/\partial y$, $\partial u/\partial y = -\partial v/\partial x$. In this proof essential use was made of the methods of the theory of functions of a real variable (in particular the theory of measure and the Lebesgue integral). As for Riemann's assertion about removing the possible singularities located on an arc of a curve on which the function is continuous and in a neighborhood of which it is analytic, it is false without additional restrictions on the arc. It is true, however, as P. Painlevé was the first to show in his paper "Sur les lignes singulières des fonctions analytiques" (*Ann. fac. sci. Toulouse*, **2** (1888)), under a condition that Riemann may have considered obvious: that the arc be rectifiable. In any case, once the necessary clarifications have been introduced into Riemann's theorem, its full value is retained for all the rest of the discussion.

In § 13 Riemann considered the case when for some point O' of the surface one knows only that for a certain natural number n the quantity $w(z - z')^n$ is infinitesimal along with $z - z'$. Then, applying the preceding theorem to $w(z - z')^{n-1}$, we find that the latter function is finite and continuous at the point O'. If its value at O' is a_{n-1}, then $w(z - z')^{n-1} - a_{n-1}$ is infinitesimal at that point and consequently $w(z - z')^{n-2} - a_{n-1}/(z - z')$ is finite and continuous at O'. Repeating this reasoning a sufficient number of times, we arrive at a rational function of the form

$$\frac{a_1}{z - z'} + \frac{a_2}{(z - z')^2} + \cdots + \frac{a_{n-1}}{(z - z')^{n-1}}.$$

Subtracting this function from w, we obtain a function that is finite and continuous (analytic) at the point O'. We see that this reasoning actually enabled Riemann to determine the *principal part of the Laurent expansion of the function* in a neighborhood of a pole.

Remarking that what had been said in §§ 12 and 13 could be applied to any (multi-sheeted) surface, in § 14 Riemann investigated how the corresponding results are changed if the point O' is a branch point of order $n - 1$. The mapping of a neighborhood of O' onto a neighborhood of the origin in the ζ-plane given by $\zeta = (z - z')^{1/n}$ naturally leads to the conclusion that the preceding results remain valid if $z - z'$ is replaced by $(z - z')^{1/n}$. As a result, instead of the rational function obtained above, one obtains an expression of the form

$$\frac{a_1}{(z - z')^{1/n}} + \cdots + \frac{a_m}{(z - z')^{m/n}}.$$

In particular, one can also assert that if w is analytic on the surface and satisfies the condition that $w(z - z')^{1/n}$ is infinitesimal in a neighborhood of a branch point O' of order $n - 1$, then w is continuous at the point O'.

In § 15 Riemann proved that a function w analytic on some surface T maps it in a one-to-one and continuous manner onto some surface S. From the point of view of later requirements Riemann should have proved first that a neighborhood of each point of the surface again maps to a neighborhood. However, he settled for just a proof that the function w cannot be constant along a curve unless it is constant on the entire surface.

The contents of §§ 16–18 played a very significant role in the history of mathematics. In these sections Riemann discussed the so-called *Dirichlet principle*, to which he gave a form suitable for applications in problems of the theory of functions. Of its origins and value Riemann spoke later, in his paper "Theorie der Abel'schen Functionen," where this principle is the main tool of investigation. Here is what he said in that paper.

As the basis for the study of a transcendent, it is first of all necessary to establish a system of conditions that are independent among themselves and sufficient to determine it. This end can be achieved in many cases, in particular in the study of integrals of algebraic functions and their inverses, using a principle that Dirichlet has applied for several years in his lectures on forces acting in inverse proportion to the square of the distance in order to solve this problem for functions of three variables satisfying Laplace's equation; he may have been led to it by a similar idea of Gauss. However for such an application to the theory of transcendents the important case is precisely one to which this principle cannot be applied in its simplest form and which has remained unexamined as a result of its completely subordinate importance. This is the case when the function must have prescribed discontinuities at known points of the domain in which it is to be defined. By this I mean that at each such point the function is required to be discontinuous in exactly the same way as a given discontinuous function or must differ from that function by a continuous function.[88]

The "simplest form" of the principle with which Riemann could have become familiar at Dirichlet's lectures says that among all the functions $u(x, y, z)$ assuming given continuous values on the boundary of a three-dimensional domain G, the minimum of the integral

$$\iiint \left[\left(\frac{\partial u}{\partial x} \right)^2 + \left(\frac{\partial u}{\partial y} \right)^2 + \left(\frac{\partial u}{\partial z} \right)^2 \right] dx\, dy\, dz$$

occurs for a function u that is harmonic in G. Thus the solution of the Dirichlet problem, the problem of finding a harmonic function given its boundary values, reduces to the corresponding problem of finding the minimum of the integral.

Here one can only conjecture the meaning of Riemann's reference to Gauss. The principle that forms the basis for the method of least squares (1821) presents an analogy with the Dirichlet principle, as does the principle used by Gauss in

---

88) *Ibid.*, S. 129.

finding a distribution of masses on a given surface for which the corresponding single-layer potential $V = \int \dfrac{m\,ds}{r}$ coincides with a given continuous function U (in particular a constant). Gauss regarded the problem of the existence of the required distribution as solved on the grounds that for such a distribution the integral $\int (V - 2U)\,ds$, which is bounded below, must have minimal value ("Allgemeine Lehrsätze in Beziehung auf die im verkehrten Verhältnisse des Quadrats der Entfernung wirkenden Anziehungs- und Abstossungskräfte. *Resultate aus den Beobachtungen des magnetischen Vereins*, Bd. 4, 1839).

Relying, like Gauss, on the conviction that for a functional (integral) that is bounded below there must exist a function that gives it a minimal value, Riemann expressed the Dirichlet principle in the specialized form that he had in mind above (more precisely, a consequence of it), as the following theorem of § 18:

If a complex-valued function [generally not analytic] $\alpha + i\beta$ of x and y is defined on a connected surface T made into a simply connected surface T^* by cuts, and the integral

$$\int \left[\left(\frac{\partial \alpha}{\partial x} - \frac{\partial \beta}{\partial y} \right)^2 + \left(\frac{\partial \alpha}{\partial y} + \frac{\partial \beta}{\partial x} \right)^2 \right] dT$$

extended over the entire surface T has a finite value, it can always be transformed in a unique manner into a function of z [analytic] by subtracting from it a function $\mu + \nu i$ of x and y satisfying the following conditions:

1) $\mu = 0$ on the boundary except at isolated points, and ν can assume an arbitrary value at one point;

2) μ can have discontinuities on the surface T and ν can have discontinuities on the surface T^* only at isolated points such that the integrals

$$\int \left[\left(\frac{\partial \mu}{\partial x} \right)^2 + \left(\frac{\partial \mu}{\partial y} \right)^2 \right] dT \text{ and } \int \left[\left(\frac{\partial \nu}{\partial x} \right)^2 + \left(\frac{\partial \nu}{\partial y} \right)^2 \right] dT$$

extended over the entire surface are finite; when crossing a cut the function ν can change only by a constant term [the same constant for all points of the cut].[89]

Here it is the function μ that must give the minimal value of the integral

$$\Omega(\alpha - \lambda) = \int \left\{ \left[\frac{\partial(\alpha - \lambda)}{\partial x} - \frac{\partial \beta}{\partial y} \right]^2 + \left[\frac{\partial(\alpha - \lambda)}{\partial y} + \frac{\partial \beta}{\partial x} \right]^2 \right\} dT.$$

This belief that such a function exists amounts to the Dirichlet principle in Riemann's interpretation. As for the function ν it is constructed from μ as the following line integral

$$\nu = \int \left[\frac{\partial \beta}{\partial x} + \frac{\partial(\alpha - \mu)}{\partial y} \right] dx + \left[\frac{\partial \beta}{\partial y} - \frac{\partial(\alpha - \mu)}{\partial x} \right] dy + C,$$

---

89) *Ibid.*, S. 66–67.

which extends from a fixed point to a variable point (x, y) over paths lying in T^*.

The criticism to which Weierstrass subjected this principle in 1870 ("Über das sogenannte Dirichlet'sche Prinzip," presented to the Berlin Academy on 14 July 1870),[90] pointing out that a functional bounded below (an integral in the simplest variational problem) does not always attain its minimal value in the class of admissible functions, cast doubt on the legitimacy of using this principle in analysis and the theory of functions. Only after the papers of Hilbert (1901, 1909) was it possible to give a rigorous justification of the correctness of the route chosen by Riemann. The important thing, however, is that in Riemann's research and that of his students the Dirichlet principle played its historical role of a powerful heuristic method.

As an example of the application of the theorem of § 18 of his paper to the theory of functions Riemann stated and proved in § 21 his famous theorem on the existence of a conformal mapping of one simply connected domain onto another.

Two given simply connected plane surfaces can always be placed in correspondence in such a way that to each point of one there corresponds a point of the other that varies continuously with it, and corresponding parts of the surfaces are similar on the infinitesimal level; this can be done in such a way that the points correspond-ing to one interior point and one boundary point can be chosen arbitrarily; the relationship will then be determined for all points.[91]

Riemann remarked that there is no loss in generality in the theorem if one of the surfaces is assumed to be a disk with center at $w = 0$ and radius 1. On the surface T he made a cut l from a point z_0 to any boundary point and constructed a function $\alpha + i\beta$ (generally non-analytic) which coincides with $\ln(z - z_0)$ in some neighborhood of the point z_0 and then is extended to the boundary in such a way that α vanishes at the boundary points. When this is done, l is a line of discontinuity for β: in passing from the negative side of the cut l to the positive side, β undergoes a jump of -2π. Applying the theorem of § 18 (the Dirichlet principle) to $\alpha + i\beta$, Riemann obtained a new function

$$t = (\alpha - \mu) + i(\beta - \nu) = m + in,$$

that is analytic in T except on the cut l, which is a line of discontinuity for n; moreover at the initial point $z = z_0$ of the cut l the function m has a logarithmic pole (i.e., it behaves like $\ln |z - z_0|$). Relying on the properties of harmonic func-tions he had established earlier, Riemann easily verified that the level lines of the function m are closed curves enclosing the point z_0, and that in traversing such a curve in the positive direction the function n increases, having jumps equal to -2π when crossing the cut l. It remained only to set $w = e^t$ in order to obtain finally a function mapping T onto the unit disk in such a way that z_0 maps to the center of the disk.

---

90) Weierstrass, K. *Mathematische Werke.* Bd. 2.

91) Riemann, B. *Werke.* S. 72.

Riemann's proof, which is conceptually simple, served as the point of departure for seeking other proofs that imposed additional simplifying hypotheses on the boundary of T but were free of the use of the dubious Dirichlet principle.

Conformal Mappings

We have seen that the Riemann mapping theorem was derived by Riemann from his "Dirichlet principle" in his 1851 dissertation. However, the absence of a convincing justification of this principle forced mathematicians to seek another proof of the theorem, and this was being done even before Weierstrass criticized the Dirichlet principle openly in 1870. An important role in this work was played by the papers of H. A. Schwarz. Karl Hermann Amandus Schwarz (1843–1921) studied at the Gewerbeinstitut and the University of Berlin in the years 1860–1866. He was a student of Weierstrass. After 1867 he taught mathematics at the University of Halle and Technische Hochschule in Zürich. After 1875 he was professor of Mathematics at the University of Göttingen and after 1892 at the University of Berlin. In his short paper "Zur Theorie der Abbildung. Programm der Eidgenössischen polytechnischen Schule in Zürich für das Schuljahr 1869–1870" Schwarz studied the problem of conformal mapping of an arbitrary bounded convex region of the plane onto the unit disk. Of fundamental importance in this paper was the fact that for the first time the study of the conformal mapping of regions (regarded as open sets) was separated from the behavior of the mappings on the boundary (the question of the correspondence of boundaries under conformal mappings). The solution of the first problem was reduced to the study of a sequence of mappings of simpler regions approximating the given region from within and a subsequent passage to the limit. Following this route, he invented a method — the Schwarz alternating method — in a paper "Grenzübergang durch alternierende Verfahren," (*Viertel-Jahr. Naturforsch. Ges. Zürich*, **15** (1870)), making it possible to solve the Dirichlet problem (the problem of constructing a harmonic function from its boundary values) under very broad hypotheses. Thus the Dirichlet principle was practically rehabilitated, to be sure in a rather narrow sense (as, for example, Poincaré later interpreted it), but with wide applications.

At that point Schwarz was able to prove specifically the possibility of a conformal mapping of any convex region (except the entire plane) onto the disk, as well as any simply connected region whose boundary consisted of a finite number of analytic arcs such that two adjacent arcs meet to form nonzero angles. In this important special case he also established that the mapping could also be continued to the boundary while remaining analytic at the interior points of the bounding arcs (i.e., the points different from the vertices of the generalized polygon that the region represented). In Schwarz' papers from 1869 and 1870 we also encounter, as an auxiliary device, the famous Schwarz lemma (which sharpens the maximum modulus principle) and the Schwarz reflection principle (the Riemann-Schwarz principle).

The next step of fundamental importance for the general theory was taken by Poincaré, who proved in his paper "Sur les groupes des équations linéaires" (*Acta*

H. A. Schwarz

Math., **4** (1884)) a uniqueness theorem that can be stated as follows: if the point of a region that is to map to the center of the disk is fixed and the direction that is to transform to the positive real axis is given, then there can exist at most one function that realizes such a mapping. To be sure, the uniqueness condition can be discerned in a different form in Riemann's dissertation (1851): Riemann had proposed choosing arbitrarily the image of one interior point and one boundary point of the region being mapped. But in his dissertation he had proved (on the basis of the Dirichlet principle) only that it was possible to satisfy these conditions; no grounds were given for asserting that there was no other mapping satisfying the same conditions.

Returning to Poincaré, we note that he is also entitled to the credit for developing an important method of solving the Dirichlet problem different from Schwarz' alternating method — the so-called method of *balayage*, which was developed in his paper "Sur les équations aux dérivées partielles de la physique mathématique" (*Amer. J. Math.*, **12** (1890)).

A proof of the existence of a conformal mapping of any simply connected domain onto the disk independent of the structure of its boundary (provided the boundary contain at least two distinct points) was first achieved in 1900 by the mathematician W. F. Osgood (1864–1943), a professor at Harvard. In the following year Hilbert gave a rigorous proof of the Dirichlet principle under minimal hypotheses (all restrictions were removed by him in 1909). Thus the Dirichlet principle was completely rehabilitated.

It followed immediately from Riemann's theorem that all simply connected regions of the plane (having more than one boundary point) are conformally equivalent, i.e., essentially indistinguishable from the point of view of conformal mapping. The situation was different for multiply connected regions. The first systematic study of the relevant questions was undertaken by F. H. Schottky (1851–1935), a student of Weierstrass and a professor in Zürich, then later at Mannheim and the University of Berlin. Schottky wrote his doctoral dissertation, "Über konforme Abbildung von mehrfach zusammenhängenden Fläche" (*J. für Math.*, **83** (1877)), and a number of other papers on this topic. As it turned out, even in the simplest case of doubly connected domains each bounded by two circles a conformal mapping of one onto the other is possible only under an additional restriction. For example, in the case of two concentric annuli, a suitable restriction is that they be similar, in other words, that the ratio of the larger radius of each annulus to the smaller be the same for both annuli. In the case of a domain bounded by p ($p \geq 2$) closed curves (analytic, though this is not essential) Schottky discovered that there exist $3p-3$ real constants that characterize the conformal class to which this region belongs. They came to be known as the *moduli* of the class. Having reduced the problem of conformal mapping to a linear differential equation with algebraic coefficients (a similar idea can be found in one of Riemann's fragments), he conjectured that a conformal mapping of one p-connected domain onto another is possible if and only if their respective moduli are equal. The first to succeed in showing that such is the case was P. Koebe (1882–1945), a student of Schwarz and professor at Leipzig University; his paper appeared in 1907. As a canonical domain on which the mapping is carried out Koebe chose a domain bounded by p pairwise disjoint circles. Hilbert obtained an equivalent result about the same time, but in his paper the role of the canonical domain was played by a half-plane from which $p - 1$ line segments parallel to the boundary line (some of which could degenerate into a point) were removed. The case of a mapping of infinitely connected regions onto one another caused particular difficulties.

We shall not go into the details of this topic, which takes us outside the era we are studying.

All the great achievements characterized above, whose subsequent development, as was shown in a number of examples, leads into the twentieth century, were of the nature of existence theorems. Meanwhile needs of a theoretical or applied character arose (dictated by the problems of physics and mechanics) to define specific functions that produce various conformal mappings.

In the last issues of Cauchy's *Exercices*, which, as we have noted, appeared in the early 1850's, one finds allusions to the mappings produced by single-valued branches of multi-valued elementary functions such as $\ln z$, $\arctan z$, and the like. But here is a curious fact, which shows that even in the early 1870's the elementary properties of fractional-linear transformations were not regarded as generally known even by the leading mathematicians: Weierstrass, in a letter to Sof'ya

Kovalevskaya of 9 June 1873,[92] sends her the following proposition: *It is always possible to map a given half-plane $\mathfrak{C}$ onto an arbitrary disk $\mathfrak{K}$ in such a way that the interior of a given disk K in $\mathfrak{C}$ corresponds to a disk $\mathfrak{K}_1$ concentric with $\mathfrak{K}$.* Weierstrass considered it necessary to add a number of clarifications to this, among them the fact that the center of $\mathfrak{K}_1$ does not correspond to the center of K. It is obvious that this question involves a simple corollary of the property that the symmetry of points with respect to a circle or a line is preserved under fractional-linear transformations. And this occurs after such questions as the reduction of Abelian integrals to elliptic integrals, problems of calculus of variations, and properties of theta functions had been discussed in previous letters. This example testifies in any case to the fact that in the early 1870's mappings by means of elementary functions were not yet part of the basic equipment of the theory of functions.

Nevertheless by that time certain individual results that were by no means trivial had been proved, which gradually turned conformal mapping into one of the important sources of new transcendental functions in analysis. And here, as in many other cases, Schwarz was in the vanguard. Thus, while still a student in the 1864 seminar conducted by Weierstrass, he obtained a formula in the form of a certain integral for mapping a half-plane onto any triangle.[93] This formula was soon independently generalized by E. B. Christoffel, a professor at the Technische Hochschule in Zürich, in his paper "Problema delle temperature stazionare e la representazione di una data superficie" (*Ann. mat.*, Ser. 2, **1** (1867), pp. 79–103), and is now called the Schwarz-Christoffel formula.[94] To be specific, if an n-gon in the z plane has angles $\alpha_1 \pi, \ldots, \alpha_n \pi$ and its vertices map respectively to the points $a_1, \ldots, a_n$ of the real axis under a conformal mapping onto the upper w-half-plane, then the mapping function is

$$z = z(w) = C_1 \int_{w_0}^{w} (t - a_1)^{\alpha_1 - 1} \cdots (t - a_n)^{\alpha_n - 1} \, dt + C_2,$$

where C_1 and C_2 are complex constants. This is the Schwarz-Christoffel formula. The main difficulty that arises in applying it when $n > 3$ is that of determining the real constants $a_1, a_2, \ldots, a_n$. Only three of these can be given arbitrarily.

We remark that the integral that occurs in the Schwarz-Christoffel formula (also called by their names) represents a significant generalization of the hyperelliptic and elliptic integrals, because the real numbers α_j, $(j = 1, \ldots, n)$ are not always rational. An elliptic integral is obtained in the case of a rectangle (then $\alpha_1 = \alpha_2 = \alpha_3 = \alpha_4 = \frac{1}{2}$).

The problem of conformal mapping of a polygon bounded by arcs of circles onto a disk or a half-plane turned out to be more complicated. In papers from 1869 and 1873 Schwarz arrived at the following theorem:[95]

---

92) Kochina, P. Ya. *Briefe von Weierstrass an Sophie Kowalewskaja, 1871–1891*. Moscow 1973, S. 24–25.

93) Schwarz, H. A. *Gesammelte Abhandlungen*. Berlin 1890, Bd. 2, S. 65–83.

94) For more details on E. B. Christoffel (1829–1900) see Chapter 1.

95) Schwarz, H. A. *Gesammelte mathematische Abhandlungen*. Bd. 2, S. 65–83, 84–101, 211–259.

Let T be a simply connected region of the z-plane bounded by arcs of circles (a circular polygon). Then the function $z(Z')$ that maps it onto the upper Z'-half-plane must satisfy the following third-order differential equation, whose left-hand side contains the so-called Schwarzian:

$$\{z, Z'\} = \frac{d^2}{dZ'^2} \ln \frac{dz}{dZ'} - \frac{1}{2}\left(\frac{d}{dZ'} \ln \frac{dz}{dZ'}\right)^2 = F(Z'),$$

and the right-hand side contains a rational function $F(z)$ defined up to several constants that are to be determined.

The function $z(Z')$ can be represented as the quotient of two linearly independent solutions of a linear second-order Fuchsian differential equation. In the case of a circular triangle this is the hypergeometric equation (Schwarz, Klein). However, to define and study the properties of function now called the Schwarz modular function Schwarz (1873) began with a triangle having zero angles formed by three equal arcs of circles orthogonal to the unit circle and defined the function $Z = \lambda(z)$ as a conformal mapping of this triangle onto the half-plane $\operatorname{Im} Z > 0$ such that the vertices of the triangle map to the points $0, 1, \infty$. The modular function is then analytically continued to the whole unit disk by the reflection principle (now known by Schwarz' name). Here the geometric method of defining (and studying the properties of) a transcendental function stood out sharply in connection with the theory of elliptic functions. However the transcendental function had been known since the time of Gauss and had been introduced and studied in a purely analytic way earlier. We remark that this function, as has already been mentioned, was one of the simplest examples of automorphic functions, the beginning of whose theory we shall discuss below. We have mentioned that Gauss' results remained unknown. The situation with respect to Riemann's results was somewhat more propitious than in the case of Gauss in regard to the study of the same function. They were included in his lectures on the hypergeometric series (1858–1859) and written down by W. Bezold, who attended the lectures.[96] However, there is no explicit evidence that Schwarz was acquainted with these lectures at the time.

It is of interest to note that by the 1880's, on the one hand there had accumulated such a wealth of material on particular conformal mappings, while on the other hand the cases in which they were applied in problems of physics and mechanics were multiplying so rapidly that a special monograph of G. Holtzmüller (1844–1914) was needed (*Einführung in die Theorie der isogonalen Verwandtschaft und der konformen Abbildungen mit Anwendung auf mathematische Physik*, Leipzig 1882).

---

96) Riemann, B. *Werke.* S. 595.

Karl Weierstrass

The final structure of the magnificent edifice of mathematical analysis whose foundations were laid by Newton and Leibniz owes perhaps more to Weierstrass than to any other mathematician of the nineteenth century. Beginning in the 1860's Weierstrass was the first to make the theory of real numbers the foundation of all of analysis, without any appeal to intuition and to propose as a universal method for constructing rigorous proofs of the principle of the existence of least upper (and greatest lower) bounds of bounded sets. All this raised analysis to a new level of rigor in comparison with what had been done by Cauchy. Weierstrass demonstrated convincingly the marvelous complexity of the behavior of a continuous function, constructing an example of a nowhere differentiable continuous function (in this he had been anticipated by B. Bolzano). On the other hand, he succeeded in showing that the simplest and most ideal function in terms of its properties — a polynomial — is capable of reproducing the values of any continuous function within an arbitrarily small error.[97]

Following Newton and Lagrange, Weierstrass made the power series, or rather a system of power series connected by the property of analytic continuation (which neither Newton nor Lagrange had discovered), the center of analytic function theory, laid the foundations of the theory of entire functions, polished the theory of elliptic functions to a high degree of perfection, and crowned the entire edifice of the theory of Abelian functions, which was the most impressive achievement of nineteenth-century analysis. As a by-product he laid the foundations of the theory of analytic functions of several complex variables. Finally, he made great achievements in the calculus of variations (the Weierstrass function, in terms of which conditions for a strong extremum can be expressed), the theory of minimal surfaces, and linear algebra (theory of elementary divisors).

Karl Theodor Wilhelm Weierstrass, whose enormous contribution to the foundations of mathematical analysis has already been discussed in Chapter 1, was born 31 October 1815 in Ostendorf (Westphalia) to the family of a civil servant. In 1834 he graduated from the Gymnasium in Paderborn, having achieved great success ("first in his class"), and in the same year, yielding to his father's insistence, he entered the University of Bonn to study law. Neglecting his university studies, which did not interest him, he studied the mathematical sciences independently. Weierstrass studied Laplace's *Mécanique céleste*, Jacobi's *Fundamenta nova*, and Abel's first publications on the theory of elliptic functions in Crelle's *Journal*. Jacobi's difficult book presupposed that the reader was familiar with the extensive treatise of Legendre on the theory of elliptic integrals, which was not available to Weierstrass at the time. Fortunately copies of Gudermann's lectures on the theory of elliptic functions came into his hands, which significantly facilitated his work. A letter from Abel to Legendre published in Crelle's *Journal* in 1829, in which Abel stated that for the function $\lambda(x)$ inverse to the integral $x = \int_0^y \dfrac{dy}{\sqrt{(1-y^2)(1-c^2y^2)}}$

---

97) Weierstrass' contribution to the foundations of analysis will be discussed in detail in another volume of this series.

K. Weierstrass

(Jacobi's function sn x), he had found a representation as the quotient of two everywhere-convergent power series, motivated Weierstrass to deduce this result independently, during the 7th semester of his studies in Bonn.[98]

After eight semesters Weierstrass ended his study in Bonn without taking his examinations and announced to his father that his calling was mathematics. Over his father's opposition he entered the Münster Academy in 1839, intending to become a high school teacher. There he attended only the lectures of Gudermann. In these lectures he was able, in particular, to become familiar with the idea of uniform convergence of series and infinite products, which first occurred in the papers of Gudermann from 1838.[99]

In the autumn of 1840 Weierstrass presented his examination paper "Über die Darstellung der modulären Funktionen" (see above). Following his teacher, he used the term *modular functions* to mean elliptic functions. The main object of attention was their representation as power series (see above). This work was judged very highly by Gudermann, but Weierstrass did not learn of the full extent

---

98) Dugac, P. "Éléments d'analyse de Karl Weierstrass," *Arch. Hist. Exact Sci.*, **10** (1973), p. 45.

99) *Ibid.*, p. 47.

of this evaluation until 1853. If he had known of it sooner, as he later wrote to Schwarz, he might have found a position in a university much earlier than he did.

This paper, along with three others written in Münster in 1841 and 1842, were not published until 1894, in the first volume of Weierstrass' *Mathematische Werke*. For that reason they could not have had any influence on the development of mathematics at the time. As it happens, they contained the entire machinery of the approach to analytic function theory associated with the name of Weierstrass. This last statement refers to the paper on the expansion of elliptic functions mentioned above. Weierstrass himself regarded his approach to elliptic functions as a prefiguration of his later interpretation of Abelian functions. In his paper "Darstellung einer analytischen Funktion einer komplexen Veränderlichen, deren absoluter Betrag zwischen zwei gegebenen Grenzen liegt" (Münster 1841),[100] which we have already mentioned, he deduced the Laurent expansion of a function in an annulus. This was two years before Laurent published the result, and was done without any direct use of the results of Cauchy (the Cauchy integral formula), and without using geometric language. It is characteristic that a complex number with absolute value r is represented not as $r(\cos\theta + i\sin\theta)$, as Cauchy and, following him, Laurent had done, but rather as a fraction $r\frac{1+\lambda i}{1-\lambda i}$, where λ is a real number. In order for such a point to describe the circle once, it is necessary for λ to vary monotonically from $-\infty$ to $+\infty$.

The next Münster paper, "Zur Theorie der Potenzreihen" (Münster, Autumn 1841)[101] used the property of uniform convergence of a power series. In the process power series in several variables were studied. The main result here is the theorem on the legitimacy of collecting like terms in the series obtained after each variable in turn is replaced by a corresponding power series.

Finally, the paper "Definition analytischer Funktionen einer Veränderlichen vermittelst algebraischer Differentialgleichungen" (Münster, Spring 1842)[102] contains three essential results. The first of these, which asserts that the solutions of a system of differential equations of the form $dx_j/dt = G_j(x_1, \ldots, x_n)$ $(j = 1, \ldots, n)$ determined by initial conditions $x_j(0) = a_j$ are analytic functions if $G(x_1, \ldots, x_n)$ is a polynomial, is not new. This proposition was contained in earlier investigations of Cauchy on his method of majorants. But the lithographed Turin paper of 1832 was not known to Weierstrass. He also was unacquainted with the reproduction of the contents of that paper in the 14th *cahier* of Cauchy's *Exercices d'analyse*, which was published in 1841. The second result, which Weierstrass later (1894) asserted could not be found in the work of Cauchy, established by use of a majorant of a power series not only the absolute (unbedingt) convergence of the series, but also its uniform (gleichförmig) convergence. Formally Weierstrass' assertion is correct. But in essence, as already stated above, the underlying idea itself of Cauchy's method of majorants meant a systematic use of the property of uniform convergence of power series, even to the extent of an explicit upper bound on the

---

100) Weierstrass, K. *Mathematische Werke*. Bd. 1, S. 52–66.

101) *Ibid.*, S. 67–74.

102) *Ibid.*, S. 75–84.

remainder terms. Finally, the third result, the description of the process of analytic continuation of power series and its application to the representation of solutions of a system of differential equations outside the original domain of convergence, which is a neighborhood of the point $t = 0$, was entirely due to Weierstrass. In this area there was no one for him to refer to, no matter how carefully he had studied the works of his predecessors and contemporaries. However (and we have also stated this above) less than a decade later V. Puiseux independently arrived at the same idea and expounded it, to be sure in connection with the special problem of representation of algebraic functions, but in a form just as general as that of Weierstrass. In a different form, without any direct use of power series, the idea of analytic continuation appeared a year after Puiseux's paper, in the dissertation of Riemann.

After one year's experience in Münster, Weierstrass received in 1842 a position as a teacher in the progymnasium at Deutsch-Krone (West Prussia). From 1848 to 1855 he was a Gymnasium teacher in Braunsberg (East Prussia). These were small isolated villages, in which he was completely alone with his intellectual pursuits. He had to teach a great variety of subjects: German language, botany, geography, history, penmanship, and gymnastics. His journey to Berlin in October 1844 had as its direct purpose the improvement of the teaching of his gymnastics. His teaching load reached 30 hours per week. Nevertheless Weierstrass reflected incessantly on mathematical problems. The center of his quest was the theory of Abelian functions, to which he remained faithful throughout his life.

The appendix to the annual report of the Braunsberg Gymnasium for 1848–1849 contained his short article "Beitrag zur Theorie der Abelschen Integrale. Beilage zum Jahresbericht über Gymnasium Braunsberg in dem Schuljahre 1848–1849,"[103] in which he stated that he had long been occupied with Abelian integrals, especially the main problem of inverting integrals of first kind, posed by Jacobi, and that he had succeeded in solving this problem completely following a route different from that of Göpel and other mathematicians. We recall that Göpel and Rosenhain had limited themselves to the inversion of hyperelliptic integrals in the case involving the square root of a polynomial of degree 5 or 6, which leads to functions of two independent variables, and that their method did not admit of immediate extension to the general case.

However, Weierstrass' 1849 paper was only a sort of announcement and remained unnoticed by contemporaries. His next two papers, which were published in volumes 47 (1854) and 52 (1856) of Crelle's *Journal*, met a very different reception. These papers bore the titles "Zur Theorie der Abelschen Funktionen"[104] and "Theorie der Abelschen Funktionen"[105] Although the first of them, according to Dirichlet's report, "gave only partial proofs of his results and lacked the

---

103) *Ibid.*, S. 111–129.

104) *Ibid.*, Bd. 2, S. 133–152.

105) *Ibid.*, Bd. 2, S. 297–355.

intermediate explanations,"[106] it was immediately appreciated. It was this paper that earned him the doctoral degree *honoris causa* awarded by the University of Königsberg (graduating from the Münster Academy did not earn Weierstrass the doctor's degree), and the title of *Oberlehrer*. Most importantly, however, on the basis of Dirichlet's report Weierstrass could now obtain a year's leave to recover his health and complete the second of the two papers mentioned above, in which the expanded proofs of the results obtained were discussed.

Weierstrass never returned to teaching in the Gymnasium. With the intervention of A. F. Humboldt and F. J. Richelot (1808–1875, a student of Jacobi and his successor at the University of Königsberg) he was named professor at the Gewerbeinstitut (later the Technische Hochschule) in Berlin in the summer of 1856. Weierstrass owed to Kummer the fact that in the autumn of that year he was also named professor extraordinarius at the University of Berlin (after 1864 he was ordinarius) and was elected a member of the Berlin Academy of Sciences.

In his *Antrittsrede* upon entering the Academy Weierstrass expounded his scientific credo in the following words:

> Ever since I first became acquainted with the theory of elliptic functions under the guidance of my teacher Gudermann this relatively new branch of mathematical analysis has exerted a powerful attraction on me, which remains the determining factor for the whole course of my mathematical development.

Further on, speaking of the periodic functions of several variables that are a generalization of elliptic functions, he said:

> To give an actual representation of these quantities of a completely new type that is unique in analysis and to study their properties more closely has now become one of the main problems of mathematics, which I have decided to investigate. However, it would be foolish if I were to try to think only about solving such a problem, without being prepared by a deep study of the methods that I am to use and without first practicing on the solution of less difficult problems.[107]

Thus from elliptic functions to Abelian functions, relying on the profound development and experimental testing of the necessary methods of analysis and the theory of functions — such is the character of the main thrust of the work of Weierstrass, as he himself gave it at the beginning of his Berlin period, which was to continue for more than 40 years, to the very end of his life. He died 19 February 1897, after a long illness, which precluded any creative work after the early 1890's.

The results of his research during the Berlin period were reflected in his oral communications at the Berlin Academy of Sciences, a few publications in journals,

---

106) Dugac, P. "Éléments d'analyse de Karl Weierstrass," *Arch. Hist. Exact Sci.*, **10** (1973) p. 52.

107) Akademische Antrittsrede. *Monatsber. Preuss. Akad. Wiss. Berlin*, 1857, S. 348–351; Weierstrass, K. *Mathematische Werke*. Bd. 1, S. 223–226.

and the courses of lectures given at the University of Berlin. Starting in 1894 his collected works (*Mathematische Werke*) began to be issued, a project that is not yet complete. Of the seven published volumes, two were published during his lifetime. The seventh volume, published in 1927, contains his course of lectures on calculus of variations given in 1879.

Of Weierstrass communications and papers the greatest value for analysis and theory of functions are the following: "Über das sogenannte Dirichletsche Prinzip" (1870); "Über eine stetige Funktion einer reellen Veränderlichen, die auf keinem Punkte ein Derivat besitzt" (1870, see Book 3); "Zur Theorie der eindeutigen analytischen Funktionen" (1876, the paper containing the theorem on the decomposition of entire functions into prime factors, which was the starting point for the general theory of entire and meromorphic functions); and "Einige sich auf die Theorie der analytischen Funktionen mehrerer Veründlichen beziehende Sätze" (1879).[108] This last paper opened with the so-called *preparation theorem*, which laid the foundations of the divisibility theory of power series for the case of several variables. The process of analytic continuation and the singular points were then studied for these series (in particular points of indeterminacy, which do not occur for functions of one variable), and a conjecture was stated (later proved by Poincaré and P. Cousin) that a single-valued function having no singularities except poles and points of indeterminacy can be represented as the quotient of two entire functions. The article concluded with the theorem on the representation of an n-fold periodic entire function of n variables by a Fourier series. The importance of this paper, which is intimately connected with Weierstrass' studies of the theory of Abelian functions, lies in the fact that it was the origin of the later theory of analytic functions of several complex variables. We should also mention his paper "Zur Funktionenlehre" (1880),[109] in which it was first revealed that a single series $\sum_{0}^{\infty} f_n(x)$, where $f_n(x)$ are analytic functions, converging uniformly inside a disconnected open set may represent different analytic functions on different components of the set. In this paper Weierstrass' posed the fundamental problem of refuting the corresponding assertion of Riemann (see above). Thus he wrote, "The purpose of this paper is to prove that the concept of a monogenic function of a complex variable does not coincide completely with the concept of dependence expressed by means of arithmetic operations on quantities...," adding in a footnote that "the contrary assertion was stated by Riemann."[110]

Weierstrass' paper "Über die analytische Darstellbarkeit der sogenannten willkürlichen Funktionen einer reellen Veränderlichen," was published in 1885 (see above). A number of Weierstrass' papers in the theory of functions published earlier or only communicated were collected and published by him as the collection *Abhandlungen aus der Funktionenlehre* (Berlin 1886).

---

108) Both of these papers were printed in 1886.

109) Weierstrass, K. *Mathematische Werke*. Bd. 2, S. 201–203.

110) *Ibid.*, S. 210.

The paper "Untersuchungen über die Flächen, deren Krümmung überall gleich Null ist),[111] in which the theory of minimal surfaces was constructed, is connected with the theory of analytic functions, as well as with calculus of variations. In its original form it was communicated to the Berlin Academy of Sciences in 1866.

But the worldwide fame of Weierstrass and his influence on the minds of mathematical youth, which grew so quickly from the late 1850's on, were not due to his communications to the Academy and journal publications alone. A large role in this phenomenon was played by his courses at the University of Berlin, discussed above, which were not published during his lifetime. He gave these courses, with interruptions caused by illness, for over 30 years. They consisted of cycles in which each course depended on its predecessors. Here are the approximate contents of one such cycle in its mature form:[112] Introduction to the analytic function theory including the theory of real numbers, Theory of elliptic functions, Applications of elliptic functions to problems of geometry and mechanics, Theory of Abelian integrals and functions, Calculus of variations.[113]

In order to reach the summit those who attended Weierstrass' lectures had to begin from the beginning. The theory of real numbers, one of whose independent creators was Weierstrass, was a novelty for the audience of 1874, when the course "Einleitung in die Theorie der analytischen Funktionen" was put together. How students reacted to it when it was repeated 10 years later was elegantly related by the Russian student M. A. Tikhomandritskiĭ, who was in Berlin in the autumn of 1884 and was interested in the theory of hyperelliptic and Abelian integrals:

> ...in Berlin I attended only a few lectures devoted to the concept of a number and the four operations on numbers with which Weierstrass always considered it necessary to begin these courses [meaning the course "Einleitung in die Theorie der analytischen Funktionen"]. This beginning, however, seemed boring to the majority of the audience, and I was witness to a phenomenon that is familiar to us: after the first lecture in the large auditorium, which did not have enough room for the whole audience, he had to move to a larger lecture hall, specially built in the garden behind the university building, which could seat more than a thousand listeners. but their number (perhaps as the result of the poor acoustic and optical properties of the hall) soon decreased, so that after several lectures, he went back to an auditorium smaller than

111) *Ibid.*, Bd. 3.

112) Biermann, K.-R. *Die Mathematik und ihre Dozenten an der Berliner Universität, 1810–1920.* S. 77.

113) All of these courses except the introduction to analytic function theory were published posthumously as the successive volumes of Weierstrass' collected works from 1902 to 1927. Volume 4 was *Theorie der Abelschen Transcendenten*, Vol. 5 was *Theorie der elliptischen Funktionen*, Vol. 6 was *Anwendungen der elliptischen Funktionen auf Probleme der Mechanik*, and Vol. 7 was *Variationsrechnung*.

the original, one that could seat at most 150–200 listeners, and even that one was far from being full.[114]

However, no matter what the attitude of part of the audience at various times to different courses of Weierstrass, it was these courses that attracted numerous students from many lands to him. Among these students, those who took up the cause of Weierstrass in analytic function theory were G. Mittag-Leffler, H. A. Schwarz, L. Fuchs (1833–1902), and F. Schottky. Schwarz in particular, whose work we have already discussed, did much to promote the propagation of the functions $\sigma(z)$, $\zeta(z)$, and $\wp(z)$ introduced into the theory of elliptic functions by Weierstrass. He published as a sort of handbook (unfortunately unfinished) a systematic survey of the contents of this theory: *Formeln und Lehrsätze zum Gebrauche der elliptischen Funktionen. Nach Vorlesungen und Aufzeichungen des Herrn K. Weierstrass...*, Göttingen 1883, 2 Ausg., Berlin 1893), which has since become the main source for the study of this theory. Among the students and disciples of Weierstrass in the area of the theory of analytic and Abelian functions was the Khar'kov professor Tikhomandritskiĭ (1844–1921). He is the author of a number of monographs written in the spirit of Weierstrass: *Inversion of Elliptic Functions* (1885), *Theory of Elliptic Integrals and Elliptic Functions* (Khar'kov 1895), *Foundations of the Theory of Abelian Integrals* (Khar'kov 1895). In this last book, in contrast to Weierstrass, he used the concept of a Riemann surface and applied largely algebraic methods; for this Tikhomandritskiĭ referred to the work of Max Noether, a professor at the Universities of Heidelberg and Erlangen, and his own work. The second, revised and enlarged, edition of this work was published in 1911 in Petersburg, written in French. Desiring to eliminate ideas and methods alien to Weierstrass from Tikhomandritskiĭ's exposition, V. P. Ermakov (1845–1922), a professor at the University of Kiev, published the *Theory of Abelian Functions without Riemann Surfaces* (Kiev 1897).

Also among Weierstrass' students were G. Cantor (1845–1918), a professor at the University of Halle, who laid the foundations of the theory of sets, and G. Frobenius (1849–1917), an outstanding algebraist, professor at the Zürich Technische Hochschule, and later at the University of Berlin. A special place belongs to S. V. Kovalevskaya, who took private lessons from Weierstrass in the years 1870–1874. and maintained close ties of friendship with him throughout her life. Her life and work will be discussed in the part of this series devoted to the theory of partial differential equations, but in the present chapter, we shall also take time to discuss the importance of the methods of analytic function theory for her investigations.

Analytic Function Theory in Russia. Yu. V. Sokhotskiĭ and the Sokhotskiĭ-Casorati-Weierstrass Theorem

Although the original accumulation of material for the theory of analytic functions of a complex variable, due to the work of Euler, was closely connected with the

---

114) Tikhomandritskiĭ, M. A. "Karl Weierstrass," *Сообщ. Харьк. Мат. Общ.* (2), **6** (1897), p. 50.

Petersburg Academy in the eighteenth century, for a long time there were no Russian mathematicians who wished to devote themselves to this new area of mathematics. Nevertheless one can name several who found the subject congenial.

In the abstract of his famous memoir "Sur les intégrales définies entre des limites imaginaires," published in the *Bulletin de Férussac* for 1825, Cauchy mentioned with approval among the scholars who had recently studied such integrals the "young Russian" M. Ostrogradskiĭ, who "has also undertaken the application of these integrals and their transformation into ordinary integrals, has given new proofs of the formulas I mentioned above, and has generalized other formulas proposed by me in the 19th issue of the *Journal de l'École Polytechnique.*"[115]

We know very little of this research of M. V. Ostrogradskiĭ. It is known that during his years in Paris he wrote his "Remarks on definite integrals" in 1824 and presented it to the Academy in 1826. In this work he derived Cauchy's formula for the residue of a function with respect to a pole of order n.[116] Subsequently he described the applications of the theory of residues and the Cauchy integral formula to the computation of definite integrals in an extensive course of public lectures given in 1858–1859.[117] However Ostrogradskiĭ's studies of analytic function theory were only a minor incident in his research, which was devoted predominantly to the development of the methods of mathematical physics, calculus of variations, integral calculus, and mechanics.

One of the most remarkable discoveries in the history of science was made in 1826 — the discovery of non-Euclidean geometry by N. I. Lobachevskiĭ. It took decades for mathematicians to realize the importance of this discovery for analytic function theory.

It can be asserted that an important facet of the extensive cycle of fundamental research on the theory of functions of a complex variable in the last decades of the nineteenth century and the first decade of the twentieth (F. Klein, H. Poincaré, P. Koebe) was the gradual explanation of the fact that hyperbolic geometry is also the geometry of analytic functions of one complex variable.[118]

A number of papers of Lobachevskiĭ having immediate importance for mathematical analysis, in particular for the theory of functions of a complex variable, belong primarily to the 1830's. His "Algebra, or the computation of finite quantities" (Kazan' 1834) contains the theory of elementary functions of a complex variable. Here $\cos x$ and $\sin x$ are defined first for x real as the real and imaginary parts of the function $e^{x\sqrt{-1}}$. From this, using now the previously established

---

115) Cauchy, A.-L. *Œuvres complètes*, sér. 2, T. 2, pp. 57–65.

116) See Yushkevich, A. P. "On the unpublished early work of M. V. Ostrogradskiĭ" [in Russian], *ИМИ*, 1965, No. 16, pp. 12–26. Ostrogradskiĭ's paper was never published and is preserved in the Archive of the Academy of Sciences in Paris.

117) A record of these courses, written by an unknown member of the audience, was published with commentaries by V. I. Antropova in the book *Mikhail Vasil'evich Ostrogradskiĭ. 1 January 1862–1 January 1962. Pedagogical Legacy. Documents on his Life and Career*. Moscow 1961, pp. 152–263.

118) See below and also the book *Essays on the History of Analytic Function Theory* [in Russian]. Moscow 1951, Essay 2.

Yu. V. Sokhotskiĭ

properties of the exponential function and the application of power series, all the basic properties of the trigonometric functions are developed. Lobachevskiĭ seems to have ascribed particular importance to such a purely algebraic construction of trigonometry, independent of Euclidean geometry.

We have already mentioned above that in 1850 I. I. Somov, a professor at the University of Petersburg (later a member of the Academy) published his *Foundations of Analytic Function Theory*, whose basis was Jacobi's *Fundamenta nova...* (see above). While this book was thus slightly out of date mathematically, the doctoral dissertation of M. E. Vashchenko-Zakharchenko "Riemann's theory of functions of a complex variable" (1866), also mentioned above, brought the Russian reader into contact with the most advanced ideas of the theory of functions at the time.

There is no doubt, however, that the first genuinely original Russian researcher in the area of the theory of analytic functions of a complex variable was Yu. V. Sokhotskiĭ.

Yulian Vasil'evich Sokhotskiĭ (1842–1927), a native of Warsaw, studied at Petersburg University, from which he graduated in 1866. In 1868 he defended his master's thesis "Theory of integral residues with certain applications" (St. Petersburg 1868). As he himself explained, he had to entreat P. L. Chebyshev for a long

time before the latter agreed to accept his thesis. This may have been a manifesta-
tion of Chebyshev's dislike of the theory of functions of a complex variable. After
the autumn of 1868 Sokhotskiĭ was Privatdocent at Petersburg University, where
he gave courses on the theory of functions of a complex variable and continued
fractions, with applications to analysis. In 1873 Sokhotskiĭ defended his doctoral
dissertation "On the definite integrals and functions used in series expansions,"
(St. Petersburg 1873), and in 1882 he became professor ordinarius at Petersburg
University. A number of later papers of Sokhotskiĭ involve the theory of elliptic
functions, as well as algebra and number theory. He is the author of the original
courses *Higher Algebra* (1882), *Number Theory* (1888), and an outstanding study
of algebraic number theory based on the work of E. I. Zolotarёv, A. A. Markov,
and himself: *Elements of the Greatest Common Divisor in Application to the Theory of Divisibility of Algebraic Numbers* [in Russian] (1903). Sokhotskiĭ died in
Leningrad in 1927.

Sokhotskiĭ's master's thesis was devoted to the applications of the theory of
residues to the inversion of a power series (the Lagrange series) and in particular
to the continued fraction expansion of analytic functions, as well as Legendre
polynomials. In the preface, noting the importance of the theory of residues for
mathemtical analysis, he pointed out that "the calculus of integral residues is
hardly being developed at all by the scholars of the present day." He continued:

> In the present argument I expound the general elements of the calulus of
> integral residues and exhibit some of its applications, namely those that I
> cannot find in any of the papers of Cauchy, or those whose exposition I have
> found to be less simple and intuitive than my own presentation.[119]

In this paper, in particular, one finds stated and proved the famous theorem on
the behavior of an analytic function in a neighborhood of an essential singularity.
Sokhotskiĭ stated it in the following old-fashioned form:

*If a given function $f(z)$ becomes infinite to infinite order at some point z_0, then
the function $f(z)$ must inevitably assume all possible values at that point.*[120]

Here an essential singularity is defined as a point at which $f(z)$ "becomes infinite
to infinite order" (what Sokhotskiĭ has in mind is the nature of the expansion of
$f(z)$ in a neighborhood of z_0), and the values of the function $f(z)$ at the point
z_0 are interpreted as the set of limiting values of the function at that point. This
is clear both from the examples he had given previously (for example, $\sin \dfrac{1}{z-b}$
"assumes all possible values at the point $z = b$"), and from the reasoning he
employed in the proof, which is essentially the same as the modern argument.
This theorem was published in the same year by Casorati in his famous course
and eight years later by Weierstrass in the paper "Zur Theorie der eindeutigen

---

119) Sokhotskiĭ, Yu. V. *Theory of Integral Residues with Certain Applications*, St. Petersburg 1868, p. 1 (unnumbered).

120) *Ibid.*, p. 17.

analytischen Funktionen" mentioned above. For that reason it could be called the Sokhotskiĭ-Casorati-Weierstrass theorem.

The fate of this theorem is both interesting and instructive.[121] It reflected the development of a number of general presentations of the theory of functions in the third quarter of the nineteenth century of fundamental importance.

As soon as Liouville's theorem became known the following theorem became almost obvious: *If an entire function $f(z)$ is not constant, then every complex number C, finite or infinite, is either among the values assumed by that function or a limit of those values.*

In the case when $C = \infty$, this is precisely Liouville's theorem. If $C \neq \infty$ and C is not among the values assumed by the function, then $[f(z) - C]^{-1}$ is an entire function, and it follows from the same theorem that C must be a limit of the values of $f(z)$. We have reproduced here the reasoning of Briot and Bouquet in their book *Théorie des fonctions doublement périodiques et, en particulier, des fonctions elliptiques* (1859). To be sure, the result was stated in a form that we find unusual: *A function that is monodromic and monogenic on the entire plane assumes all possible values.* Here the point is that the authors, like many of their contemporaries, made no distinction between the values of a function and limits of those values.[122] In its only possible interpretation the theorem of Briot and Bouquet is correct and was rigorously proved by them (this theorem contains in particular the so-called fundamental theorem of algebra). For convenience we shall call it the "preliminary" theorem (on the route to the Sokhotskiĭ-Casorati-Weierstrass theorem).

Apparently guided by the analogy with the properties of elliptic functions, Briot and Bouquet quickly drew the following conclusion from the "preliminary" theorem:

Two functions that are monodromic and monogenic and have the same zeros and infinities [i.e., poles], are equal up to a constant factor.

This conclusion, of course, is wrong: the quotient of the two functions does not *asssume* the value zero or infinity at finite points of the plane, but nothing prevents these numbers from occurring as *limiting values* in complete consistency with the "preliminary" theorem.

The mistake of Briot and Bouquet was noticed by their contemporaries, in particular Casorati.[123] In the second enlarged edition of the book of Briot and Bouquet (1875) not only is the incorrect corollary missing, but so is the correct theorem from which it was deduced.

---

121) The facts in the case were discussed in detail on the basis of archival materials in a recent paper of E. Neuenschwander: "The Casorati-Weierstrass theorem," *Hist. Math.*, **5** (1978), pp. 139–166.

122) This is why one should not regard this proposition as "an incorrect statement of what was to become Picard's theorem" (*Abrégé d'histoire des mathématiques. 1700–1900. Sous la direction de J. Dieudonné.* Paris 1978, T. 1, p. 149). Picard's theorem, after all, speaks only of the values actually assumed by the function!

123) See the article of Neuenschwander mentioned above.

To pass from the "preliminary" theorem to the complete theorem it was necessary to take two more important steps.

First, it was necessary to find the analogue of Liouville's theorem for the case when the function $f(z)$ is single-valued and analytic only in some neighborhood of the point z_0 (except perhaps at this point). Such an analogue can be deduced from Riemann's doctoral dissertation (1851) in the following form: *If* $\lim\limits_{z \to z_0} f(z)(z - z_0) = 0$, *then* z_0 *cannot be a singular point of* $f(z)$. It is obvious that Riemann's condition holds when $f(z)$ is bounded in a given neighborhood of the point z_0.

Second, it was necessary to make a distinction between a pole and an essential singularity, which were not distinguished (for $z_0 = \infty$) in the statement of the "preliminary" theorem. To do this it suffices, for example, to show that the condition $\lim\limits_{z \to z_0} f(z) = \infty$ holds if and only if there exists a natural number n — the order of the pole — and a function $\varphi(z)$ that is analytic at the point z_0 and does not vanish on it such that $f(z) = (z - z_0)^{-n}\varphi(z)$. Using Riemann's theorem and the Taylor series, this was established by several mathematicians without any use of the Laurent expansion.

After that it remained only to assume that z_0 is "an infinity of infinite order" (i.e., an essential singularity of the function) and repeat the reasoning that earlier led to the "preliminary" theorem without any changes for this case. As a result the Sokhotskiĭ-Casorati-Weierstrass theorem was obtained this time. All these steps were taken and published in 1868, by both Sokhotskiĭ and Casorati.

It is interesting to note that the textbook *Eléments de la théorie des fonctions* of Durège (1864), which we have already mentioned, marks a sort of intermediate stage on the route from Briot and Bouquet (which Durège could have been familiar with) and Sokhotskiĭ and Casorati. To be specific, Durège's book contains the theorem of Riemann, correctly stated and proved using the Cauchy integral theorem, and also the necessary characterization of a pole.[124] However, he did not go beyond Briot and Bouquet in the direction we are interested in, confining himself to the "preliminary" theorem, which establishes, as he expresses it, a harmony in the theory of functions that disappears if complex values of the variable are excluded. Finally, like Briot and Bouquet, he gives the same erroneous corollary of this theorem. As a result, the priority in publication of this theorem in its full form belongs to Sokhotskiĭ and Casorati. Independently of each other they chose the same line of reasoning, which, however, had been in some measure suggested by their predecessors.

As for Weierstrass, although there is some information that this theorem occurred in his lectures in 1863 and 1866,[125] he still did not publish it until 1876,

---

124) In this connection we cannot agree with Neuenschwander's claim (on p. 156 of the article cited above) that Weierstrass was the first to extend the theorem of Liouville (in 1876!) to the case of functions that are single-valued and analytic in a neighborhood of an essential singularity. This had been done earlier by Riemann and later by Durège, Sokhotskiĭ, and Casorati.

125) See the article of Neuenschwander, p. 161.

in a paper devoted mostly to the analytic representation of functions having a finite number of essential singularities. It was stated at the very end of the paper. The route to the proof suggested by Weierstrass was by no means direct. To establish the necessary analogue of Liouville's theorem, he referred to his theorem on the expansion of an entire function as a product, which was obtained in the same paper.

Sokhotskiĭ's doctoral dissertation (1873) was remarkable in that it laid down the foundations of the theory of singular integral equations containing the Cauchy principal values of integrals. In the very first sentence of the preface Sokhotskiĭ spoke of the problem of solving an integral equation containing an integral of Cauchy type:

> In this work I try to direct the reader's attention to the solution of the equation $\varphi(x) = \int_a^b \frac{f(t)\,dt}{t-x}$, i.e., to the determination of the function $f(x)$ from a given function $\varphi(x)$.

Continuing, after referring to integral equations in the work of Abel, Cauchy, and Murphy,

> One can exhibit several other cases of the determination of the function under an integral sign: one of the simplest is the one we have already pointed out. It merits particular attention because solving it leads us to the fundamental elements of the theory of functions of a complex variable.

The first section of this paper is of fundamental interest from the point of view of the theory of functions, since the concept of an integral of Cauchy type in full form was introduced here for the first time: $\int_z^b \frac{f(t)\,dt}{t-x} = \varphi(x)$, where a and b are two arbitrary complex numbers. The integral was assumed to be taken over some curve ("trajectory") joining a and b. The function $f(t)$ "may be discontinuous between the limits of integration." As for the integral $\varphi(x)$ itself, at each point x of the trajectory it assumed two values, which were regarded as corresponding to two points x_1 and x_2, on opposite sides of the trajectory. As a result, the latter was regarded as a line of discontinuity of the function $\varphi(x)$. Sokhotskiĭ went on to prove a number of theorems, of which the most important for applications were the following:

If $\varphi(x_1)$ and $\varphi(x_2)$ are two values of the function $\varphi(x)$ corresponding to two adjacent points of the line of discontinuity, then $(\varphi(x_1) - \varphi(x_2))/2\pi i = f(x + 0) + f(x - 0)/2$, so that if the function $f(x)$ is continuous at the point x, then $(\varphi(x_1) - \varphi(x_2))/2\pi i = f(x)$ (Theorem 1).

If the function $f(x)$ is continuous at the point x on the line of discontinuity, so that for infinitely small h

$$f(x + h) - f(x - h) = \theta h^\alpha,$$

where α is a finite positive quantity and θ assumes a finite positive value for $h = 0$, then

$$\varphi(x_1) = \int_a^b \frac{f(t)\,dt}{t-x} + \pi i f(x),$$

where the integral on the right-hand side denotes the limit of the sum

$$\lim \left[\int_a^{x-h} \frac{f(t)\,dt}{t-x} + \int_{x+h}^b \frac{f(t)\,dt}{t-x} \right]$$

for $h = 0$ (Theorem 2).

It was in this last theorem that the *Cauchy principal value of the integral* $\int_a^b \frac{f(t)\,dt}{t-x}$ appears, along with a *singular integral equation* (with respect to $f(x)$). Here a Lipschitz condition is imposed on the function $f(x)$, which under the assumption (implicitly understood) that there exists a tangent to the trajectory at a point, does indeed guarantee the existence of the principal value of the integral.

These and certain other results of Sokhotskiĭ on the boundary values of an integral of Cauchy type were rediscovered 35 years later by J. Plemelj ("Ein Ergänzungssatz zur Cauchyschen Integraldarstellung analytischer Funktionen, Randwerte betreffend," *Monatshefte für Math. und Phys.*, **19** (1908), S. 205–210), and were known only by his name for a long time. In the mid-twentieth century they received widespread application in papers on the theory of elasticity by the prominent Georgian mathematician Academician N. I. Muskhelishvili and his school.

In speaking of the development of analytic function theory, one must not leave out the research of S. V. Kovalevskaya, although its main importance lies outside this theory. In her paper "Zur Theorie der partiellen Differentialgleichungen" (*J. für Math.*, **80** (1875)), Kovalevskaya proved a very general existence theorem for analytic solutions of a system of partial differential equations, in which the derivatives of highest order of the unknown function are expressed as analytic functions of the independent variables, the unknown function, and its derivatives of lower orders (the normal form of the system).

As is known, Kovalevskaya's most important creative work was her research on the theory of motion of a rigid body, especially her paper "Sur le problème de la rotation d'un corps solide autour d'un point fixe," (*Acta Math.*, **12** (1889)), which won a prize in a competition of the Paris Academy of Sciences. Her success was due to her completely new formulation of the problem in terms of analytic functions. Regarding time t as a complex variable, Kovalevskaya decided to find in general all the cases when the parameters of the motion could be expressed as meromorphic functions of t containing five arbitrary constants. As a result it was revealed that in her formulation of the problem, besides the known cases of Euler and Lagrange, there exists one and only one more case (the Kovalevskaya case) characterized by the fact that two of the principal moments of inertia are equal to twice the third principal moment, and the center of gravity of the body lies in the plane of the two equal moments.

This result was later supplemented by Academician A. M. Lyapunov, who showed that the cases of Euler, Lagrange, and Kovalevskaya are the only ones in which the parameters determined by the differential equations of motion are single-valued functions of time for all possible initial conditions ("On a property of the differential equations of the problem of motion of a heavy rigid body having a fixed point" [in Russian], *Сообщ. Харьк. Мат. Общ.* (2), **4** (1894, 1895)). Kovalevskaya's research served as the point of departure for numerous papers devoted to finding and studying cases of the motion of a rigid body under various special assumptions on the analytic nature of the corresponding functions.

One of the areas of application of analytic function theory in mechanics is related to the problem of stability of oscillations.

A complex-valued function of a real variable

$$e^{(\sigma+i\omega)t} = e^{\sigma t}\cos\omega t + ie^{\sigma t}\sin\omega t,$$

where σ and ω are real constants, can be interpreted as the complex notation for an elementary oscillation with frequency ω and amplitude $e^{\sigma t}$. This oscillation will be stable if its amplitude does not become arbitrarily large as time passes, i.e., if $\sigma \le 0$. The coefficient $p = \sigma + i\omega$ of t in the exponent can be found as the root of the equation $f(p) = 0$, where $f(p)$ is an entire function of the complex variable p or, in the simplest case, a polynomial. Therefore in practice it is very important to have a criterion for deciding in which cases all of the zeros of this polynomial or entire function lie in the left half-plane, i.e., their real parts are nonpositive (the criterion for stability). This problem was posed in 1868 by the famous physicist J. C. Maxwell (1831–1879). In 1877 E. J. Routh (1831–1907) gave an analytic solution of it for the case of a polynomial, which, however, was soon forgotten. In 1895 A. Hurwitz proposed a new solution, based on the theory of residues, and the problem is now called the Routh-Hurwitz problem.

An original geometric method of solving the problem was proposed simultaneously with Routh by the Russian scholar I. A. Vyshnegradskiĭ (1831–1895), who was professor at the Petersburg Technical Institute after 1862 and its director after 1875, in a paper "On the regulators of direct action" (*Изв. Спб. Практ. Техн. Инст.*, 1877). His idea was to include the polynomial being studied in a family

$$P(p)\xi + Q(p)\eta + R(p),$$

where P, Q, and R are certain polynomials, and ξ and η are real parameters.

Suppose the polynomial being studied is obtained with parameter values ξ_0 and η_0. Vyshnegradskiĭ studied the plane of the complex variable $\omega = \xi + i\eta$ and obtained a partition of this plane into separate regions, in each of which the polynomials of the family have a definite number of roots with positive and negative real parts (and have no purely imaginary roots). The region for whose points the number of zeros with negative real part is k and those with positive real part is $n - k$ will be denoted by $D(k, n - k)$. The boundaries of the regions can be established without the need to find the zeros themselves. It then remains only to determine precisely which one of them contains the point $\omega_0 = \xi_0 + i\eta_0$ corresponding to the polynomial in question. In particular, the condition for stability is that ω_0 belong to the domain $D(n, 0)$ (the domain of stability).

Vyshnegradskiĭ discussed his method only through the example of polynomials of degree three with real coefficients. The extension of the Vyshnegradskiĭ's method to the case of arbitrary polynomials is due to Soviet mathematicians.[126]

Entire and Meromorphic Functions. Picard's Theorem

Cauchy's 1844 theorem (known as Liouville's theorem), which asserts that a bounded entire function is a constant, can be considered the first significant result of the general theory of entire functions. For a long time this theorem played the role of a subsidiary device in the general theory of elliptic functions (Liouville, Briot and Bouquet). But it implied immediately that the maximum modulus $M(r)$ of an entire function in the disk $|z| < r$ must increase without bound as r increases, and moreover (by a slight generalization of Cauchy's reasoning) that for a transcendental entire function (i.e., an entire function that is not a polynomial) it increases faster than any power of r. This was one step toward the classification of entire functions according to the rate of growth of their maximum modulus. But this step was taken only at the end of the century by É. Borel.

The point of departure in the study of entire functions was the theorem on expansion of an entire function in an infinite product popularized by Weierstrass in 1876 ("Zur Theorie der eindeutigen analytischen Funktionen," *Berl. Abh.*, 1876, S. 11–60).[127] It asserts that for any sequence of complex numbers $\{z_n\}$ with non-decreasing absolute values tending to infinity one can construct an entire function having zeros at these points and only these points, and that any entire function with these zeros can be represented by the formula

$$f(z) = e^{g(z)} z^\lambda \prod_{1}^{\infty} \left(1 - \frac{z}{z_n}\right) \exp\left(\frac{z}{z_n} + \frac{z^2}{2z_n} + \cdots + \frac{z^n}{nz_n}\right).$$

Here $g(z)$ is an entire function (if $f(z)$ is given, it is determined by this formula up to a term of the form $2k\pi i$, where k is an integer), λ is the multiplicity of the zero of $f(z)$ at the origin, and the z_n are assumed to be all nonzero.

One of the first expansions of an entire function in a product was obtained by Euler for $\sin \pi z$ in 1734–1735. It seemed to him to be a natural analogue of the expansion of a polynomial in factors, not requiring any special justification. Given that the equation $\sin \pi z = 0$ has the roots $0, \pm 1, \pm 2, \ldots$ and no others, it follows that $\sin \pi z = Cz(1 - z/1)(1 + z/1) \cdots$; the value 1 was easily obtained for the constant (see HM, T. 3, pp. 328–329).

---

126) Pontryagin, L. S. *Ordinary Differential Equations*, Reading, MA 1962.
127) Weierstrass, K. *Mathematische Werke*. Bd. 2.

J. Bouquet

However, in one of the earliest definitions of the gamma function, contained in a letter of Euler to Goldbach (October 1729) we see the expansion $\dfrac{1}{\Gamma(z)} = \lim\limits_{n\to\infty} \dfrac{z(z+1)\cdots(z+n)}{n^z n!}$, which differs only slightly from the formula that led Weierstrass to the general result

$$\frac{1}{\Gamma(z)} = ze^{\pi z}\left[\left(1+\frac{z}{1}e^{-z/1}\right)\right]\left[\left(1+\frac{z}{2}\right)e^{-z/2}\right]\cdots .$$

Although the roots of the equation $1/\Gamma(z) = 0$ here do reduce to $0, -1, -2, \ldots$, the product $z(1+z/1)(1+z/2)\ldots$ diverges. To obtain a true result, the auxiliary factors $e^{-z/1}, e^{-z/2}, \ldots$, which have no zeros, are needed to guarantee convergence. It is these factors that Weierstrass introduced in the general form $\exp(z/z_n + z^2/2z_n^2 + \cdots + z^n/nz_n^n)$.

However, he informed Kovalevskaya (and Schwarz) of the expansion theorem in December 1874. This theorem enabled him to prove immediately that every meromorphic function of one variable is the quotient of two entire functions $f = g/h$. The term *meromorphic* was introduced by Briot and Bouquet; Weierstrass himself used the rather cumbersome description *a single-valued analytic function*

$f(z)$ *having the character of a rational function for each finite value of z.* To this theorem Weierstrass added that one can require the numerator and denominator of the fraction to have no common zeros. However he did not succeed in obtaining the analogous theorem for functions of several variables, although he pursued it, so to speak, in connection with his perfection of the theory of Abelian functions. Success fell to the lot of H. Poincaré (1883), but only for the case of two variables, after which his student P. Cousin (1867–1933), who worked first at the Lycée in Cannes, then as docent and professor at the University of Bordeaux, studied the general case (1895). The two proofs use different methods, which of course differ from the method of Weierstrass (for one variable). One might say that the properties specific to the theory of functions of several complex variables come to light here for the first time.

While Weierstrass' theorem was a generalization of the theorem on the decomposition of a polynomial into linear factors, the question naturally arose of generalization of the theorem on the representation of a rational function as a sum of partial fractions to the case of any meromorphic function. The affirmative answer was obtained and published the following year by one of Weierstrass' students, the Swedish scholar Gösta Mittag-Leffler (1846–1927), who later became a professor at the University of Stockholm and founded the journal *Acta Mathematica*. Mittag-Leffler published the result in his paper "Om den analytiska framställningen af en funktion af rationel karakter med en godtyckligt vald gränspunkt" (*Öfversigt Kgl. Vet. Akad. Förhandl.*, **34** (1877)). Mittag-Leffler's theorem makes the following assertion:

It is always possible to construct a meromorphic function with prescribed principal parts $G_n(z)$ (partial fractions) of the Laurent expansion at prescribed poles $\{z_n\}$ provided the condition $\lim_{n\to\infty} z_n = \infty$ holds; any such function $f(z)$ can be represented as

$$f(z) = \varphi(z) + \sum_1^{\infty} [G_n(z) + P_n(z)],$$

where $\varphi(z)$ is a certain entire function and $\{P_n(z)\}$ are polynomials that guarantee the convergence of the expansion.

Once Mittag-Leffler's theorem was proved Weierstrass' theorem could be obtained from it as a simple corollary (but not conversely!).

The theorem of Weierstrass just discussed was naturally the point of departure for the subsequent construction of the theory of entire functions. But a much deeper result is Picard's theorem, from which the general theory of meromorphic (and entire) functions grew in the 1920's — the theory of distribution of values. É. Picard (1856–1941) was a professor at the University of Toulouse, and after 1889 at the Sorbonne and the École Normale Supérieure; after 1917 he was the permanent secretary of the Paris Academy of Sciences. His theorem was published in a note "Sur une propriété des fonctions entières," (*Comptes Rendus*, **88** (1879)), pp. 1024–1027). According to Picard's theorem, for each nonconstant entire function $f(z)$ the equation $f(z) = A$ has roots, where A is any complex number, with

É. Picard

at most one exceptional value of A. This value, which depends on the function, is called the Picard exceptional value (for example, $A = 0$ is the exceptional value for the function $f(z) = e^z$). As a means of proving the theorem Picard used the infinitely many-valued function inverse to the so-called Schwarz modular function (see above).

In the same year of 1879, in a note "Sur les fonctions analytiques uniformes dans le voisinage d'un pointe singulier essentiel" (*Comptes Rendus*, **89** (1879)), by a slight complication of the proof, Picard obtained the so-called big Picard theorem (in contrast to the preceding one, called the little Picard theorem): *In any neighborhood of an essential singularity z_0 of the function $f(z)$ the equation $f(z) = A$ has infinitely many solutions for any complex number A, finite or infinite, with at most two exceptions, depending only on the function.* (Here the function $f(z)$ is allowed to have poles in any neighborhood of z_0.) An example with $z_0 = \infty$ occurs in the case of the function $\tan z$, for the numbers $z = \pm i$: the equation $\tan z = \pm i$ has no roots.

It is natural to regard Picard's big theorem as a significant generalization of the Sokhotskiĭ-Casorati-Weierstrass theorem. From the latter it follows only that the equation $f(z) = A$ can be satisfied approximately, not exactly, in a neighborhood of an essential singularity, i.e., one can find a sequence $\{z_n\} \to z_0$

such that $\{f(z_n)\} \to A$. However, for entire transcendental functions Picard's theorem also admits another interpretation that also resembles the fundamental theorem of algebra.

In fact, if we regard an entire transcendental function $f(z)$ as a sort of polynomial of infinite degree, it turns out that the equation $f(z) = A$ has a number of roots equal to the degree of $f(z)$ (to be sure, there may be an exceptional value A). The development of this analogy required the establishment of a quantitative distinction between two kinds of "infinities," understood as the degree of the equation, and as the number of roots. The ideas of G. Cantor, which were new at the time, on the existence of different infinite cardinal and ordinal numbers were inapplicable here, but one can assume that they did not pass unnoticed; these results will be discussed below.

However that may be, the attempt to classify different transcendental entire functions according to their "degrees" was at first connected only with the form of their expansions into factors according to Weierstrass' formula, and the first idea in this area was proposed by E. N. Laguerre (1834–1886), a professor at the École Polytechnique and the Collège de France and a member of the Paris Academy of Sciences. In his note "Sur quelques équations transcendantes" (*Comptes Rendus*, **94** (1882), pp. 160–163) Laguerre introduced the concept of the genus of an entire function. Let $\{z_n\}$, $(z_n \neq 0)$ be the sequence of zeros of the entire function $f(z)$, and suppose there exists a nonnegative integer k such that the series $\sum_{n=1}^{\infty} \left| \dfrac{1}{z_n} \right|^k$ diverges while the series $\sum_{n=1}^{\infty} \left| \dfrac{1}{z_n} \right|^{k+1}$ converges (for the function $\sin z$, for example, $k = 1$).

It then follows from Weierstrass' proof that the expansion of $f(z)$ into factors assumes the simpler form

$$f(z) = e^{g(z)} z^\lambda \prod_{1}^{\infty} \left(1 - \frac{z}{z_n}\right) e^{\frac{z}{z_n} + \cdots + \frac{z^k}{k z_n^k}}$$

(the exponential factor in the product is replaced by 1 if $k = 0$). If in addition, it is discovered that the entire function $g(z)$ is a polynomial of degree m, then the larger of the two numbers m and k is called the *genus* p of the function $f(z)$. The genus is set equal to infinity when at least one of the two conditions does not hold. It is easy to verify that for the functions $\sin z$, $(\sin \sqrt{z})/\sqrt{z}$, and $\sin(\sin z)$ the genus is 1, 0, and ∞ respectively.

Laguerre was particularly interested in the analogy between the genus of an entire function and the degree of a polynomial in the case of real coefficients. Thus, it follows immediately from Rolle's theorem that if such a polynomial has only real zeros, all the zeros of its derivative are also real. In the 1882 note just mentioned Laguerre stated a theorem that was proved completely by Borel in his monograph *Leçons sur les fonctions entières* (Paris 1900): *In the case of an entire function of genus p having q complex zeros in addition to its real zeros, the derivative can*

H. Poincaré

have (at most) $p + q$ complex zeros, in addition to the real zeros whose existence follows from the multiplicity of the zeros of the function and Rolle's theorem.

Poincaré took up the search for relations between the genus of an entire function on the one hand and the growth of its maximum modulus $M(r)$ and the asymptotic behavior of the sequence of coefficients of its Taylor expansion $f(z) = \sum_{0}^{\infty} c_n z^n$ on the other. In particular in his paper "Sur les fonctions entières" (*Bull. soc. math. France*, **11** (1883), pp. 136–144) he established that for a function of genus p and any $\varepsilon > 0$ the following relations hold:

$$\lim_{z \to \infty} e^{-\varepsilon|z|^{p+1}} |f(z)| = 0 \quad \text{and} \quad \lim_{n \to \infty} c_n \Gamma\left(\frac{n + \varepsilon + 1}{p + 1}\right) = 0,$$

where $\Gamma(z)$ is the gamma function. Ten years later the converses of these theorems were proved by J. Hadamard in his paper "Étude sur les propriétés des fonctions entières et en particulier d'une fonction considérée par Riemann," (*J. math. pures et appl.* (4), **9** (1893), pp. 171–215).

However it followed already from these early papers that the concept of genus was not a happy find, if one may say so. Thus, for example, for two entire functions $f(z)$ and $g(z)$ of genus not higher than p one cannot conclude that the genus of

$f(z) + g(z)$ is not higher than p; on the contrary it is possible that it exceeds p (but not by more than 1).

The concept of order of an entire function, which was introduced by Borel in 1896[128] and plays the same role for entire functions that degree plays for polynomials, had enormous significance. Years were to pass before it occupied the decisive place, having supplanted the concept of genus. Not without reason did Borel accompany it by the cautious epithet "apparent" (*ordre apparent*).

Since the relation $\varlimsup\limits_{r\to\infty} \dfrac{\ln M(r)}{\ln r} = \infty$ holds for an entire transcendental function $f(z)$ ($M(r)$ increases faster than any power of r), it is natural to define the order (of growth) as the quantity $\varlimsup\limits_{r\to\infty} \dfrac{\ln\ln M(r)}{\ln r} = \rho$. It turns out to be equal to 1, for example, for e^z or $\sin z$ and $1/2$ for $(\sin\sqrt{z})/\sqrt{z}$, to the natural number n for e^{z^n}, and to ∞ for e^{e^z}.

A relation between the coefficients and the order

$$\rho = \varlimsup_{n\to\infty} \frac{\ln n}{\ln \sqrt[n]{1/|a_n|}},$$

which is analogous to the Cauchy-Hadamard formula, follows from combining the results of Hadamard (1893) and Borel (1896).

While the order ρ characterizes the maximum possible rate of growth of a transcendental entire function, $\varkappa$ — the exponent of convergence of the sequence of zeros $\{z_n\}$, $(z_n \neq 0)$ — gives one of the possible characterizations of the maximal growth of $|z_n|$ as a function of n, and consequently the greater or lesser abundance of zeros in the plane. It is defined by the formula

$$\varkappa = \varlimsup_{n\to\infty} \frac{\ln n}{\ln |z_n|}$$

and can also be characterized as the lower bound of the numbers N ($0 \leq N \leq +\infty$) for which the series $\sum\limits_{n=1}^{\infty} \dfrac{1}{|z_n|^N}$ converges; for the zeros of $\sin z$, for example, $\varkappa = 1$.

Hadamard established that the relation

$$\varkappa \leq \rho$$

always holds.

This result, together with a series of other remarkable theorems, was obtained by J. Hadamard in the paper "Étude des propriétés des fonctions entières et, en particulier, d'une fonction considérée par Riemann" (1893) mentioned above.

At this point we must return to the classical paper of Riemann "Über die Anzahl der Primzahlen, die unter einer gegebenen Grösse liegt" (1859) mentioned above. Although it belongs number theory, since its announced purpose (which, as

---

128) See, for example, the monograph of Borel mentioned above.

it happens, was not achieved) was to study the asymptotic behavior of the function $\pi(n)$, the number of primes not greater than n (see Book 1, Chapt. II), its value goes beyond number theory and is very important for the history of mathematics, especially for analytic function theory. Attempts to fill the gaps in Riemann's reasoning provided the motivation for other papers, which have continued to appear down to the present time. Among such papers was the paper of Hadamard just mentioned, as shown by from its title.

Riemann's great merit was that he reintroduced into mathematics (number theory and theory of functions) the function $\zeta(s)$, which had been known to Euler (see HM, T. 3, p. 109). For a real $s > 1$ it could be defined by either of the following formulas:

$$\zeta(s) = \prod \frac{1}{1 - 1/p^s} = \sum \frac{1}{n^s},$$

where p ranges over the primes and n over all natural numbers. Riemann's brilliant idea was to regard $\zeta(s)$ as a function of a complex variable s. By ingenious transformations he proved that it can be analytically continued to the entire complex plane as a single-valued meromorphic function and that it has a simple pole at the point $s = 1$ and simple zeros at the points $-2, -4, -6, \ldots$ (the so-called trivial zeros). After establishing for $\zeta(s)$ a functional equation that was also known to Euler in 1749 (the product $\Gamma(s/2)\pi^{-s/2}\zeta(s)$ does not change if s is replaced by $1 - s$ (see HM, T. 3, p. 338)), Riemann introduced the auxiliary entire function

$$\xi(t) = \Gamma(s/2 + 1)(s - 1)\pi^{-s/2}\zeta(s).$$

Here $s = 1/2 + it$ so that $\xi(t)$ is an even function by virtue of the theorem just mentioned. Due to the factor $\Gamma(s/2 + 1)$, none of the trivial zeros of $\zeta(s)$ (i.e., the numbers $-2, -4, \ldots$) is a zero of $\xi(t)$. Therefore the numbers α — the zeros of $\xi(t)$ — are generated only by the nontrivial zeros ρ of the function $\zeta(s)$, to which they are related by the equation $\alpha = (2\rho - 1)/2i$. In regard to them, Riemann remarked that they all lie in the strip $|\operatorname{Im} t| < 1/2$ and stated a number of propositions about them without proof, one of which — the Riemann conjecture — remains neither proved nor disproved as of the present: *All the zeros of $\xi(t)$ are real numbers, i.e., all the non-trivial zeros of $\zeta(s)$ lie on the line* $\operatorname{Re} s = 1/2$. Riemann reduced the study of the asymptotic behavior of the function $\pi(n)$ to the study of the distribution of these zeros (see Book 1, Chapt. III).

In the 1893 paper mentioned above Hadamard confirmed some of Riemann's assertions. In doing so he used the general dependence between the coefficients of a power series, the rate of growth of an entire function, and the distribution of its zeros, in the development of which he had taken such an active part. In particular Hadamard established that for the nontrivial zeros $\{\rho\}$ of the zeta function the series $\sum \dfrac{1}{|\rho|}$ diverges (consequently there are infinitely many of them, as Riemann had claimed), while the series $\sum \dfrac{1}{|\rho|^2}$ converges.

Hadamard went on to obtain the expansion of the function $\xi(t)$ in an infinite product of the form[129]

$$\xi(t) = \xi(0) \prod \left(1 - \frac{t^2}{\alpha^2}\right),$$

from which the corresponding representation of $\xi(s)$ of course follows.

Three years later, in 1896, this result was used by Hadamard himself and simultaneously by the Belgian mathematician Charles de la Vallée-Poussin to give a complete proof of the asymptotic law of distribution of the prime numbers (Book 1, p. 172).

We now return to the general theory of entire functions as it was in the late nineteenth century. By the successive efforts of Poincaré (1883), Hadamard (1893), and Borel (1897, 1900) results were obtained that established a rather complete dependence between the (finite) order ρ of an entire transcendental function $f(z)$ having zeros $\{z_n\}$ and the exponent of convergence $\varkappa \le \rho$. (Here $z_n \ne 0$, and there may in addition be a λ-fold zero at the origin.)

Let k be the smallest nonnegative integer among the numbers N for which the series $\displaystyle\sum_{n=1}^{\infty} \frac{1}{|z_n|^N}$ converges ($[\varkappa] \le k < [\varkappa] + 1$); then

$$f(z) = e^{g(z)} z^\lambda \prod_{1}^{\infty} \left(1 - \frac{z}{z_n}\right) e^{\frac{z}{z_n} + \cdots + \frac{z^k}{k z_n^k}}$$

(here the exponential factor in the product is replaced by 1 if $k = 0$); the entire function $g(z)$ must necessarily be a polynomial of degree m not larger than $[\rho]$, and $\rho = \max\{m, \varkappa\}$. The converse is also true: if the entire function $f(z)$ admits an expansion of this form, where $g(z)$ is a polynomial of degree m and the exponent of convergence of the zeros $\{z_n\}$ is a finite number, with $[\varkappa] \le k < [\varkappa] + 1$, then $f(z)$ is a function of finite order ρ with $\rho = \max\{m, \varkappa\}$.

These theorems are connected with a remarkable sharpening of Picard's theorem for a function of finite order ρ. If A and B are two distinct complex numbers and $\{z_n(A)\}$ and $\{z_n(B)\}$ are respectively the sequences of roots of the equation $f(z) = A$ and $f(z) = B$, then Hadamard's theorem immediately implies the inequalities $\varkappa(A) \le \rho$ and $\varkappa(B) \le \rho$. Borel proved, however, that with only one possible exception (depending only on the function) the equalities $\varkappa(A) = \varkappa(B) = \varkappa$ hold, i.e., independently of the value of A the points at which the function assumes this value are have the same frequency distribution on the plane. In the case when ρ is not an integer there are no exceptional values in this sense.

One can say that these results capped off the basic achievements of the theory of entire functions in the nineteenth century. These results were the contents of Borel's *Leçons sur les fonctions entières* (1900), which was the first book in the famous series of monographs by different authors on the theory of functions of

---

129) This product is given in altered form in Book 1, p. 171.

a real and complex variable, which was later published over several decades by Gauthier-Villars, edited by Borel himself.

As for the general theory of meromorphic functions, it was still in an embryonic state. Here the analogue of the concept of maximum modulus $M(r)$ was lacking. However Jensen's formula had already appeared ("Sur un nouvel et important théorème de la théorie des fonctions," *Acta Math.*, **22** (1899), p. 359). It had the simple form

$$\ln|f(0)| = \frac{1}{2\pi} \int_0^{2\pi} \ln|f(re^{i\varphi})|\, d\varphi - \sum_{|a_j|<r} \ln \frac{r}{|a_j|} + \sum_{|b_k|<r} \ln \frac{r}{|b_k|},$$

where a_j are the zeros and b_j the poles of the function $f(z)$. The point of departure for the later general theory of distribution of values of meromorphic functions was already implicitly contained in this formula. But a quarter-century was still to pass before this theory was established.

Abelian Functions

As already pointed out, Weierstrass' two papers from 1854 and 1856 contained the results of his studies of the problem of inverting hyperelliptic integrals of first kind as stated by Jacobi.

The second of these papers was an expanded exposition of the first. However, it remained unfinished. Here is what Tikhomandritskiĭ wrote about it:

... the paper entitled "Theorie der Abelschen Funktionen" was written under difficult circumstances and unfortunately left unfinished due to the loss of the manuscript, as I heard from Weierstrass himself. He broke off at the beginning of the second section, entitled "Einige allgemeine Betrachtungen über die Darstellung eindeutiger analytischen Funktionen durch Reihen," which contained a digression on his research in the area of elliptic functions and formed the subject of the first paper of the present volume [Volume 1 of his collected works].[130]

However, this incompleteness had been pointed out earlier by Riemann, who, in characterizing the first paper as a survey of "extremely profound" papers of Weierstrass, "crowned with beautiful results," added:

However up to the present time a complete reproduction has followed (*J. für Math*, **52** (1856), p. 285) only of the part of these papers that develops §§ 1, 2, and the first half of § 3 (on elliptic functions) in the paper in question.[131]

Here is the principal content of these papers of Weierstrass.

Let us set $R(x) = A_0(x - a_1) \cdots (x - a_{2\rho+1})$, where the a_j are any complex numbers with $a_j \neq a_k$ for $j \neq k$, and let $P(x) = (x - a_1) \cdots (x - a_\rho)$.

---

130) Tikhomandritskiĭ, M. A. "Karl Weierstrass," *Сообщ. Харьк. Мат. Общ.* [in Russian] (2), **6** (1897), p. 41.

131) Riemann, B. *Werke.* p. 133–134.

Then the Jacobi inversion problem for the corresponding hyperelliptic integrals of first kind can be represented as follows: *find solutions $x_j = x_j(u_1, \ldots, u_\rho)$ of the system of differential equations*

$$du_m = \sum_{j=1}^{\rho} \frac{P(x_j)}{x_j - a_m} \frac{dx_j}{\sqrt{R(x_j)}} \quad (m = 1, \ldots, \rho),$$

satisfying the initial conditions $x_j(0, \ldots, 0) = a_j$ $(j = 1, \ldots, \rho)$. (Jacobi's formulation of the problem can be obtained by termwise integration of each of the equations of the system.) As Jacobi himself had foreseen, there must exist single-valued analytic functions $P_1, \ldots, P_\rho$ of the variables $u_1, \ldots, u_\rho$ such that $x_1, \ldots, x_\rho$ can be regarded as the roots of the equation $x^\rho + P_1 x^{\rho-1} + \cdots + P_\rho = 0$. Analogously, added Weierstrass, there exist single-valued analytic functions $Q_1, \ldots, Q_\rho$ of the same variables such that the polynomial $Q_1 x^{\rho-1} + \cdots + Q_\rho$ in x gives the corresponding values of $\sqrt{R(x_j)}$ for the values of $x = x_j$ $(j = 1, \ldots, \rho)$.

It should follow from this theorem that all rational and symmetric functions of $x_1, \ldots, x_\rho$ and $\sqrt{R(x_1)}, \ldots, \sqrt{R(x_\rho)}$ are single-valued analytic functions of $u_1, \ldots, u_\rho$. In particular, setting $\varphi(x) = (x - x_1) \cdots (x - x_\rho)$, one can assert that $\sqrt{h_1 \varphi(a_1)}, \ldots, \sqrt{h_{2\rho+1} \varphi(a_{2\rho+1})}$, where $h_1, \ldots, h_{2\rho+1}$ are constants, are single-valued analytic functions of $u_1, \ldots, u_\rho$; the functions $P_1, \ldots, P_\rho$ introduced above can be expressed in terms of them. Weierstrass denoted the functions $\sqrt{h_k \varphi(a_k)}$ with a suitable set of constants h_k as $\mathrm{al}(u_1, \ldots, u_\rho)_k$ (see above; $k = 1, \ldots, 2\rho+1$). These functions are distinguished, said Weierstrass, by being completely analogous to the elliptic functions $\sin \mathrm{am}\, u$, $\cos \mathrm{am}\, u$, and $\Delta \mathrm{am}\, u$, to which they reduce when $\rho = 1$. It is these functions and a number of others connected with them that one should reasonably call Abelian functions. The author's purpose was to study them. Weierstrass considered his main achievement to have been that he had succeeded in extending the method applied in his very first paper from 1840 on elliptic functions ("Über die Darstellung der modulären Funktionen") to Abelian functions and establishing that all the functions can be represented as fractions with a common denominator, and with numerator and denominator expandable in series of positive integer powers of the variables $u_1, \ldots, u_\rho$ that are absolutely convergent. This result, the author asserted, had been obtained by him in 1847, although he had not published the proof.

In this work Weierstrass was relying on the following property of Abelian functions (which he called the principal property: If u_j is replaced by $u_j + v_j$, the values of the Abelian functions thereby obtained can be expressed rationally in terms of

$$\mathrm{al}\,(u_1, \ldots, u_\rho)_k \text{ and } \mathrm{al}\,(v_1, \ldots, v_\rho)_k \quad (k = 1, \ldots, 2\rho+1)$$

and their first-order partial derivatives. It was therefore a question of an algebraic addition theorem for them.[132]

---

132) Actually one can get by without the partial derivatives here. See, for example, Siegel, K. *Automorphe Funktionen in mehreren Variabeln*, Göttingen 1955, pp. 160–162, (Remark 1).

As for the method, which, as already stated, was originally developed by Weierstrass using the example of elliptic functions, in its main outline it consists of the following: It was proved using general theorems on differential equations that the solutions $x_j(u_1, \ldots, u_\rho)$ of the initial system, like the functions $R(x_j)$ $(j = 1, \ldots, \rho)$ are single-valued analytic functions of the variables $u_1, \ldots, u_\rho$ in a neighborhood of the origin. Then a suitable application of Abel's theorem on integrals of first kind enabled Weierstrass to pass from the expansion in a neighborhood of the origin to the analogous expansions in a neighborhood of the point $u_k = b_k$ $(k = 1, \ldots, \rho)$ arbitrarily far from the origin (in the case $\rho = 1$, i.e., when the integrals are elliptic, it suffices to appeal to the addition theorem for elliptic integrals). As a result it was established that the rational symmetric functions of $x_1, \ldots, x_\rho$ (and $\sqrt{R(x_1)}, \ldots, \sqrt{R(x_\rho)}$), in particular Abelian functions, have the character of rational functions in some neighborhood of any point of the space of the variables $u_1, \ldots, u_\rho$, i.e., can be represented as the quotient of two power series (generally depending on the neighborhood). The difficulty that Weierstrass could not completely overcome in its general form was to derive the existence of a representation of each Abelian function as the quotient of two everywhere-convergent power series, i.e., entire functions, from what had been said. Here is what Weierstrass himself said about this:

> Here we encounter a problem that, as far as I know, has not yet been studied in its general form, but is nevertheless of great importance (von besonderer Wichtigkeit ist) for the theory of functions.[133]

Weierstrass made an attempt to solve this problem under the additional hypothesis that the function satisfies an algebraic differential equation, but on this route he succeeded in reaching the required result only in the case of a function of one variable. We attribute particular importance to the general problem stated above, to which Weierstrass returned many times in almost the same language in his papers, lectures, and letters. His attempts to solve it led him, in particular, to the theorem on the decomposition of an entire function into factors, from which it followed that the problem has a positive answer in the case of one variable (see above).

In communicating this last result to Kovalevskaya in a letter of 16 December 1874, Weierstrass noted

> To this we can add the next theorem, which *is regarded as unproved in my theory of Abelian functions* [our emphasis]. Every single-valued analytic function of x that has the character of a rational function for each value of this quantity can always be represented as the quotient of two ordinary everywhere-convergent power series, moreover in such a way that the numerator and denominator have no common zeros.[134]

It is curious that on the same day he sent a letter to Schwarz in which, after noting that he needed this theorem to develop the theory of single-valued functions, since

133) Weierstrass, K. *Mathematische Werke.* Bd. 1, S. 347.
134) Kochina, P. Ya. *Briefe von Weierstrass an Sophie Kowalewskaja. 1871–1891.* S. 193.

that theory was lacking a foundation, he added, "Up to now I have been seeking the solution of this problem on the wrong route."[135]

In our opinion P. Dugac is not completely correct when he concludes that "this letter also shows the extent to which Weierstrass' thought was directed toward the unifying theorems of analysis."[136] For Weierstrass this was a matter of routes for the search that had not been successful, not providing a link in his construction of the general theory of Abelian functions.

This problem was the motivation for studying general properties of functions of several variables, in particular constructing the divisibility theory of multiple power series and characterizing rational functions as functions having rational character everywhere including infinity. However, the main problem, as already mentioned, was not completely solved by him; and that, in our opinion, is one of the reasons the theory of Abelian functions did not receive a complete exposition in his hands. The solution of this problem was obtained only in the 1880's by Poincaré and Cousin (see below).

Returning to the paper of Weierstrass on Abelian functions published in Crelle's *Journal* in 1856, one must note another essential reason why the author did not publish the expected continuation of this paper. That reason was the publication of Riemann's fundamental paper "Theorie der Abel'schen Functionen" in 1857, in the same volume of Crelle's *Journal*. In this paper the general problem was treated from the point of view of Riemann's doctoral dissertation and in essence amounted to an extensive field of application of the "Dirichlet principle" to the construction of the foundations of the theory of algebraic functions and their integrals. Although the principle itself was not in Weierstrass' style and seemed to him to be an illegitimate application of "the transcendental" to establish fundamental algebraic truths (see his letter to Schwarz of 3 October 1875, cited below), nevertheless Riemann's paper so far surpassed in breadth and profundity anything that Weierstrass could have given at the time (and in doing so got by with minimal computations, in contrast to Weierstrass' paper) that Weierstrass decided to withdraw, at least for a time.

We shall give a resume of the results of Riemann's paper below. At the present juncture we note only that Weierstrass continued to study the questions of the theory of Abelian functions to the end of his life, gradually broadening and deepening his research. He devoted several courses of lectures, letters to colleagues, and published communications to this basic problem. In all these writings he discussed only particular, but fundamentally important questions of the general theory, for example, the general properties of $2p$-fold periodic single-valued functions of p complex variables. Judging from all this material, he succeeded in going far from the statements and results of the papers from the 1840's and 1850's studied above. But even in the huge fourth volume of his *Mathematische Werke* (1902) — more than 600 pages — in which his lectures on the theory of Abelian functions from

---

135) Dugac, P. "Eléments d'analyse de Karl Weierstrass," *Arch. Hist. Exact Sci.*, **16** (1973).

136) *Ibid.*

the winter semester of 1875–1876 and the summer semester of 1876 are printed in full, two-thirds of the work is devoted to the theory of algebraic functions and Abelian integrals, and only one-third treats Jacobi inversion problems for Abelian integrals (to be sure, in contrast to his earlier papers, he did not confine himself to hyperelliptic integrals here). With good reason the German mathematicians Hettner and Knoblauch, who prepared the text of these lectures for print in 1889, wrote in the preface that the theory of Abelian functions in the proper sense of the term is sketched only briefly (nur kurz skizziert ist), while the fundamental algebraic research and the theory of Abelian integrals are presented with precision. The general theory of Abelian funcions, which we shall discuss further in connection with the work of Poincaré and Picard, remains untouched here.

Abelian Functions (Continuation)

Nevertheless, let us return to Riemann's paper "Theorie der Abel'schen Functionen" (1857). It should be said at the outset that, despite the fact that the papers of Riemann and Weierstrass both bore the same title, the problems these two distinguished authors posed were not the same. The fact is that Weierstrass, as we have seen, defined Abelian functions to be the single-valued analytic functions of several variables to which his solution of the Jacobi problem led. Riemann, however, in accordance with the tradition of the time (which also originated with Jacobi), understood Abelian functions to be simply Abelian integrals, and the heart of his paper was the study of algebraic functions and their integrals, as well as theta functions of several variables. The solution of the Jacobi problem was obtained more or less in passing, as a by-product of the entire study.

Riemann preceded his paper with a brief and clear exposition of the basic propositions of his doctoral dissertation of 1851. In particular he emphasized the concept of a Riemann surface, the topological properties of a surface (order of connectivity), and the Dirichlet principle. The theory of Abelian functions proper (as he understood it) was divided into two parts. The first part contained the general theory of algebraic functions and Abelian integrals on a surface of any genus p. From our point of view, it is this part that is of most importance for the subsequent development of the theory. The second part contained the theory of theta functions of p complex variables, and in particular, without exhibiting the corresponding result as an explicitly stated theorem, the solution of the Jacobi inversion problem (not just for hyperelliptic integrals, but for any Abelian integrals).

We shall concentrate here on a survey of the contents of the first part, confining ourselves to just a few brief remarks on the second.

The purpose of the first part, as already noted, was to construct a general theory of algebraic functions and their integrals for the case of an arbitrary closed Riemann surface of any genus.

The case $p = 0$ is almost trivial. (Riemann does not even mention it; the corresponding surface is the Riemann sphere, which as it happens, is not found in

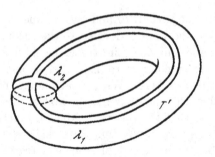

Fig. 16

his published works, though he did use it in his lectures.) At the same time this case is quite instructive for the understanding of the general scheme of the paper. In this case all the rational functions (i.e., single-valued algebraic functions of a point of the sphere) are completely characterized by the fact that they are single-valued on the entire sphere and have only a finite number of singularities of the form $B(z-z_0)^{-1}+\cdots+K(z-z_0)^{-k}$ (if $z_0 = \infty$, the difference $z-z_0$ is replaced, as usual, by z^{-1}). Here the following relation from the theory of residues holds: $\sum B = 0$. Similarly the integrals of rational functions, which are in general multi-valued (the trivial case of Abelian integrals), are characterized by having only a finite number of singularities of the more general form $A\ln(z-z_0)+B(z-z_0)^{-1}+\cdots+K(z-z_0)^{-k}$ with the condition $\sum A = 0$.

If each point ε_ν for which $A_\nu \neq 0$ is joined to a fixed point of the sphere by some path l_ν (the various l_ν must not intersect themselves or one another), single-valued branches of the integral, differing only by additive constants, can be defined on the sphere cut along these paths. Each branch undergoes a jump of $-2\pi i A_\nu$ in passing from one edge of the cut l_ν (the negative edge) to the other (the positive edge).

Riemann was able to show that these elementary regularities, with a suitable modification to fit the topological properties of the surface, can also be extended to any closed surface of arbitrary genus $p > 0$.

In accordance with the principles of his dissertation he regarded such a surface T as a multi-sheeted covering of the plane (sphere). If any point is singled out on this surface as a hypothetical boundary point, then starting from that point one could make successive closed cuts, distinguishing the two edges of each (positive and negative). These cuts, linked together, so to speak, are performed until the remaining portion T' of the surface T becomes simply connected. The number of such cuts is always even $(2p)$, and is a topological invariant of the surface. As already mentioned above, in 1865 Clebsch proposed calling half of this number the *genus* of the surface.

A topological model of a closed surface of genus p is provided by a sphere with p handles. Figure 16 uses the example of a torus $(p = 1)$ to illustrate the process described above for converting the original surface T into a simply connected surface T'. Here by means of an obvious topological transformation the region T' can be made into a rectangle whose opposite sides represent the two edges of the same cut.

Since the surface T of genus $p > 0$ covers the sphere, which is a surface of genus 0, it covers it with several layers, say n, and a finite number of branch points arise on it. If ω is the number of simple branch points (i.e., branch points of order 1; Riemann regarded a branch point of higher order $\nu > 1$ as the coalescence of ν simple branch points), the relation $\omega - 2n = 2(p - 1)$ holds among n, ω, and p.

A problem entirely analogous to the elementary problem posed above for the sphere is to give a complete characterization of the class of functions that are analytic, in general multi-valued, on T, and have only a finite number of singularities, each characterized by an expression of the form

$$A \ln(z - z_0) + B(z - z_0)^{-1/\nu} + \cdots + K(z - z_0)^{-k/\nu},$$

where the nonnegative integer $\nu - 1$ is the order of the branch point above z_0 (if the point of the surface on which the singularity lies is not a branch point, then $\nu = 1$). The only restriction on the coefficients of these expressions is that $\sum A = 0$.

If each singular point ε_ν for which the corresponding coefficient A is not zero is joined to some fixed point of the simply connected surface T' by a path l_ν (the l_ν must not intersect themselves or one another), the single-valued branches of the function with these singularities on the remaining part of T' will, first of all, undergo a jump equal to $-2\pi i A_\nu$ when crossing l_ν and, second, undergo jumps h_j in passing from one edge of any of the original cuts λ_j to the other (we must not forget that different branches of this function on T' differ only by additive constants, since their derivatives are identically equal here).

Relying on the Dirichlet principle, Riemann deduced that there always exists an analytic function ω on T (in general multi-valued) having prescribed singularities at given points (i.e., the difference between ω and the expressions written out above is a function that is analytic in some neighborhood of the corresponding point), and that, in addition, for its single-valued branches on T' the complex numbers h_h that express the jumps undergone in passing across the discontinuities λ_j, $j = 1, \ldots, 2p$, (called by Riemann *moduli of periodicity*) can have any preassigned (arbitrary) real parts. This proposition can be regarded as the basis for the entire Riemann theory of algebraic functions and their integrals. At the same time it is a broad generalization of a regularity that, until Riemann, had been known for the sphere ($p = 0$) and could have been stated for the torus ($p = 1$) as a deduction from the theory of elliptic functions and their integrals, which was quite well developed at the time.

Within the class he was studying Riemann distinguished three simple classes of functions such that any function ω could be expressed as linear combinations of them (and their derivatives):

I. Abelian integrals of first kind, defined as functions that are finite everywhere on T, but in general not single-valued. For these functions all the coefficients $A, B, C, \ldots, K$ are zero. These functions are determined up to an additive constant if the real parts of their moduli of periodicity h_h $(j = 1, \ldots, 2p)$ are prescribed arbitrarily. Riemann established that the number of linearly independent integrals of first kind coincides with the genus p of the surface. He denoted any definite system of such linearly independent integrals as $w_1, \ldots, w_p$ (or $u_1, \ldots, u_p$).

II. Abelian integrals of second kind, defined as those that have a unique simple pole ε located above a certain point z_0 of the z-plane. In the corresponding expansion all the coefficients except B are zero. If $t^0(\varepsilon)$ is one of these integrals, any other having the same pole is of the form

$$t(\varepsilon) = \beta t^0(\varepsilon) + \alpha_1 w_1 + \cdots + \alpha_p w_p + \text{const},$$

where the complex numbers $\beta, \alpha_1, \ldots, \alpha_p$ are chosen so that the real parts of the $2p$ moduli of periodicity h_j have given values.

III. Abelian integrals of third kind, which have one logarithmic pole at each of two distinct points of T, say ε_1 and ε_2. In the corresponding expressions that characterize the singularities of the integral, all the coefficients except A are 0, and the values of A at the two points ε_1 and ε_2 differ only in sign. If one of these integrals is denoted by $\tilde{\omega}^0(\varepsilon_1, \varepsilon_2)$, and we set $A_2 = -A_1 = 1$, then any other with the same two logarithmic poles can be represented as

$$\tilde{\omega}(\varepsilon_1, \varepsilon_2) = \tilde{\omega}^0(\varepsilon_1, \varepsilon_2) + \alpha_1 w_1 + \cdots + \alpha_p w_p + \text{const}.$$

Assuming for simplicity that ε_1 and ε_2 are not branch points and are distinct from ∞, and denoting by z the projection onto the complex plane of a point of T in a neighborhood of ε_1, Riemann remarked that the derivative of $\tilde{\omega}(\varepsilon_1, \varepsilon_2)$ with respect to z (with the real parts of the moduli of periodicity fixed) gives an integral of second kind $t(\varepsilon_1)$ having a simple pole at the point ε_1. Differentiating $\tilde{\omega}(\varepsilon_1, \varepsilon_2)$ with respect to z a sufficient number of times, we obtain an Abelian integral having at ε_1 a pole of any prescribed order.

It follows from all this that any function ω of the class being studied, having prescribed singularities and arbitrarily given real parts for its moduli of periodicity, can be represented as a linear combination of integrals of first kind $w_1, w_2, \ldots, w_p$, integrals of third kind $\tilde{\omega}$, and their derivatives; the function is determined up to an additive constant.

Riemann went on to distinguish the subclass of the class of all functions ω consisting of algebraic functions with the same branching on the surface T (i.e., branching only at the branch points of T with orders not exceeding the orders of those branch points). These had to be single-valued on T and have no singularities except poles. All such functions are encompassed in the formula

$$s = \beta_1 t_1 + \cdots + \beta_m t_m + \alpha_1 w_1 + \alpha_2 w_2 + \cdots + \alpha_p w_p + \text{const}.$$

The poles ε_i of the functions t_i are assumed to be, distinct (simple) in general. However poles of multiplicity ρ ($1 < \rho \le m$) at certain points η are also possible; then the ρ functions t_i corresponding to this point are replaced by a function $t(\eta)$, and its first $\rho - 1$ derivatives with respect to z. The constants β and α can be chosen so that s is single-valued on T, i.e., takes the same value on both edges of any cut l_j. The number of arbitrary constants that the function s depends on (s has poles of first order only at the m prescribed points of T and is continuous at all other points of the surface) is in all cases equal to $2m - p + 1$ (consequently m cannot be less than a certain number depending only on the genus p of the surface). Riemann proved that a function of this type is indeed algebraic in the ordinary sense (a similar result had been obtained before him in a different, less precise form, by Puiseux). To be specific there exists an algebraic equation $F(\overset{n}{s}, \overset{m}{z}) = 0$, where $F(\overset{n}{s}, \overset{m}{z})$ is a polynomial of degree n with respect to s and degree m with respect to z, which is either irreducible or a power of an irreducible polynomial, which $s = s(z)$ satisfies identically.

Throughout the construction just described Riemann started from a closed Riemann surface T and arrived at an algebraic function on it. He went on to devote one section of his paper to the opposite line of thought: he started from an irreducible algebraic equation $F(\overset{n}{s}, \overset{m}{z}) = 0$ and arrived at a corresponding n-sheeted Riemann surface with certain branch points and an algebraic function $s = s(z)$ as a single-valued function on T. In doing this he essentially obtained the same results as Puiseux, but without citing him anywhere.

Returning to the original formulation of the problem, Riemann sought a more general expression for functions s' having the same branching as s and poles of first order at m' prescribed points of the surface T and no others. He established that there exists a family of rational functions of s and z that depend linearly on $m - p + 1$ arbitrary constants and satisfy these conditions. Of course $s' = R(s, z)$ are also algebraic functions on T and have the same branch points, but they satisfy different algebraic equations.

Without going into the details of the first part of Riemann's paper, we note only that he combined irreducible algebraic equations into a single class if and only if for any pair of them $F(s, z)$ and $F_1(s_1, z_1)$, it was possible to express s and z rationally in terms of s_1 and z_1 so that the first equation became the second and, conversely it was possible to express s_1 and z_1 rationally in terms of s and z so that the second equation became the first. In this case the corresponding Riemann surfaces T and T_1 admit a one-to-one conformal mapping onto each other, i.e., they are conformally equivalent and consequently have the same genus p, although the degrees of the equations $F(s, z) = 0$ and $F_1(s_1, z_1) = 0$ might be different in both variables. This enabled him to speak of one class of algebraic functions with the same branching. Such a class can be realized on any of the corresponding conformally equivalent surfaces T: the functions of this class can be represented as the set of rational functions of the corresponding variables s and z, regarded as functions of any one of them. In this way the problem of the minimal degree of the equation $F(s, z) = 0$ in the given class (which depends on the genus p) arose and was solved.

The second part of Riemann's paper was devoted to the study of theta functions of p complex variables, defined by series of the form

$$\theta(v_1, \ldots, v_p) = \sum_{-\infty}^{\infty} \cdots \sum_{-\infty}^{\infty} \exp\left(\sum_{\mu=1}^{p} \sum_{\mu'=1}^{p} a_{\mu\mu'} m_\mu m_{\mu'} + 2\sum_{\mu=1}^{p} v_\mu m_\mu\right),$$

where the outside sums extend over all ordered sets of integers $(m_1, m_2, \ldots, m_p)$. Riemann gave a condition for convergence of such series in the special case when $p = 2$ encountered in the work of Göpel and Rosenhain: the real part of the quadratic form $\sum_{\mu=1}^{p} \sum_{\mu'=1}^{p} a_{\mu\mu'} m_\mu m_{\mu'}$ must be negative for all choices of integers $m_1, \ldots, m_p$, not all zero. This is a condition on the coefficients of the form $a_{\mu\mu'}$.

The most important property of functions of this class is that they are entire functions having $2p$ systems of moduli of periodicity: First, $(0, \ldots, \pi i, \ldots, 0)$ (here πi occurs in the jth place and all the other places are zero, $j = 1, 2, \ldots, p$); second, $(a_{1\mu}, a_{2\mu}, \ldots, a_{p\mu})$, $\mu = 1, \ldots, p$. If the numbers of one of the first systems are added to $v_1, \ldots, v_p$ respectively (i.e., v_μ is replaced by $v_\mu + \pi i$, all other v being left unchanged), then $\theta(v_1, \ldots, v_p)$ retains its value. If the numbers $a_{1\mu}, \ldots, a_{p\mu}$ respectively are added to $v_1, \ldots, v_p$ for some μ ($\mu = 1, \ldots, p$), then $\theta(v_1, \ldots, v_p)$ acquires an exponential coefficient $e^{2v_\mu + a_{\mu\mu}}$.

The theta functions can be made into an auxiliary tool for studying Abelian functions due to the fact that the values of p linearly independent integrals $u_1, \ldots, u_p$ can be taken in place of $v_1, \ldots, v_p$, and each of these integrals is an Abelian integral of first kind of a certain rational function of s and z on a closed Riemann surface of genus p. Then $\ln \theta$ becomes a function of one variable z that can change only by a linear function of the quantity u as the points s and z return to their original values after varying continuously. Here it is necessary to choose a system of $2p$ closed cuts $a_1, \ldots, a_p$, $b_1, \ldots, b_p$ of the surface T (which make it into a simply connected surface T') in a certain special way, namely so that the moduli of periodicity of the integral of first kind u_μ is πi on the cut a_μ and 0 on all the others. As for the modulus of periodicity $a_{\mu\nu}$ of the same integral u_μ on the cuts b_ν ($\nu = 1, \ldots, p$): first, the equalities $a_{\mu\nu} = a_{\nu\mu}$ must hold (*Riemann's equalities*); second, the real part of the quadratic form $\sum_{\mu=1}^{p} \sum_{\mu'=1}^{p} a_{\mu\mu'} m_\mu m_{\mu'}$ must be negative for any set of integers $(m_1, \ldots, m_p)$ not all zero (*Riemann's inequality*, which guarantees, as shown above, that the series that defines the theta-function converges). Riemann proved that all these conditions can always be satisfied.

We note, finally, that Riemann solved the Jacobi inversion problem along the way in the process of studying the theta-functions. For some reason, however, he did not state this important result as a special theorem. Moreover he is lacking (at least in explicit form) the representation of any Abelian function (as a symmetric function of the limits of integration — the result of the inversion) as a ratio of two entire functions. Riemann does not even mention that an Abelian function has rational character, i.e., can be represented in a neighborhood of any ordered set of values $(u_1^{(0)}, u_2^{(0)}, \ldots, u_p^{(0)})$ as the quotient of two series of powers of the differences $u_1 - u_1^{(0)}, \ldots, u_p - u_p^{(0)}$. We know the importance that Weierstrass assigned to this fact.

Already in his paper "Über die allgemeinsten eindeutigen und $2n$-fach periodischen Funktionen von n Veränderlichen" (*Monatsber. Dtsch. Akad. Wiss. Berlin*, 1869)[137] he had called attention to the fact that the theta functions considered by Riemann are not the most general, since they depend on $p(p+1)/2$ essential constants (the moduli of periodicity $a_{11}, \ldots, a_{pp}, a_{12} = a_{21}, \ldots, a_{ij} = a_{ji}, \ldots$) subject to certain conditions (the Riemann equations). Weierstrass sought a generalization of the Jacobi problem that would lead to the most general theta functions. Moreover he took his time in replacing the variables in the expression $\theta(v_1, \ldots, v_p)$ by integrals of first kind on a given closed Riemann surface, thereby reducing the function of p independent variables to a function of one variable z (and an algebraic function s of it). Weierstrass was interested in the function $\theta(v_1, \ldots, v_n)$ primarily as an entire function of p independent complex variables with certain properties of generalized periodicity, and the ratio of two such functions with the same "moduli" as a single-valued $2p$-fold periodic function of p complex variables.

However, Riemann, like Weierstrass had long been interested in the general properties of such $2p$-fold periodic functions independently of the way in which they are generated. As early as 1859, when Riemann visited Berlin, he seems to have raised the question of the existence of such functions in a conversation with Weierstrass. In any case, a letter to Weierstrass of 26 October 1859 (published by Weierstrass after Riemann's death) gives a complete proof of the theorem that a single-valued analytic function of n variables cannot have more than $2n$ systems of linearly independent periods (moduli of periodicity, in the terminology of Riemann).[138] We know that the special case of this theorem for $n = 1$ had been proved by Jacobi as early as 1834. Weierstrass published many notes on such functions. The third volume of his *Mathematische Werke*, published posthumously, contains his long paper "Allgemeine Untersuchungen über $2n$-fach periodische Funktionen von n Veränderlichen," written after 1870.[139] The discussion is carried out independently of the Jacobi inversion problem on the basis of a general construction of the theory of theta functions of n variables as entire functions satisfying conditions of the type

$$\theta(u_1 + w_1, \ldots, u_n + w_n) = \theta(u_1, \ldots, u_n)e^{w_1 u_1 + \cdots + w_n u_n + w_{n+1}}$$

for $2n$ sets of linearly independent values $(w_1 \ldots, w_n)$; these last form a system of $2n$ fundamental generalized periods of the function (each such period is a vector

137) Weierstrass, K. *Mathematische Werke*. Bd. 2.

138) Riemann, B. *Werke*. S. 326–329.

139) Weierstrass, K. *Mathematische Werke*. Bd. 3, S. 53–114.

with n complex components). They can be used to solve the problem of the existence of $2n$-periodic single-valued functions. A question arises at this point: Can all of them be expressed in terms of theta functions? The goal of the paper is to prove that each single-valued analytic function of n variables having rational character in a neighborhood of each finite point and having a system of $2n$ joint linearly independent periods can be formed of theta functions of the same variables.

The proof relies on the properties of Abelian integrals of first kind. Weierstrass was forced to resort to them because the functions are assumed only to be locally meromorphic, but the result required it to be established that they are meromorphic functions, i.e., quotients of entire functions.

The existence theorem for general Abelian functions of p complex variables having a given matrix of $2p$ periods

$$\Omega = \left\| \begin{array}{cccc} \omega_1^{(1)} & \omega_1^{(2)} & \cdots & \omega_1^{(2p)} \\ \cdots\cdots\cdots\cdots\cdots\cdots\cdots \\ \omega_p^{(1)} & \omega_p^{(2)} & \cdots & \omega_p^{(2p)} \end{array} \right\|$$

was stated independently by Riemann and Weierstrass (under certain conditions imposed on the periods expressed by the Riemann equations and inequality). However a complete proof of the theorem was not published until 1883, in a joint paper of Poincaré and Picard ("Théorème de Riemann relatif à fonctions de n variables, admettant $2n$ systèmes de périodes," *Comptes Rendus*, **97** (1883)). In this paper the authors relied on a proposition of Weierstrass (1862) that had been stated in print earlier but also had not been proved, to the effect that there must be an algebraic relation among any $p + 1$ Abelian functions with the same periods. The first proof of this fact was published by Poincaré, but not until 1897, in his paper "Sur les fonctions Abéliennes" (*Comptes Rendus*, **24** (1897)).

The line of thought of Riemann and Picard in the 1883 paper was approximately as follows. The established that the equations $y_1 = f_1(u), \ldots, y_{p+1} = f_{p+1}(u)$, where $f_j(u)$ are Abelian functions of a complex vector $u = \begin{pmatrix} u_1 \\ \cdots \\ u_p \end{pmatrix}$, determine a p-dimensional algebraic (Picard) variety V_p whose points are in one-to-one correspondence with the points of the parallelotope Π_0 (constructed on the fundamental periods $\omega^{(1)}, \ldots, \omega^{(p)}$; each of them is a vector in the p-dimensional complex space $\mathbb{C}^p$). On V_p the variables $u_1, \ldots, u_p$ are represented as everywhere-finite algebraic integrals of total differentials (Picard integrals of first kind, which for $p = 1$ become Abelian integrals of first kind). Among them there are certain relations of equality and inequality that express the conditions first exhibited by Riemann as necessary for the existence of the Abelian functions themselves. The fact that these conditions are not only necessary, as Riemann knew already, but also sufficient, was proved by W. Wirtinger using the theory of general theta functions of p variables in the papers "Zur Theorie der $2n$-fach periodischen Funktionen" (*J. Math.*, **6** (1895); **7** (1896)) and in his book *Untersuchungen über Thetafunktionen* (Leipzig 1895).

One of the strongest results of Poincaré, whose significance goes beyond the theory of Abelian functions, was a theorem originally proved by him in 1883 for the case of two variables ("Sur les fonctions de deux variables," *Acta Math.*, **2** (1883)). As already mentioned, one could say that Weierstrass sought a proof of it throughout his life. Here is the theorem:

If the function $F(X, Y)$ has the character of a rational function in a neighborhood of each finite point (X, Y), it can be represented as the quotient of two entire functions of the same variables.

The original proof of Poincaré was rather complicated, and did not admit of extension to the case of functions of a larger number of variables. A comparatively simple proof of the theorem for any number of variables was first obtained by P. Cousin in his doctoral dissertation ("Sur les fonctions de n variables complexes," *Acta Math.*, **19** (1895)). For that reason the corresponding general theorem, which is the basis of the theory of meromorphic functions of several complex variables, is known as the Poincaré-Cousin theorem. The result published by Weierstrass in 1876 is only the simplest special case of it.

Of the other numerous papers of the late nineteenth century on Abelian functions we mention here a paper of G. Frobenius "Über die Grundlagen der Theorie der Jacobischen Funktionen" (*J. für Math.*, **97** (1884)) devoted to the systematic study of theta functions of p independent variables and also slightly more general functions, called *Jacobian* functions by Klein (Briot and Bouquet called them *intermediate* functions). They were introduced independently of the Jacobi inversion problem. In particular, Frobenius succeeded in proving that the existence of the Jacobian functions implied the same Riemann inequality and equalities that Riemann had derived from the moduli of periodicity of the special classes of theta functions studied by him, introduced in connection with the problem of inverting Abelian integrals.

Using these results of Frobenius, P. Appel ("Sur les fonctions périodiques de deux variables," *J. math. pures et appl.* (4), **7** (1891)) constructed for $p = 2$ a complete theory of Abelian functions defined as single-valued $2p$-periodic functions of p complex variables, without any explicit connection with the Jacobi problem that had generated the whole topic. In the twentieth century a fully developed theory of Abelian functions was constructed for any number of variables in this general interpretation through the efforts of a number of mathematicians, among them Italians (Scorza, Conforto) and Americans (Lefschetz and others). For $p = 1$ this theory degenerates to the theory of elliptic functions. It has found many applications in the algebraic geometry of p-dimensional varieties (cf., for example, the excellent monograph of Conforto).[140]

Automorphic Functions. Uniformization

We have already said that the modular function first considered by Gauss in papers not published at the time was the first nontrivial example of the extensive class

---

140) Conforto, F. *Abelsche Funktionen und algebraische Geometrie*, Berlin 1956.

of automorphic functions. Its importance for the theory of elliptic functions was a direct consequence of the fact that this function expressed the ratio K'/K of the periods of the Jacobi elliptic functions in terms of the square of the modulus k^2 of the corresponding elliptic integral of first kind in the Legendre normal form; this is the reason for the name *modular function.* We have also noted that Schwarz later gave it a purely geometric definition independent of elliptic functions in connection with the conformal mapping of a regular circular triangle with zero angles inscribed in the unit disk onto the half-plane, normalized so that the vertices of the triangle correspond to the points 0, 1, and ∞. Even this simple example of an automorphic function found remarkable applications at the time outside the theory of elliptic functions, such as the solution of the general algebraic equation of degree 5 (C. Hermite, *Théorie des équations modulaires et la résolution de l'équation du 5e degré*, Paris 1859) and the proof of Picard's theorem.

The task of developing the general theory of automorphic functions fell to the mathematicians of the final decades of the nineteenth century (Schottky, Klein, Poincaré). It led to results of even higher theoretical importance, especially as a result of the work of Poincaré during the years 1880–1884.[141] The concept on which these papers are based can nowadays be described in the following terms. Let $S = \{T_n(z)\}$ be a group of fractional-linear transformations, where $T_0(z) = z$, and
$$T_n(z) = \frac{a_n z + b_n}{c_n z + d_n}, \quad a_n d_n - b_n c_n = 1.$$
When $c_n \neq 0$, we call the circle $|c_n z + d_n| = 1$ the *isometric circle* of the transformation. It is completely characterized by the fact that the lengths and areas of all arcs and regions lying inside it are increased under the mapping, while for those outside it lengths and areas are decreased. In the case when the group S is infinite, the closure of the set of centers of the isometric circles is denoted by H. We shall consider only properly discontinuous groups S, i.e., those for which there exist a point z_0 and a positive number r such that for each $T \in S$, $T \not\equiv z$, the condition $|T(z_0) - z_0| \geq r$ holds. Groups of this type contain either a finite or countable set of transformations and, following Poincaré, we can divide them into three classes: 1) elementary groups containing all finite groups and those infinite groups for which H contains at most two points; 2) Fuchsian groups, or groups with a boundary circle, for which there exists a fixed disk (or half-plane) that maps into itself under all $T \in S$; 3) Schottky-Klein groups, encompassing all groups not belonging to either of the other two classes.

Now let $S = \{T_n(z)\}$ be a group (properly discontinuous), let G be a region of the complex plane such that $T_n(G) \subset G$ for all n, and let $w(z)$ be a single-valued analytic function in G. This function is said to be *automorphic* with respect to the group S if $w(T_n(z)) = w$ for all $T_n \in S$. In particular, when S is a Kleinian group, the function $w(z)$ is also called Kleinian (all the concepts and terms given

---

141) Although these and other papers of Poincaré on analytic function theory are being discussed in the present chapter, his work as a whole goes far beyond the time frame that forms the main subject matter of *Mathematics of the Nineteenth Century*, and his biography will be given in one of the subsequent volumes of this series. This applies to a number of scholars mentioned in this chapter: P. Painlevé, É. Borel, J. Hadamard, and others.

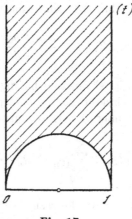

Fig. 17

here — properly discontinuous group, Fuchsian and Kleinian groups — are due to Poincaré).

Trivial examples of automorphic functions (with respect to elementary groups) are e^z, $\sin z$, $\cos z$, $\tan z$, and the elliptic functions (of Jacobi or Weierstrass). In the case of e^z, for example, G is the finite plane, and the group S consists of translations of the form $T_n(z) = z + 2\pi in$, $(n = 0, \pm 1, \pm 2, \ldots)$; in the case of the Weierstrass elliptic function $\wp(z)$ with fundamental periods $2\omega_1$ and $2\omega_2$ (Im $(\omega_2/\omega_1) > 0$) G is also the finite plane and S consists of all translations of the form $T(z) = z + 2\omega_1 m + 2\omega_2 n$, where $m = 0, \pm 1, \pm 2, \ldots$, $n = 0, \pm 1, \pm 2, \ldots$.

In order to give the simplest possible characterization of the Schwarz modular function $\lambda(z)$ as a Kleinian automorphic function we first transform the unit disk $|z| < 1$ into the upper half-plane Im $t > 0$ by a fractional-linear transformation $z = L(t)$ so that the vertices of an equilateral triangle with zero angles map respectively to 0, 1, and ∞. When this is done, the triangle assumes the form of the shaded region in Fig. 17, as had been known to Gauss. As for the function $\lambda(z)$, it becomes $\lambda[L(t)]$; we shall continue to denote it by $\lambda(t)$ and call it the modular function. In this case the region G is the upper half-plane Im $t > 0$, and the group S with respect to which $\lambda(t)$ is automorphic is obtained by taking all possible products of integer powers of the two transformations $T_2(t) = t + 2$ and $T_2(t) = t/(2t + 1)$. All this can easily be verified using the Schwarz reflection principle and taking account of the fact that $w = \lambda(t)$ maps the region of Fig. 17 conformally onto the upper half-plane Im $w > 0$ so that the vertices map respectively to the points 0, 1, and ∞.

In his paper "Sur les fonctions fuchsiennes," (*Acta Math.*, **1** (1882), pp. 193–294) Poincaré proposed a simple method of representing a function automorphic with respect to a given properly discontinuous group S as the quotient of two series (the so-called Poincaré theta series, or, as he himself called them, the thetafuchsian functions).

Suppose, as before, that the transformations of the group S have the form

$$T_0(z) = z, \quad T_n(z) = \frac{a_n z + b_n}{c_n z + d_n} \quad (a_n d_n - b_n c_n = 1),$$

and let $R(z)$ be any rational function having no poles on H (i.e., either at the centers of the isometric circles or at the limit points of these centers). Then, after choosing any natural number $m > 1$, we construct the (Poincaré theta) series

$$\theta(z) = \sum_{n=0}^{\infty} (c_n z + d_n)^{-2m} R(T_n(z)).$$

Remarking that $T_n T_k = T_\nu$, we can easily see that

$$\theta[T_k(z)] = (c_k z + d_k)^{2m} \theta(z).$$

Constructing two distinct functions $\theta_1(z)$ and $\theta_2(z)$ of this form for the same m (and different $R(z)$), we obtain a single-valued analytic function

$$w(z) = \theta_1(z)/\theta_2(z),$$

that is automorphic with respect to the given group S.

The source of the theory of automorphic functions in Poincaré's construction was provided by his reflections on the theory of differential equations. We shall relate briefly what the situation was, following his *Analyse de ses travaux par lui-même*, which was written in 1901 but not published until 1921. The part of this work that interests us bears the title "Fonctions fuchsiennes" and was placed in the first volume of the French version of Poincaré's collected works.[142]

In giving a clear exposition of the evolution of his ideas Poincaré emphasized many times that for him the "thread of Ariadne" was the classical theory of elliptic functions and along with it hyperbolic non-Euclidean geometry, in regard to which, "it seemed at first that it was simply a mind game, of interest only to a philosopher, incapable of being of any use for the mathematician."

In making the original formulation of the problem more precise Poincaré considered a second-order linear differential equation with rational (or more broadly, algebraic) coefficients. Suppose the independent variable is x. It is known that the integral can be represented in a neighborhood of any point by means of a certain series. But Poincaré was interested in the global study of the integral. Here he advanced a brilliant idea similar to those that Abel and Jacobi (and even earlier Gauss) had once proposed: instead of studying the integral as a function of x (which is too complicated), study the variable x itself as a function of the integral, not in isolation, however, but rather the ratio z of two linearly independent integrals of the equation: $x = x(z)$. One might conjecture that he was led to this

---

142) Poincaré, H. Œuvres (nouveau tirage). 1951, T. 1.

idea by the representation of the Schwarz modular function as the quotient of two integrals of the hypergeometric equation.

The new formulation of the problem, as Poincaré showed, has two important advantages: first, the function $x = x(z)$ is single-valued; second, for it there exists a group S of fractional-linear transformations with respect to which $x(z)$ is invariant, i.e., $x(z)$ is automorphic with respect to S. Assuming that the coefficients of the transformations $T(z) = \dfrac{az+b}{cz+d} \in S$ are real, Poincaré obtained a group of Fuchsian type (it maps the half-plane into itself, which immediately reduces to the case of a disk, which he called fundamental).

Poincaré went on to determine the conditions under which S is a properly discontinuous group and discovered that in the case of a Fuchsian function, which he was studying, the polygons bounded by arcs of circles orthogonal to the fundamental circle played a role analogous to the period parallelograms for elliptic functions; at first he placed their vertices on the unit circle, but later, using an idea of Klein, began to choose them inside the circle as well. To verify that these polygons cover the interior of the unit disk without any gaps or overlap, he invoked hyperbolic geometry, since all of its theorems are true if one imagines an arc of a circle in the unit disk orthogonal to the boundary circle whenever Lobachevskiĭ spoke of a line.

To obtain a complete analogy of the Fuchsian functions with elliptic functions it was still necessary to obtain series analogous to the theta functions such that a Fuchsian function can be represented as a quotient of two such functions. He actually constructed such series, as indicated above, calling them theta-fuchsian series (Poincaré series). We confine ourselves to this account here in order to note that *the original reason for introducing the class of automorphic functions into mathematics was the theory of differential equations* and to emphasize at the same time that *Poincaré regarded his theory as a far-reaching analogue of the theory of elliptic functions for whose construction he had to invoke hyperbolic geometry instead of Euclidean geometry.*

But this was not the only application of the theory of automorphic functions to the theory of differential equations in Poincaré's hands, and perhaps not even the most important argument in favor of introducing a large new class of transcendental functions into the subject. Another application of these functions, of fundamental importance, was the solution of the so-called *uniformization problem*, obtained in its main outline by the same great mathematician in the same years.

The problem is to find a complex parameter t for a given multi-valued analytic function $y = F(x)$ such that x and y can be expressed as two single-valued functions of t in some region of the t-plane: $x = \varphi(t)$, $y = \psi(t)$. Such a uniformization had been known for a long time for the elementary cases of multi-valued algebraic functions. Thus, for example, in the case of the algebraic equation $x^2 + y^2 = 1$, the uniformization can be obtained either as $x = \cos t$, $y = \sin t$ or using rational functions $x = \dfrac{1-t^2}{1+t^2}$, $y = \dfrac{2t}{1+t^2}$. We shall now let Poincaré himself have the floor to give the description of his achievements in the study of multi-valued functions that he wrote in 1901.

The theory of multi-valued functions is far less advanced than the theory of single-valued functions. I first showed ("Sur une propriété des fonctions analytiques," 1888) that the number of values that they have, if infinite, has the cardinality of the first infinity in the sense of Cantor.

Although the character of the behavior of these functions in a neighborhood of a given point is well known, and although the introduction of Riemann surfaces threw much light on the still-mysterious parts of their theory, much remains to be done before we reach their main properties. For that reason I was inspired by the desire to reduce the study of them to single-valued tran-scendental [functions]. The theory of Fuchsian functions had already brought me close to my goal; in fact I had proved that if $f(x, y) = 0$ is the equation of any algebraic function, one can choose the parameter z so that x and y are single-valued functions of that parameter. I had thereby solved the problem for the simplest class of multi-valued functions, namely algebraic functions.

Naturally I asked whether this was a special property of algebraic functions or one that could be extended to any multi-valued function ("Sur un théorème de la théorie générale des fonctions," 1883). I was able to answer that question and prove the following theorem: *Let the analytic function $y = f(x)$ be multi-valued. One can always find a variable z such that x and y are both single-valued functions of z.*

My point of departure was the proof of the Dirichlet principle proposed by Schwarz. But this principle alone did not enable me to overcome all the difficulties that resulted from the great generality of the theorem to be proved. It was necessary first to define an infinite-sheeted Riemann surface, which I attempted to map conformally onto a region of the plane. I chose this domain so that it would be simply connected, so that none of the singular points belonged to the surface proper and all of them were, so to speak, on its boundary, and finally, so that the function y could not assume two distinct values at the same point of the surface [the *universal covering*].

I selected a finite part R of this surface and mapped it conformally onto a disk, which enabled me to invoke the Schwarz theorem. This mapping was done by means of a certain function u. Then, finally, forcing the region R to increase to infinity, we obtain a conformal mapping of larger and larger parts of our Riemann surface. I then needed to show that the analytic function u that I had mentioned above, tends to a definite finite limit. After this was done, only the first difficulties were overcome. In fact it remained to show that the limit of the function u was itself an analytic function.

For this it was necessary that the analytic function u tend uniformly (gle-ichmässig) to its limit, which I succeeded in proving.

Thus the study of multi-valued functions had been reduced in all possible cases to the much simpler study of single-valued functions.[143]

---

143) *Ibid.*, T. 4.

We have quoted this long excerpt because it conveys, so to speak, the whole pathos of the problem from the original source. Still, some brief explanations are needed.

The theorem of Poincaré mentioned at the beginning, which asserts that the set of values of a multi-valued function is always at most countable, indicates that the set of sheets of a Riemann surface is at most countable. This important theorem was published by the Italian mathematician V. Volterra in the same year of 1888, independently of Poincaré. For that reason it is called the Poincaré-Volterra theorem.

Poincaré went on to tell about his earliest and most impressive results, namely the possibility of uniformizing any algebraic function by means of a some Fuchsian automorphic function. Up to that time the possibility of uniformizing algebraic functions had been known only for curves of genus 0 (unicursal curves, which in particular encompass all conic sections — for this last case the uniformization could always be achieved using rational functions, as shown by the example of the circle given above) or genus 1, which contains all curves of order 3 or 4 — for this case the uniformization can always be achieved using elliptic functions (this theorem was established by Clebsch). Poincaré had thus provided an infinite extension of the previously known possibilities. But in regard to the ideas of the proofs of both this theorem and its extension to the general case of multi-valued analytic functions of any nature (1883), the affair required an initial push, in the form of a letter from Klein to Poincaré (1882) giving one of the considerations stated orally by Schwarz. It was a question of the purely topological problem of constructing a universal covering (Universelle Überlagerungs-Fläche) for any closed Riemann surface. It was this surface for the case of any Riemann surface that Poincaré had in mind in the passage quoted above, where he enumerated the properties that characterize it. One can say that the construction of this infinite-sheeted surface $\mathfrak{T}$ solves the topological part of the uniformization problem, since any functions that are analytic on the Riemann surface F originally given and have only algebraic singularities at its branch points (i.e., are in general not single-valued on the surface) can be explicitly regarded as single-valued functions of the point $\tau \in \mathfrak{T}$ that projects to the point in question on F. The goal now was to establish that the universal covering surface is simply connected and map it one-to-one and conformally onto some region of the t-plane, say a circular polygon whose sides are orthogonal to the unit circle. When this is done, the vertices may lie either on this circle or inside it. The points of the disk t will then represent a uniformizing parameter, and the function $\tau = \tau(t)$ will represent a Fuchsian automorphic function. However, it later turned out that for full generality one had to assume that the mapping of the universal covering can represent the whole finite plane, or even the extended plane.

Poincaré's reasoning in the paper of 1883 to which he referred in his resume of his own work contained other problems that were not completely solved (Hilbert pointed out several in 1900). In complete and final form the general uniformization theorem was established only in 1907 by Poincaré himself ("Sur l'uniformisation des fonctions analytiques," *Acta Math.*, **31** (1907)) and independently by P. Koebe ("Über die Uniformisierung beliebiger analytischer Kurven," *Gött. Nachr.* (1907))

in the form of the following theorem, sometimes called the theorem with boundary circle:

On every universal covering surface $\mathfrak{T}$ there exists a function $t = t(\tau)$ $(\tau \in \mathfrak{T})$ that maps $\mathfrak{T}$ conformally onto either the entire extended plane (elliptic case) or the entire finite plane (parabolic case) or the unit disk (hyperbolic case).

The complex variable t, which ranges over one of these three regions is a uniformizing parameter for all of the multi-valued functions whose Riemann surfaces have the universal covering surface $\mathfrak{T}$. In the hyperbolic case (the most important) the function $\tau = \tau(t)$ is a Kleinian automorphic function in the unit disk that cannot be extended beyond its boundary (in a very special case, the modular function $\lambda(t)$).

The uniformization theorem stated and proved by Poincaré in 1883, to be sure with certain difficulties that he himself analyzed at the beginning of his paper "Sur l'uniformisation des fonctions analytiques" can jusitifiably be considered the summit of the entire development of the theory of analytic functions of a complex variable in the nineteenth century.

Sequences and Series of Analytic Functions

The fundamental result that the limit of a sequence of analytic functions that converges uniformly inside some region (i.e., in each closed disk belonging to the region) is an analytic function was established by Weierstrass in his paper "Zur Theorie der Potenzenreihen," written in Münster in 1841. Actually Weierstrass discussed the possibility of gathering like terms in a series whose terms were power series when the basic series converges absolutely and uniformly. Here the theorem was stated for the case of any number of (complex) variables. As already noted earlier, this paper was not published until it appeared in the first volume of Weierstrass' *Mathematische Werke* in 1894. However, Weierstrass introduced it in his courses on analytic function theory at the University of Berlin and, in addition, published it as a lemma in his paper "Zur Funktionenlehre" (1880), in slightly more general form. To be specific, he proved that it was possible to gather like terms in a series $\sum_{0}^{\infty} P_\nu(z)$ that converges uniformly inside an annulus provided each term of the series was represented by some Laurent series.

However, the main problem of the paper "Zur Funktionenlehre" was, from our point of view, to refute one of the propositions of Riemann's dissertation, which Weierstrass interpreted too literally. Riemann had claimed that the definition of an analytic function he had adopted, i.e., a function satisfying the equation $\partial f/\partial y = i\partial f/\partial x$ at each point of its domain (a Riemann surface) leads to the same class that would be obtained by defining functions by means of the operations of addition, multiplication, and passage to the limit.

Of course the requirement that the region be connected was assumed here by Riemann himself. He had discussed connectivity in § 2 of the dissertation. It was also assumed at the beginning of his 1857 paper on Abelian functions, where Riemann reproduced briefly the contents of his 1851 dissertation. Here, in particular, he discussed the uniqueness of continuation of a function along various strips belonging to the portion of the plane under consideration, assuming the equation $i\partial w/\partial x = \partial w/\partial y$. To be sure, in the conjecture just mentioned Riemann lacked the requirement that the limiting passage be uniform. But one can surmise that Riemann took it for granted.

In this paper, however, Weierstrass considered that he had refuted Riemann by constructing a rather cumbersome example of a sequence of analytic functions that converges uniformly inside the union of several pairwise disjoint regions of two half- planes; the limit of the sequence represented different analytic functions in different regions, functions that were not analytic continuations of one another. Thus the example was constructed under the condition that the set of convergence lacked connectedness — a condition that Riemann explicitly assumed, as we have seen. For that reason Weierstrass' assertion in a footnote, "The contrary was asserted by Riemann," cannot, we believe, be considered fair to Riemann.

We wish to emphasize, therefore, that this paper of Weierstrass is a continuation of the old rivalry between Weierstrass and Riemann, caused mainly by the development of the theory of Abelian functions, a rivalry whose echoes we discern — on an entirely different plane, to be sure — in a fragment of a letter from Weierstrass to Schwarz of 3 October 1875; in this letter Weierstrass wrote that the more he studied the principles of the theory of functions, the more he was convinced that "they must be constructed on the basis of algebraic truth, and therefore one is on the wrong path when the "transcendental" is applied to prove simple and fundamental algebraic theorems; one can see this at a glance, for example, in the reasoning by which Riemann discovered so many important properties of algebraic functions... ."[144] Of course here he had in mind primarily Riemann's paper "Theorie der Abel'schen Functionen," and the theoretical expression of Weierstrass' protest against the "transcendental" was his criticism of Riemann's Dirichlet principle, which preceded this letter (1870). To what has been said here about Weierstrass' jealous (if not worse) attitude to Riemann's ideas and methods one can add that in his correspondence with Kovalevskaya,[145] where Riemann's name is mentioned at least seven times, nearly every time Weierstrass accompanies it by critical remarks. Either he quotes a fragment of a letter of Richelot to himself in which a decisive preference is expressed for the route chosen by Weierstrass in the theory of Abelian functions as opposed to the route of Riemann, Clebsch, and Gordan (20 August 1873) or he mentions "one of the principal defects of his theory of linear differential equations" (19 November 1873), or he shares with Kovalevskaya his intention to write a paper on Abelian functions as a series of letters to Richelot, in which he will be able "to point out the uniqueness of my method without hesitation and get into a criticism of Riemann and Clebsch" (12 January 1875), or he refutes a remark of Hermite to the effect that Riemann

144) Weierstrass, K. *Mathematische Werke*. Bd. 2, S. 225.

145) Kochina, P. Ya. *Briefe von Weierstrass an Sophie Kowalewskie, 1871–1891*. Moscow 1973.

had derived a necessary condition for the so-called Riemann matrix in the general case ("...there is nothing about this in Riemann's papers," he writes on 14 June 1882), or, finally, he criticizes Riemann's definition of the integral as inadequate (16 May 1885).

The 1880 paper we are discussing here, was a continuation of his polemic with Riemann's ideas, but on a different topic, the topic of the very concept of an analytic function, where Weierstrass invariably placed power series as the starting point, in opposition to Riemann's idea.

However that may be, the result of this paper was unexpected both for Weierstrass himself and for his contemporaries. J. Tannery, almost immediately after Weierstrass, remarked that to exhibit the fact that a sequence could converge uniformly to different functions one need not resort to cumbersome computations with elliptic functions, as Weierstrass had done. It sufficed to consider the sequence of rational functions $\dfrac{1+z^n}{1-z^n}$, which obviously converges to $+1$ inside the unit disk and to -1 outside it. Weierstrass discussed this example in detail in a special printed appendix to his basic paper (despite his antipathy to publishing!) entitled "Zur Funktionenlehre. Nachtrag" (*Monatsber.*, 21 febr. 1881).[146] In making detailed comments on Tannery's example, Weierstrass mentioned him, along with Mittag-Leffler and Picard, as one of the young mathematicians producing interesting research related to his papers.

This paper of Weierstrass was also of great interest to his contemporaries because its last section contained an example of a power series $\sum\limits_{0}^{\infty} a_n x^{b^n}$ ($a_n > 0$, b a natural number larger than 1), which could not be continued beyond the boundary of the disk of convergence. This example was connected with an example of a nondifferentiable continuous function of a real variable constructed earlier by Weierstrass, represented by a trigonometric series $\sum\limits_{0}^{\infty} b^{\nu} \cos a^{\nu} t$, $ab > 1 + \frac{3}{2}\pi$ (see below).

Weierstrass' example belongs to the class of so-called lacunary power series. This is the name given to series of the form $\sum\limits_{0}^{\infty} a_n x^n$ in which groups of arbitrarily many successive zero coefficients occur infinitely often. Retaining only the nonzero coefficients, one can write a lacunary power series as $\sum\limits_{0}^{\infty} a_n z^{\lambda_n}$, where $\varlimsup\limits_{n\to\infty} (\lambda_{n+1} - \lambda_n) = \infty$. A theorem of Hadamard ("Essai sur l'étude des fonctions données par leurs développement de Taylor," *J. math. pures et appl.* (4), **8** (1892), 101–186) on lacunary series of the form $\sum\limits_{0}^{\infty} a_n z^{\lambda_n}$ under the assumption $\varlimsup\limits_{n\to\infty} \sqrt[\lambda_n]{|a_n|} = 1$

(i.e., the radius of convergence is 1) asserts that the series cannot be continued if $\lim_{n \to \infty} \frac{\lambda_{n+1}}{\lambda_n} > 1$ (in Weierstrass' example $\lambda_{n+1}/\lambda_n = b > 1$).

An even stronger result is due to the French mathematician E. Fabry (1856–1944), who proved in a paper "Sur les points singuliers d'une fonction donnée par son développement en série et l'impossibilité du prolongement analytique dans les cas très généraux" (*Ann. Ec. Norm.* (3), **13** (1896), pp. 367–399) that, in the same notation, the series cannot be continued if $\lim_{n \to \infty} \frac{\lambda_n}{n} = \infty$ (for example $f(z) = \sum_1^\infty \frac{z^{n^2}}{n}$, where $\lambda_n = n^2$). Obviously this condition holds whenever Hadamard's condition does (for in Hadamard's condition the λ_n increase at least as fast as the terms of a geometric progression with ratio larger than 1); the converse is obviously false.

Starting from the example of a noncontinuable power series, Poincaré, in his survey of his own papers written in 1901 and published in 1921, introduced a peculiar classification of all analytic functions whatsoever. Poincaré called the single-valued analytic functions that cannot be continued to the whole (finite) plane *fonctions uniformes à l'espaces lacunaires* and accordingly distinguished three classes of functions: 1) single-valued functions in the entire plane; 2) single-valued functions with lacunary spaces; 3) multi-valued functions.

He studied the functions of the second class in a number of papers. These include, first of all, his studies of the classes of automorphic functions that he named after Fuchs and Klein (1883). These functions in general cannot be continued beyond a certain disk or more complicated region. Second, in a number of papers he studied functions that are sums of series of the form $\sum_1^\infty \frac{A_n}{z - b_n}$. Assuming that the series $\sum_1^\infty |A_n|$ converges, and the points b_n are everywhere dense on the boundary of some region D, this sum cannot be extended beyond the region D. Poincaré conjectured (1889) that replacing power series by series of this form might lead to a broader concept of analytic continuation in comparison with the Weierstrassian continuation using power series. É. Borel came close to this interpretation in his doctoral dissertation ("Sur quelques points de la théorie des fonctions." Paris 1894; also published in *Ann. Éc. Norm.*, 1895). His ideas on sums of series $\sum_1^\infty \frac{A_n}{z - b_n}$, where the points b_n are dense in some region in which the sum of the series is studied, later gave rise to the theory of functions monogenic in the sense of Borel (1917). These functions have many properties of analytic functions, but in general are not analytic at any point of the plane. Important and interesting studies of the properties of the functions $\sum_1^\infty \frac{A_n}{z - b_n}$ under various assumptions on the coefficients A_n and the points b_n were carried out by A. Denjoy, and quite recently by the Soviet mathematician A. A. Gonchar.

Many papers of the late nineteenth century were devoted to attempts to obtain expansions of analytic functions in series in regions other than disks and annuli. The most general results were obtained by K. Runge (1856–1927), who started from the Cauchy integral formula ("Zur Theorie der eindeutigen analytischen Funktionen," *Acta Math.*, **6** (1885), S. 229–244). He established first of all that in any multiply connected region each analytic function can be represented as the limit of a sequence of rational functions that converges uniformly inside that region. If, however, the region is a simply connected region of the finite plane, one can speak of a sequence of polynomials, i.e., any function that is analytic in a simply connected region can be represented as a series of polynomials that converges uniformly inside the region. Thus for analytic functions of a complex variable an analogue was obtained for Weierstrass' theorem on the representation of any continuous function of a real variable as the sum of a uniformly convergent series of polynomials (1885). Hilbert obtained the same result in 1897 ("Über die Entwicklung einer beliebigen analytischen Funktion in eine unendliche nach ganzen rationalen Funktionen fortschreitende Reihe," *Gött. Nachr.* (1897), S. 63–70) by another method (using approximation of the region from within by regions bounded by lemniscatic curves, i.e., curves of the form $|P(z)| = $ const, where $P(z)$ is a polynomial). Finally, Runge gave an example of a sequence of analytic functions that converged nonuniformly inside a certain region to a function that was analytic in that region ("Zur Theorie der analytischen Funktionen," *Acta Math.*, **6** (1889), 245–248). Later (1901) W. F. Osgood remarked that in such cases each subregion must contain a disk in which the sequence converges uniformly ("A note on the functions defined by infinite series whose terms are analytic functions of a complex variable, with corresponding theorems for definite integrals," *Ann. Math.*, (2) **3** (1901), pp. 25–34).

Runge's theorems gave no effective method of obtaining the corresponding series. For that reason there was naturally some interest in a cycle of papers in which it was shown how, starting from the power series expansion of a function, say in a neighborhood of the origin, one could obtain expansions of the function in series of polynomials that converge uniformly inside a region that is larger than the disk of convergence (provided, of course, the function can be continued beyond the disk). Here also the point of departure was the Cauchy integral, which made it possible to reduce the general problem to the expansion of a function of the form $1/(1-u)$ (the Cauchy kernel) in a series of polynomials ($u = z/s$). The most interesting results in this area were obtained by Borel (1895), who proposed a new method of summing divergent series for the purpose (this method bears his name), and by Mittag-Leffler ("Fullständig analytisk framställning af hvarje entydig monogen funktion hvars singulära ställen utgöra en värdemängd af första slaget," *Öfversigt. Kg. Vet. Akad. Förhandl.*, **39** (1882), 11–45; "Om den analytiska framställningen af en allmän monogen funktion," *Öfversigt. Kg. Vet. Akad. Förhandl.*, **55** (1898), 247–282, 375–385). Mittag-Leffler's problem was to construct a series of polynomials (whose coefficients could be expressed in a certain way independent of the given function in terms of the coefficients of the original power series) representing the function in is so-called principal Mittag-Leffler star. The latter is obtained as the set of all rays originating at the center of the circle of convergence of the power series and extending to the first singular point encountered in analytic continuation of the function along the ray.

A number of important theorems established by various methods made it possible to deduce the uniform convergence of a sequence of analytic (or harmonic) functions inside a region T as a corollary of its convergence on some subset $\overline{T}$. Thus, Runge established in the paper cited above that if a sequence of functions $\{f_n(z)\}$ that are continuous in a bounded closed region T and analytic inside the region converges uniformly on the boundary of T, then it also converges uniformly inside T; consequently its limit is an analytic function in that region. To be sure, his method, which was based on the use of the Cauchy integral over the boundary of the region, assumed that the boundary was rectifiable. But it sufficed to combine the maximum modulus principle with the fundamental theorem of Weierstrass on uniformly convergent sequences of analytic functions to verify that this restriction was superfluous.

A completely analogous theorem for a sequence of harmonic functions was published by C. G. A. Harnack (1851–1888), a professor at the Technische Hochschule in Dresden (*Grundlagen der Theorie des logarithmischen Potentials und der eindeutigen Potentialfunktion in der Ebene*, Leipzig 1887). In this book he also obtained another theorem of a more special character and at first glance much more remarkable: *If the functions $\{u_n\}$, harmonic in a bounded region T, form a nondecreasing sequence, i.e., $u_n \leq u_{n+1}$ at all points of T, and the sequence converges at even one point of the region, then it converges uniformly inside the region, and consequently its limit is also a harmonic function in T.*

It is not possible to pass directly from this result to sequences of analytic functions. But here T. J. Stieltjes (1856–1894), a Dutch mathematician, professor in Groningen and later at the University of Toulouse, succeeded in proving in a posthumously published paper "Recherches sur les fractions continues" (*Ann. fac. sci. Univ. Toulouse* (8), **7** (1894), p. 56) that if a sequence of analytic function $\{f_n(z)\}$ is uniformly bounded in absolute value inside a region T (i.e., in each closed disk contained in T) and converges uniformly in some subregion of T, then it converges uniformly inside T. This theorem of Stieltjes was generalized by Vitali ("Sulle serie di funzioni analitiche," *Rend. Ist. Lombardo* (2), **36** (1903), pp. 771–774) and M. B. Porter ("Concerning series of analytic functions," *Ann. Math.* (2), **6** (1904), pp. 190–192), who showed independently of each other that the conclusion remains valid if convergence holds on some subset having at least one limit point in the region.

All these facts were placed in a clear light in the theory of compact (normal) families of analytic functions developed by P. Montel outside the period covered in the present book. It turned out, in particular that in the Vitali-Porter theorem the condition of uniform convergence on a subset of the region having at least one limit point in the region can be weakened further by assuming only simple *pointwise* convergence.

G. Mittag-Leffler

As for the ideas of the best approximation of analytic functions by polynomials and more generally by rational functions, which received so much development for functions of a real variable in the last quarter of the nineteenth century due to P. L. Chebyshev and his followers, for analytic functions of a complex variable these ideas blossomed only in the twentieth century, mostly due to the work of Faber, Tonelli, Bernshtein, Walsh, and others. It led to a new area of analytic function theory — the so-called constructive function theory.

Conclusion

Analytic function theory arose in the nineteenth century. The research of the first half of the nineteenth century, most of all the work of Cauchy, played a major role in its development. The theory of elliptic functions appeared and began to develop in parallel with it in the work of Abel, Jacobi, Liouville, and Eisenstein. These functions arose in the problem of inverting elliptic integrals and were not recognized as representatives of the class of doubly periodic meromorphic functions until the 1840's. The attempt to solve the inversion problem for hyperelliptic integrals as well encountered difficulties connected with multi-valued functions. Jacobi pronounced "absurd" any function of one variable having more than two independent periods and modified the inversion problem so that its output would be single-valued multiply periodic functions of several variables (Abelian functions). The first successes in this area were achieved by Göpel and Rosenhain.

An important step in the study of multi-valued functions and their integrals was made by Puiseux in a paper on algebraic functions. He showed that the methods invented by Cauchy in his theory of integration and series expansions of single-valued functions could also be used for multi-valued functions. But the decisive contribution in this area was Riemann's doctoral dissertation, with its idea of a Riemann surface, which treated an analytic function as a conformal mapping of one surface onto another and used the Dirichlet principle. This principle had great heuristic power in the subsequent construction of the theory of algebraic functions and their integrals. It opened the second period in the development of analytic function theory, which lasted until the end of the 1870's.

At that time Weierstrass appeared on the historical scene, having worked in quiet and solitude since the early 1840's to prepare his methods of working out the problems of analytic function theory. However after Riemann's memoir on Abelian functions appeared, he halted publication of his own research on that subject. Weierstrass' activity at the University of Berlin expressed itself primarily in the development of carefully prepared courses, including an introduction to analytic function theory, elliptic functions with their applications, and Abelian integrals. Although these courses were not published, they exerted a serious influence on the development of mathematics, embracing an ever-growing audience of highly qualified students.

In the second half of the 1870's Weierstrass' theorem on the infinite-product expansion of an entire function and Mittag-Leffler's theorem on the representation of a meromorphic function in terms of its principal parts appeared. At the same time Weierstrass was laying the foundations of the theory of analytic functions of several variables with his famous "preparation theorem," on the basis of which he constructed, in particular, the theory of divisibility of power series.

Somewhat peripheral to the basic analytic tendency of Weierstrass was the further development of the geometric theory of functions based on the work of Riemann. Here mostly through the efforts of Schwarz, Schottky, and Klein the theory of conformal mapping and the theory of modular functions were developed. It was the Schwarz modular function that, in the hands of Picard, served as the tool for proving his remarkable theorem. From this theorem grew the theory of the distribution of values of meromorphic functions, which reached maturity only in the 1920's in the work of R. Nevanlinna.

In the early 1880's even such an essential part of the theory of functions as the theory of algebraic functions began to diverge from the main theory. This occurred because Dedekind and H. Weber succeeded in giving a new construction of this theory, guided by the analogy of algebraic functions with algebraic numbers (Bk. 1, pp. 125–130). But even with this reduction in its content the history of analytic function theory in the last two decades of the nineteenth century was marked by notable achievements. These achievements opened with Poincaré's research on the theory of automorphic functions and uniformization. His theorem on the possibility of a parametric representation of any multi-valued function using automorphic functions with a boundary circle is one of the highest achievements of all nineteenth-century mathematics.

The stock of facts and concepts relating to functions of several variables grew noticeably. Poincaré and Cousin proved a theorem that had resisted many years of efforts on the part of Weierstrass, to the effect that any function of several complex variables that can be represented locally as the quotient of two power series can be represented as the quotient of two (relatively prime) entire functions. Picard and Appel laid the foundations of a general theory of Abelian functions as $2p$-periodic meromorphic functions of p complex variables. In doing so they used a result of Wirtinger on theta functions of several variables. Thus the accumulation of material of the theory of analytic functions of several variables continued, and in the mid-twentieth century began to supplant the classical theory of functions of one variable. But before this could happen the latter theory was fated to undergo a long and fruitful development along paths containing difficult, still-unsolved problems such as the Riemann problem on the zeros of the zeta-function or the less-well-known coefficient problem in the theory of schlicht functions.

Returning to the late nineteenth century, we note the appearance and development after the mid-1880's of the general theory of series of polynomials and, more generally, rational functions, whose foundation was laid by Runge's theorems; we also note the study of the mechanics of uniform convergence of series of analytic functions, which opened with Stieltjes' theorem. It was from here that the route to the further study of spaces of analytic functions began.

But the clearest expression of the achievements of classical analytic function theory in the late nineteenth century was the proof of the asymptotic law of distribution of prime numbers, obtained simultaneously by Hadamard and de la Vallée-Poussin. This proof was made possible at that time due to the profound development of the theory of entire functions, in which the main participants were Poincaré, Hadamard, and Borel. The last-named author summed up all this work in the first monograph specially devoted to entire functions, published in 1900.

Literature

Here *ИМИ* denotes the journal *Историко-Математические Исследования* (*Historico-Mathematical Research*). Articles in journals bearing a Russian name are in Russian; the titles of the articles are translated in this bibliography.

General Works

Ball, W. W. Rouse. *A Short Account of the History of Mathematics*, 4th ed. (1908); reprint New York 1960.

Bell, E. T. *The Development of Mathematics*, 2nd ed. New York-London 1945.

Bell, E. T. *Men of Mathematics*. New York 1962. (Among others, biographies of Monge, Poncelet, Gauss, Cauchy, Lobachevskiĭ, Jacobi, Hamilton, Weierstrass, Riemann, and Poincaré).

Bourbaki, N., *Éléments d'histoire des mathématiques*. Paris 1984.

Boyer, C. B. *A History of Mathematics*. New York 1968.

Cajori, F. *A History of Mathematical Notations*, Vols. 1–2. London 1928–1929.

Dieudonné, J., ed. *Abrégé d'histoire des mathématiques. 1700–1900*, T. 1–2. Paris 1978.

Enzyklopädie der mathematischen Wissenschaften und ihrer Anwendungen, 2. Aufl., Bd. 1–6. Leipzig 1898–1934, 1952–1968.

Gillispie, Ch. C., ed. *Dictionary of Scientific Biography*, Vols. 1–15. New York 1970–1976.

History of Mathematics in the Countries of the Soviet Union [in Russian], T. 1–4. Kiev 1966–1970.

Klein, F. *Vorlesungen über die Geschichte der Mathematik im 19. Jahrhundert*. New York 1956.

Kline, M. *Mathematical Thought from Ancient to Modern Times*. New York 1972.

Loria, G. *Storia delle matematiche dall' alba civiltà al tramonto del secolo XIX*. Milano 1950.

Mathematics, its Contents, Methods, and Meaning, T. 1–3. Providence, 1963.

May, K. O. *Bibliography and Research Manual of the History of Mathematics*, Toronto-Buffalo 1973.

Rybnikov, K. A. *History of Mathematics* [in Russian], 2nd ed. Moscow 1974.

Scienziati e technologi contemporei, Vol. 1–3. Milano 1974–1976.

Scienziati e technologi dalla origini al 1875, Vol. 1–3. Milano 1975–1976.

Wieleitner, H., *Geschichte der Mathematik*, Bd. 2. Berlin 1939.

Wussing, H., Arnold, W. *Biographien bedeutender Mathematiker*, Berlin 1975. (Among others biographies of Lagrange, Monge, Laplace, Gauss, Bolzano, Cauchy, Möbius, Lobachevskiĭ, Abel, Jacobi, Galois, Weierstrass, Chebyshev, Kronecker, Riemann, Cantor, Klein, Kovalevskaya, Hilbert, and E. Noether).

Yanovskaya, S. A. *Methodological Problems of Science* [in Russian], Moscow 1972.

Yushkevich, A. P., ed. *Chrestomathy of the History of Mathematics* [in Russian], T. 1–2, Moscow 1976–1977.

Yushkevich, A. P. *History of Mathematics in Russia up to 1917* [in Russian], Moscow 1968.

Collected Works and Other Original Sources

Abel, N. H. *Œuvres complètes*, T. 1–2. Christiania 1881.

Argand, J. R. *Essai sur une manière de représenter les quantités imaginaires dans les constructions géométriques*, 2. éd. Paris 1874.

Beltrami, E. *Opere matematiche*, Vol. 1–4. Milano 1902–1910.

Betti, E. *Opere matematiche*, Vol. 1–2. Milano 1903–1914.

Betti, E. "La teorica delle funzioni ellitiche," *Tortolini Annalici*, **III** (1860), 65–159, 298–310.

Bólyai, J. "Appendix" in: *Geometrische Untersuchungen* by F. Bólyai. Leipzig-Berlin 1913.

Borel, E. *Leçons sur les fonctions entières*. Paris 1900.

Borel, E. "Sur quelques points de la théorie des fonctions," *Ann. Éc. Norm. Sup.*, Sér. 3, **12** (1895), 9–55.

Briot, C., Bouquet, J. "Étude des fonctions d'une variable imaginaire," *J. Éc. Polyt.*, **21** (1856), 85–131.

Briot, C., Bouquet, J. *Théorie des fonctions doublement périodiques et, en particulier, des fonctions elliptiques*. Paris 1859.

Briot, C., Bouquet, J. *Théorie des fonctions elliptiques*. Paris 1875.

Casorati, F. *Teorica delle funzioni di variabili complesse*. Pavia 1868.

Casorati, F. *Opere*. Roma 1951.

Cauchy, A.-L. *Cours d'analyse de l' École Polytechnique. Première partie. Analyse algébrique.* in: *Œuvres*, sér. 2, T. 3.

Cauchy, A.-L. *Leçons de calcul différential et de calcul intégral.* Paris 1840–1861. (Edited by Abbé Moigno.)

Cauchy, A.-L. *Œuvres complètes,* T. 1–27 (2. sér.). Paris 1882–1974.

Cayley, A. *Collected Mathematical Papers,* Vols. 1–14. Cambridge 1889–1898.

Chasles, M. *Aperçu historique sur l'origine et le développement des méthodes en géométrie, particulièrement de celles qui se rapportent à la géométrie moderne, suivi d'un mémoire de géométrie sur deux principes généraux de la science, la dualité et l'homographie,* 2. éd. Paris 1875.

Chasles, M. *Traité de géométrie supérieure,* Paris 1852.

Chebyshev, P. L. *Collected Works* [in Russian], T. 1–5. Moscow-Leningrad 1944–1951.

Christoffel, E. B. *Gesammelte mathematische Abhandlungen.* Leipzig-Berlin 1910.

Clebsch, A., Lindemann, F. *Vorlesungen über Geometrie,* Bd. 1–2. Leipzig 1875–1876.

Clifford, W. K. *The Common Sense of the Exact Sciences.* New York 1885.

Clifford, W. K. *Lectures and Essays,* Vols. 1–2. London 1901.

Clifford, W. K. *Mathematical Papers,* London 1882; reprint New York 1968.

Cremona, L. *Sulle trasformazioni geometriche delle figure plane.* Bologna 1863.

Darboux, G. *Sur une classe remarquable de courbes algébriques et sur la théorie des imaginaires.* Bordeaux 1873.

Darboux, G. *Leçons sur la théorie générale des surfaces et les applications géométriques du calcul infinitésimal,* 2 éd., T. 1–4. Paris 1914–1925.

Dupin, C. *Développement de géométrie.* Paris 1813.

Dupin, C. *Applications de géométrie et de mécanique.* Paris 1822.

Durège, H. *Elemente der Theorie der Funktionen einer complexen veränderlichen Grösse, mit besonderer Berücksichtigung der Schöpfungen Riemanns bearbeitet.* Leipzig 1864.

Eisenstein, F. G. M. *Mathematische Abhandlungen.* Berlin 1874.

Euler, L. *Introductio in analysin infinitorum.* Lausannae 1748.

Fuchs, L. *Gesammelte mathematische Werke,* Bd. 1–3. Berlin 1904–1908.

Gauss, C. F. *Werke,* Bd. 1–12. Göttingen 1863–1929. Reprint Hildesheim-New York 1973.

Gauss, C. F. *Disquisitiones Arithmeticae.* Gottingae 1801; in: *Werke,* Bd. 1. Göttingen 1863.

Göpel, A. "Entwurf einer Theorie der Abel'schen Transcendenten erster Ordnung," in: *Ostwalds Klassiker der exakten Wissenschaften,* No. 67. Leipzig 1895.

Grassmann, H. *Gesammelte mathematische und physikalische Werke,* Bd. 1–3. Leipzig 1894–1911.

Hadamard, J. "Étude sur les fonctions entières et en particulier d'une fonction considérée par Riemann," *J. math. pures et appl.*, sér. 4, **9** (1893), 170–215.

Hamilton, W. R. *Elements of Quaternions*. New York 1969.

Hamilton, W. R. *Lectures on Quaternions*. Dublin 1853.

Hamilton, W. R. *The Mathematical Papers*, Vols. 1–3. Cambridge 1931–1967.

Harnack, A. *Die Grundlagen der Theorie des logarithmischen Potentials und der eindeutigen Potentialfunktion in der Ebene*. Leipzig 1887.

Helmholtz, H. *Wissenschaftliche Abhandlungen*, Bd. 1–3, Leipzig 1887.

Hilbert, D. *Grundlagen der Geometrie*. Leipzig 1903.

Holtzmüller, H. *Einführung in die Theorie der isogonalen Verwandtschaft und der konformen Abbildungen mit Anwendung auf mathematische Physik*. Leipzig 1882.

Jacobi, C. G. J. *Gesammelte Werke*, Bd. 1–7. Berlin 1881–1891.

Jordan, C. *Traité d'analyse de l'École Polytechnique*, 3 éd., T. 1–3. Paris 1909–1915.

Jordan, C. *Traité des substitutions et des équations algébriques*, Paris 1870; Nouveau tirage, Paris 1957.

Jordan, C. *Œuvres*, T. 1–4, Paris 1961–1964.

Klein, F. *Gesammelte mathematische Abhandlungen*, Bd. 1–3. Berlin 1921–1923.

Klein, F. *Vorlesungen über das Ikosaeder und die Auflösung der Gleichungen von fünften Grades*. Leipzig 1884.

Klein, F. *Vorlesungen über höhere Geometrie*, 3. Aufl. bearb. und hrsg. von W. Blaschke. New York 1949.

Klein, F. *Vorlesungen über Nicht-euklidische Geometrie*. New York 1960.

Kotel'nikov, A. P. *The Cross-product Calculus and Certain of its Applications to Geometry and Mechanics* [in Russian]. Kazan' 1895.

Kovalevskaya, S. V. *Scientific Works* [in Russian]. Moscow 1948.

Kummer, E. "Allgemeine Theorie der geradlinigen Strahlensysteme," *J. für reine und angew. Math.*, **35** (1874), 319–326.

Lagrange, J.-L. *Théorie des fonctions analytiques contenant les principes du calcul différentiel dégagé de toute considération d'infiniment petits, d'évanouissants, de limites, et de fluxions, et réduites à l'analyse algébrique des quantités finis*. Paris 1797; *Œuvres*. Paris 1881, T. 9, 1–428.

Laplace P.-S. *Œuvres complètes*, T. 1–14. Paris 1878–1912.

L'Huilier, S. "Mémoire sur les polyédrométries," *Ann. math. pures et appl.*, **3** (1812–1813), 169–192.

Lie, S. *Gesammelte Abhandlungen*, Bd. 1–10. Leipzig-Oslo 1934–1960.

Liouville, J. *Leçons sur les fonctions doublement périodiques (faites en 1847)*. Paris 1880.

Listing, J. B. "Der Census räumlicher Complexe, oder Verallgemeinerung des Euler'schen Satzes von den Polyedern," *Abh. Königl. Ges. Wiss. Göttingen*, **10** (1862), 97–180.

Listing, J. B. *Vorstudien zur Topologie*. Göttingen 1848.

Lobachevskiĭ, N. I. *Complete Works* [in Russian], T. 1–5. Moscow-Leningrad 1946–1951.

Lobachevskiĭ, N. I. *Scientific-pedagogical Legacy. Leadership of Kazan' University. Fragments. Letters* [in Russian]. Moscow 1976.

Möbius, A. F. *Gesammelte Werke*, Bd. 1–4. Leipzig 1885–1887.

Monge, G. *Application de l'analyse à la géométrie*. Paris 1807.

Neumann, C. *Vorlesungen über Riemann's Theorie der Abel'schen Integrale*. Leipzig 1865.

On the Foundations of Geometry. A Collection of Classic Works on the Geometry of Lobachevskiĭ and the Development of his Ideas [in Russian]. Moscow 1956. (Contains works of Lobachevskiĭ, Bólyai, Gauss, Minding, Beltrami, Cayley, Klein, Poincaré, Riemann, Helmholtz, Hilbert, V. F. Kagan, and Cartan.)

Peterson, C. *Über Curven und Flächen*. Moskau-Leipzig 1868.

Peterson, K. M., "On Bending of Surfaces," *ИМИ*, **5** (1952), 87–133.

Picard, E. *Traité d'analyse*, T. 2. Paris 1893.

Plücker, J. *Analytisch-geometrische Entwicklungen*. Bd. 1–2. Essen 1821–1831.

Plücker, J. *System der analytischen Geometrie*. Berlin 1835.

Plücker, J. *Theorie der algebraischen Curven*. Bonn 1839.

Plücker, J. *System der Geometrie des Raumes in neuer analytischer Behandlungsweise*. Bonn 1846.

Plücker, J. *Neue Geometrie des Raumes, gegründet auf Betrachtung der geraden Linie als Raumelement*, Bd. 1–2. Leipzig 1868–1869.

Plücker, J. *Gesammelte wissenschaftliche Abhandlungen*, Bd. 1–2. Leipzig 1895–1896.

Poincaré, H. *Œuvres*, T. 1–11. Paris 1928–1956.

Poisson, S. D. "Sur les intégrales des fonctions qui passent par l'infini entre des limites de l'intégration, et sur l'usage des imaginaires dans la détermination des intégrales définies," *J. Éc. Polyt.*, **11** (1820), No. 8, 295–341.

Poncelet, J. V. *Traité des propriétés projectives des figures*. Paris 1822.

Poncelet, J. V. *Applications d'analyse et de géométrie, qui ont servi, en 1822, de principal fondement au* Traité des propriétés des figures, T. 1–2. Paris 1862–1864.

Puiseux, V. "Recherches sur les fonctions algébriques," *J. math. pures et appl.*, **15** (1850), 365–480.

Riemann, B. *Gesammelte mathematische Werke und wissenschaftlicher Nachlass*, Berlin 1990.

Salmon, G. *A Treatise on Conic Sections.* Dublin 1848.

Salmon, G. *A Treatise on the Analytic Geometry of Three Dimensions.* Dublin 1862.

Salmon, G. *Higher Plane Curves.* Dublin 1862.

Schläfli, L. *Gesammelte mathematische Abhandlungen,* Bd. 1–2. Berlin 1902–1909.

Schubert, H. "Die n-dimensionalen Verallgemeinerungen der fundamentalen Abzahlen unseren Raumes," *Math. Ann.,* **26** (1866), 26–51.

Schubert, H. *Kalkül der abzählenden Geometrie,* Leipzig 1879.

Schwarz, H. A. *Gesammelte mathematische Abhandlungen.* Bd. 1–2. Berlin 1890.

Sokhotskiĭ, Yu. V. *The Definite Integrals and Functions Used in Series Expansions* [in Russian]. St. Petersburg 1873.

Sokhotskiĭ, Yu. V. *Elements of the Greatest Common Divisor in Application to the Theory of Divisibility of Algebraic Numbers* [in Russian]. St. Petersburg 1893.

Somov, O. I. *Foundations of the Theory of Elliptic Functions* [in Russian]. St. Petersburg 1850.

Somov, O. I. "On Higher-order Accelerations," *Зап. Акад. Наук,* **5** (1864), Appendix No. 5.

Somov, O. I. "A direct method of expressing the first- and second-order differential parameters and curvature of a surface in any coordinates, rectilinear or curvilinear," *Зап. Акад. Наук,* **8** (1865), Appendix No. 4.

Staudt, Ch. von. *Beiträge zur Geometrie der Lage,* H. 1–3, Nürnberg 1856–1860.

Staudt, Ch. von. *Geometrie der Lage,* Nürnberg 1847.

Steiner, J. *Gesammelte Werke,* Bd. 1–2. Berlin 1881–1882.

Stieltjes, T. J. "Recherches sur les fractions continues," *Mémoires présentés par divers savants à l'Académie des Sciences* (2e Sér.), **32**. Paris 1909.

Stieltjes, T. J. *Œuvres complètes,* T. 1–2. Grøningen 1914–1918.

Study, E. *Geometrie der Dynamen: Die Zusammensetzung von Kräften und verwandte Gegenstände der Geometrie.* Leipzig 1903.

Suvorov, F. M. *On the Characterization of Three-dimensional Systems* [in Russian]. Kazan' 1871.

Tikhomandritskiĭ, M. A. *Inversion of Hyperelliptic Integrals* [in Russian]. Khar'kov 1885.

Tikhomandritskiĭ, M. A. *Theory of Elliptic Integrals and Elliptic Functions* [in Russian]. Khar'kov 1885.

Vyshnegradskiĭ, I. A. "On the Regularity of Direct Action," *Изв. Спб. Практ. Техн. Инст.,* 1861.

Weierstrass, K. *Abhandlungen aus der Funktionentheorie.* Berlin 1894.

Weierstrass, K. *Mathematische Werke,* Bd. 1–7. Berlin 1894–1927.

Wessel, K. "Om directionens analytiske betegning, et forsøg, anvendt fornemmling til plane og sphaeriske polygoners opløsning," *Nye samling af det Kong. Danske Vid. Selsk. Skr.* Ser. 2, **5** (1799), 496–518.

Auxiliary Literature to Chapter 1

Bonola, R. *Non-Euclidean Geometry; a Critical and Historical Study of its Development.* New York 1955.

Boyer, C. G. *A History of Analytic Geometry.* New York 1956,

Coolidge, M. J. *A History of Geometrical Methods.* Oxford 1940.

Crowe, M. J. *A History of Vector Analysis. The Evolution of the Idea of a Vectorial System.* London 1967,

Depman, I. Ya. "Karl Mikhailovich Peterson and his candidate dissertation," *ИМИ,* **6** (1952), 134–164.

Efimov, N. V. *Higher Geometry* [in Russian]. Moscow-Leningrad 1946.

Engel, F. *Urkunden zur Geschichte der nichteuklidischen Geometrie,* Bd. 1–2. Leipzig 1898–1913.

Galchenkova, R. I., Lumiste, Yu. G., Ozhigova, E. P., Pogrebenskiĭ, I. B. *Ferdinand Minding, 1806–1880* [in Russian]. Moscow 1970.

Gerasimova, V. I. *Index to the Literature on Lobachevskiĭ's Geometry and the Development of his Ideas* [in Russian]. Moscow 1952.

Kagan, V. F. *Foundations of the Theory of Surfaces* [in Russian], T. 1–2. Moscow-Leningrad 1947–1949.

Kagan, V. F. *Lobachevskiĭ* [in Russian], 2nd ed. Moscow-Leningrad 1948.

Kagan, V. F. *Foundations of Geometry. A Study of the Foundations of Geometry in the Course of its Historical Development* [in Russian], Pt. 1–2. Moscow 1949–1956.

Kagan, V. F. *Essays on Geometry* [in Russian]. Moscow 1963.

Khil'kevich, E. K. "From the history of the propagation and evolution of the ideas of N. I. Lobachevskiĭ in the 1860's and 1870's," *ИМИ,* **2** (1940), 168–230.

Kötter, E. "Die Entwicklung der synthetischen Geometrie," *Jahresber. Dtsch. Math.-Verein.,* **5** (1901), 1–486.

Kramar, F. D. "Vector calculus of the late eighteenth and early nineteenth centuries," *ИМИ,* **19** (1963), 225–290.

Kramar, F. D., Milyukov, I. D. *Iosef Ivanovich Somov (1815–1876). Mathematician, Mechanician, Pedagogue* [in Russian]. Alma-Ata 1956.

Laptev, B. L. "Theory of parallel lines in the early work of N. I. Lobachevskiĭ," *ИМИ,* **4** (1951), 201–229.

Loria, G. *Curve piani speciali algebraiche e transcendenti. Teoria e storia,* Vol. 1–2. Milan 1930.

Loria, G. *Il passato e il presente delle principale teorie geometriche* Milano 1931.

Modzalevskiĭ, L. B. *Materials for a Biography of N. I. Lobachevskiĭ* [in Russian]. Moscow-Leningrad 1948.

Norden, A. P. "Gauss and Lobachevskiĭ," *ИМИ*, **9** (1956), 145–168.

Norden, A. P. "Foundational questions of geometry in the work of N. I. Lobachevskiĭ," *ИМИ*, **11** (1958), 97–132.

Olonichev, P. M. "The Kazan' geometer Fedor Matveevich Suvorov," *ИМИ*, **9** (1956), 271–316.

Pogorelov, A. B. *Foundations of Geometry* [in Russian]. Moscow 1979.

Pont, J. C. *La topologie algébrique des origines à Poincaré.* Paris 1974.

Pringsheim, A. "Zur Geschichte des Taylorschen Lehrsatzes," *Bibl. mat.*, Ser. 3, **1** (1900), 433–479.

Rashevskiĭ, P. K. *Riemannian Geometry and Tensor Analysis* [in Russian], 3 ed. Moscow 1967.

Reich, K. "Die Geschichte der Differentialgeometrie von Gauss bis Riemann (1828–1868)," *Arch. Hist. Exact Sci.*, **5** (1973), 273–372.

Rozenfel'd, B. A. *History of Non-Euclidean Geometry* [in Russian]. Moscow 1976.

Rozenfel'd, B. A. *Multi-dimensional Spaces* [in Russian]. Moscow 1966.

Rozenfel'd, B. A. *Non-Euclidean Geometries* [in Russian]. Moscow 1955.

Rozenfel'd, B. A. *Non-Euclidean Spaces* [in Russian]. Moscow 1969.

Rybkin, G. F. "Materialism was a fundamental characteristic of the world-view of N. I. Lobachevskiĭ," *ИМИ*, **3** (1950), 9–29.

Simon, M. *Über die Entwicklung der Elementar-Geometrie im XIX Jahrhundert.* Leipzig 1906.

Sommerville, D. M. Y. *Bibliography of Non-Euclidean Geometry.* London 1911.

Stäckel, P., Engel, F. *Die Theorie der Parallellinien von Euklid bis auf Gauss: Eine Urkundensammlung zur Vorgeschichte der nichteuklidischen Geometrie.* Leipzig 1895.

Struik, D. *An Essay on the History of Differential Geometry* [in Russian]. Moscow-Leningrad 1941.

Taton, R. *Gaspard Monge.* Basel 1950.

Vizgin, V. P. "On the history of Klein's 'Erlanger Programm'," *ИМИ*, **18** (1973), 218–248.

Auxiliary Literature to Chapter 2

Bachelard, S. *La représentation paramétrique des quantités imaginaires au début du XIX siècle.* Paris 1966.

Bashmakova, I. G. "On the proof of the fundamental theorem of algebra," *ИМИ*, **10** (1957), 257–304.

Belozerov, S. E. *The Principal Stages of the Evolution of The General Theory of Analytic Functions* [in Russian]. Rostov 1962.

Biermann, K.-R. *Die Mathematik und ihre Dozenten an der Berliner Universität, 1810-1920.* Berlin 1968.

Bottazzini, U. "Riemann's Einfluss auf E. Betti und F. Casorati," *Arch. Hist. Exact Sci.*, **18** (1978), 27-37.

Briefwechsel zwischen Gauss und Bessel. Leipzig 1880.

Brill, A, Noether, M. "Die Entwicklung der Theorie der algebraischen Funktionen in älterer und neuer Zeit," *Jahresber. Dtsch. Math. Verein.*, **3** (1892-1893), 107-566.

Conforto, F. *Abelsche Funktionen und algebraische Geometrie.* Berlin 1956.

Dugac, P. "Éléments d'analyse de Karl Weierstrass," *Arch. Hist. Exact Sci.*, **10** (1973), 41-176.

Enneper, A. *Elliptische Funktionen: Theorie und Geschichte.* Halle 1876.

Ermakov, V. P. *Theory of Abelian Functions without Riemann Surfaces* [in Russian]. Kiev 1897.

Festschrift zur Gedächtnisfeier für Karl Weierstrass. 1815-1965. Köln 1966.

Hurwitz, A. *Vorlesungen über allgemeine Funktionentheorie und elliptische Funktionen.* Berlin 1922.

Hurwitz, C. F. "Über die Entwicklung der allgemeinen Theorie der analytischen Funktionen," in: *Verhandlungen des ersten intern. Mathematiker-Kongr. in Zürich.* Leipzig 1898, S. 91-112.

Jourdain, P. E. B. "The theory of functions with Cauchy and Gauss," *Bibl. mat.*, Ser. 3, **6** (1905), 190-207.

Kochina, P. Ya. *Briefe von Weierstrass an Sophie Kowalewskaja, 1871-1891,* Moscow 1973.

Markushevich, A. I. *Elements of Analytic Function Theory* [in Russian]. Moscow 1944.

Markushevich, A. I. "Yu. V. Sokhotskiĭ's contribution to the general theory of analytic functions," *ИМИ*, **3** (1950), 309-406.

Markushevich, A. I. *Essays on the History of Analytic Function Theory* [in Russian]. Moscow-Leningrad 1951.

Markushevich, A. I. "Gauss' work on mathematical analysis," in: *Karl Friedrich Gauss* [in Russian]. Moscow 1956, pp. 146-216.

Markushevich, A. I. "Fundamental Concepts of Mathematical Analysis and Theory of Functions in the Work of Euler," in: *Leonhard Euler* [in Russian]. Moscow 1958, pp. 98-132.

Markushevich, A. I. *Remarkable Sines* [in Russian], 2 ed., Moscow 1975.

Markushevich, A. I. *Introduction to the Classical Theory of Abelian Functions* [in Russian]. Moscow 1979.

Markushevich, A. I. "Some questions in the history of analytic function theory," *ИМИ*, **23** (1980), 52-70.

Montferrier, A. S. de. *Dictionnaire des sciences mathématiques pures et appliquées*, T. 1–3. Paris 1836–1840.

Nalbandyan, M. B. "Analytic function theory and its applications in the work of Russian mathematicians of the nineteenth and early twentieth centuries," *ИМИ*, **17** (1966), 361–369.

Neuenschwander, E. "The Casorati-Weierstrass theorem," *Hist. Math.*, **5** (1978), 139–166.

Petrova, S. S. "The Dirichlet principle in the work of Riemann," *ИМИ*, **16** (1965), 295–310.

Petrova, S. S. "On the Dirichlet principle," in: *History and Methodology of the Natural Sciences* [in Russian]. Moscow 1966, N 5, pp. 200–218.

Petrova, S. S. "From the history of analytic proofs of the fundamental theorem of algebra," in: *History and Methodology of the Natural Sciences* [in Russian]. Moscow 1973, N 14, pp. 167–172.

Pokrovskiĭ, P. M. *Historical Essay on the Theory of Ultra-elliptic and Abelian Functions* [in Russian]. Moscow 1886.

Pontriagin, L. S. *Ordinary Differential Equations*. Reading, MA 1962.

Pringsheim, A., Molk, J. "Principes fondamentaux de la théorie des fonctions," in: *Encyclopédie des sciences mathématiques*, T. 2, vol. 1. Paris 1909, pp. 1–112.

Sokhotskiĭ, Yu. V. *Higher Algebra* [in Russian]. St. Petersburg 1882.

Sokhotskiĭ, Yu. V. *Number Theory* [in Russian]. St. Petersburg 1888.

Tikhomandritskiĭ, M. A. "Karl Weierstrass," *Сообщ. Харьк. Мат. Общ.* (2), **6** (1899), 35–56.

Timchenko, I. Yu. "Foundations of analytic function theory. Pt. 1. Historical information and development of the concepts and methods that underlie analytic function theory," *Зап. Мат. Отдел. Новороссийск. Общ. Естествоиспыт.*, **12** (1892), 1–236; **16** (1896), 1–216, 257–472; **19** (1899), 1–183, 473–655. (Separate edition under the same name, Odessa 1899.)

Vashchenko-Zakharchenko, M. E. *The Riemannian Theory of Functions of a Complex Variable* [in Russian]. Kiev 1866.

Weil, A. *Elliptic Functions According to Eisenstein and Kronecker*. Berlin 1976.

Index of Names

Printed in the United States
By Bookmasters